Cadastre: Geo-Information Innovations in Land Administration

Tahsin Yomralioglu • John McLaughlin
Editors

Cadastre: Geo-Information Innovations in Land Administration

Springer

Editors
Tahsin Yomralioglu
Department of Geomatics Engineering
Istanbul Technical University
Istanbul, Turkey

John McLaughlin
Department of Geomatics
University of New Brunswick
Fredericton, NB, Canada

Co-published by Springer International Publishing, Cham, Switzerland, with Capital Publishing Company, New Delhi, India.

Sold and distributed in North, Central and South America by Springer, 233 Spring Street, New York 10013, USA.

In all other countries, except SAARC countries—Afghanistan, Bangladesh, Bhutan, India, Maldives, Nepal, Pakistan and Sri Lanka—sold and distributed by Springer, Tiergartenstr. 15, 69121 Heidelberg, Germany.

In SAARC countries—Afghanistan, Bangladesh, Bhutan, India, Maldives, Nepal, Pakistan and Sri Lanka—printed book sold and distributed by Capital Publishing Company, 7/28, Mahaveer Street, Ansari Road, Daryaganj, New Delhi 110 002, India.

ISBN 978-3-319-51215-0 ISBN 978-3-319-51216-7 (eBook)
DOI 10.1007/978-3-319-51216-7

Library of Congress Control Number: 2016961941

Printed on acid-free paper

This Springer imprint is published by Springer Nature
The registered company is Springer International Publishing AG
The registered company address is: Gewerbestrasse 11, 6330 Cham, Switzerland

Preface

Land had represented wealth and power from the first settlement to the end of the1700s. As a result of the increase in the capital after the Industrial Revolution, land became a good that can be traded rather than a source of wealth. After 1945, effective environmental planning need emerged especially in Europe's urban areas, and land was considered as a scarce source after the population growth as a result of the reconstruction works. In the 1970s, due to the insufficient food production and scarcity in the sources, effective management of rural areas emerged as an increasing need. Thus, land was seen as a "socially" scarce source, and the need to manage this source effectively was frequently mentioned in the international community.

The relation between land knowledge, policy, administration, management and use has a dynamic structure just as the relation between humanity and land does. Countries should regularly revise their land knowledge administration regimes, land policies, land administration and management systems and land use in order to keep up with this dynamism. Land Administration Systems (LAS) dynamics are as follows: these systems are directly affected by rapid technological developments; they are the main source of land information, including land policymaking, that all public institutions and private sector need; and the government wants to provide more effective service by using these systems. "Cadastre" is one of the main components of LAS.

A cadastre is normally a parcel-based and up-to-date land information system containing a record of interests in land (e.g. rights, restrictions and responsibilities). It usually includes a geometric description of land parcels linked to other records describing the nature of the interests, the ownership or control of those interests and often the value of the parcel and its improvements. It may be established for fiscal purposes (e.g. valuation and equitable taxation) and legal purposes (conveyancing), to assist in the management of land and land use (e.g. for planning and other administrative purposes) and enable sustainable development and environmental protection. A well-functioning cadastre guarantees property, reduces land conflict, supports real estate taxation, ensures loan security, protects land sources, monitors environment and ameliorates urban planning and infrastructural development. Today, the most important duty of cadastre is to support sustainable development.

"Cadastre" is a universal concept, and it is defined as "the work of officially mapping and systemically registering the areas, borders and values of all kind of land and property". Therefore, "cadastre" is an important public inventory documenting the records of ownership, bordering and responsibility regarding the land with "title deed" on parcel and answering the questions of "whose, where and how much". That is why "cadastre" has an utmost importance for development of countries and the future of the individuals in our globalized world.

Today, of course, many articles and scientific works can be found on the cadastral works, but to find well-bounded and updated books on cadastre is limited in the international context. Therefore, a need to publish a well-designed book on cadastre is very important. In that case, taking the opportunities of "The World Cadastre Summit" (wcadastre.org) which was held between April 20 and 24, 2015, in Istanbul, some of the selected papers presented by the authorities have been published as a new book with a professional style.

This book observes the range of interdisciplinary topics in a variety of research and application fields from different parts of the world with cases. These cases provide an insight into present-day issues, challenges and opportunities and highlight the key features, principles and new methods that need to be considered in future efforts to make the land a liveable place. It provides a contemporary view of current research and development in cadastre, including surveying, land management, remote sensing and geoinformation sciences. Authors from multiple continents, in association with national and international organizations and societies, bring together the most comprehensive forum for cadastre.

This book explains the higher level of significance of cadastre in order to create a better land management system and establish a platform for further debate, discussion and research on all land-related matters. It can offer land policy matters in adopting reform agenda for achieving national land use and development goals. The materials included in the book can be imparted in higher study courses of the universities and related training institutions worldwide as it would have enough potentials to enhance and increase a significant level of knowledge and understanding in the scholarly fields of cadastre, land tenure, property registration and management, surveying and mapping, land and geoinformation management, land governance, land taxation, public administration, etc. The book contents may influence all land-related academics, researcher, policymakers, business and stakeholders around the world.

As editors, we would like to thank all the contributors to this book and take this opportunity to acknowledge all our colleagues for their time-consuming efforts to review the manuscripts of the chapters. Their efforts with high-quality review of the manuscripts contributed significantly to keep the highly scientific contents of the book. The editors also would like to thank collaborators and research scholars for supporting their research activities that helped in bringing out this work productively. Lastly, the editors wish to thank the publishers for bringing out this book successfully.

Istanbul, Turkey — Tahsin Yomralioglu
Fredericton, NB, Canada — John McLaughlin

Contents

About the Editors

Prof. Tahsin Yomralioglu graduated from the Department of Surveying Engineering of Karadeniz Technical University (KTU), Trabzon, Turkey, in 1985. He worked on Land Information Systems at the University of New Brunswick (UNB) in Fredericton, Canada. In 1993, he obtained his PhD from the University of Newcastle upon Tyne, England. First, he was appointed as a full-time professor at KTU in 2001 and then at Istanbul Technical University (ITU) in 2009. He has been the deputy chairman of the Department and the head of both Land Management and Cartography Divisions at KTU. He was also appointed as the general secretary of KTU and an adviser to the rector of KTU and the rector of ITU. He has served as a member on various commissions and also worked as a project manager and consultant in several public and private institutions. As a supervisor, he guided over 30 MSc and PhD theses. He established the ITU-GeoIT graduate programme and the first national GIS R&D innovation centre of Turkey. He has published many scientific research publications in the field of GIS, SDI, cadastre, land management, etc.

Prof. John McLaughlin is president emeritus at the University of New Brunswick (UNB), Canada. He served as chair of the Department of Surveying Engineering, vice president and 17th president of the University and vice-chancellor at UNB. He introduced and developed the first land administration programme at UNB and the first programme in land information management to be taught anywhere in the world. He has been a leader in building the North American geomatics industry and has worked extensively overseas, in more than 40 countries, on the development of property systems with the World Bank, UNDP and other international agencies. He has authored or co-authored more than 200 publications, including three books and a series of monographs published by the US National Academy of Sciences. McLaughlin has supervised or co-supervised more than a hundred graduate students and has had academic appointments at a dozen universities worldwide.

Part I
Towards Cadastre

Chapter 1
Towards a Fourth Wave of Property Reform

John McLaughlin and Robin McLaren

This was one of my prayers: for a parcel of land not so very large, which should have a garden and a spring of ever-flowing water near the house, and a bit of woodland as well as these.

(Horace, 1st century B.C.)

Introduction

These notes have been prepared as background material for a presentation scheduled for The World Cadastre Summit in Istanbul. They are built on an earlier presentation given to the FIG International Congress in Sydney (McLaughlin 2010).

The Sydney presentation provided a simple framework for reviewing the post-war history of property reform and the concomitant history of developments in the field of land management. That history was divided up into three overlapping chapters, or waves, and built upon a narrative of emerging interests in the importance of property to development, the subsequent investments in major initiatives, some early and important success stories, followed by a growing awareness of the challenges and limits to this aspect of development. Along the way, an unfortunate side effect has been that the rich elite have too often high-jacked the administration services to secure land assets at the expense of the poor and the most vulnerable in society.

The framework was, and is, overly simplified and requires careful attention to the significant caveats and limits associated with such an overarching narrative. However, it was generally well received at the time and has provided a useful platform for discussing the impact of property reform on both economic and social development and for assessing lessons learned.

In this presentation, we'll briefly rehearse this history, with a focus primarily on the third chapter of reform (roughly from the late 1970s). The discussion will examine some of the key drivers (especially technology advances, systems develop-

J. McLaughlin (✉)
Department of Geomatics, University of New Brunswick, Fredericton, NB, Canada
e-mail: jdm@unb.ca

R. McLaren
Know Edge Ltd., Edinburgh, UK

T. Yomralioglu, J. McLaughlin (eds.), *Cadastre: Geo-Information Innovations in Land Administration*, DOI 10.1007/978-3-319-51216-7_1

ments in a number of Western countries, coupled with a renewed interest in the role of property and land reform within the international development community).

After a brief historical tutorial, the presentation will then review some of the major lessons learned over the past 30 years and the current consensus (to the extent that this is possible or indeed desirable) on the importance and direction of property reform. For example, the Western hasn't transplanted well into the dynamic environments of developing countries with high levels of urbanisation, leaving citizens exposed to eviction and land grabbing.

Next, and perhaps most importantly, we will attempt to make the case that we are now witnessing the movement towards a fourth wave of reform, based in part on lessons learned, new technologies, and new development priorities, but also with both the intellectual and professional leadership increasingly coming from a group of so-called middle income countries. Put simply, we in the West are now passing the torch!

Finally, and very briefly, we also want to discuss the role of the surveying profession throughout this history. In the past, both authors have periodically expressed their concerns about the profession, but (subject to plenty of caveats) in this presentation we will advance the argument that surveyors are actually rather well placed to make a significant contribution going forward.

Framing the Historical Narrative

The Sydney presentation provided a fairly detailed review of the post-war land and property reform story, primarily from an active practitioner perspective. It began (no surprise) with the reforms immediately after World War II and especially in South Asia, primarily Japan, Taiwan and South Korea. While these reforms, if remembered at all, are often seen as part of a policy to contain communism and weaken local elites, rather than as one of the most important measures taken for market economies to flourish.

Indeed, over the years the Asian examples have featured in the core narrative about the importance of fundamental land reform to the economic development agenda. And with good reason. But we have also come to appreciate the flaws in these reform programmes, programmes imposed from without (a la Douglas MacArthur in Japan) with limited sensitivity to the social and cultural norms within which they were implemented, and embedded with institutional flaws that in some cases only emerged decades later. These early success stories largely failed to be replicated elsewhere, the Swynnerton Plan in Kenya often cited as a cautionary example, and this first wave of reform gradually grew to a close.

The second wave of land reform, which featured prominently in the 1960s and 1970s, had a very strong Latin American dimension (McLaughlin, for example, was a graduate student at the University of Wisconsin during the time when the Land Tenure Centre was actively involved with Allende's government in Chile) and was very much ideologically driven. As one would imagine, this chapter has proven to be very complicated and controversial. Although dated, a great introduction to that period for anyone interested is Russell King's *Land Reform: A World Survey* (1977). For a detailed critique of this period, see Dasgupta (2010).

Whatever successes may be attributed to this era, the overall impact was to sow doubt on the motives and concerns of property reform as part of the development agenda, and there was a wholesale retreat from the field by the funding agencies. But not for long. By the late 1970s a new generation of technocrats and development specialists were rediscovering the fundamental importance of land and property (and more often than not were trained by a distinguished cadre of pragmatic specialists who came of age in the immediate post-war era). Thus began a third wave of reform which we discuss in somewhat more detail in this paper.

The Third Wave

This most recent chapter in the property reform narrative has a large number of strands, from the major World Bank investment in property projects such as Thailand, to the East European reforms in the post-Soviet era, to Latin American jurisdictions such as Peru (with its language of formalization), to South Africa and elsewhere.

Early drivers in this chapter included:

- a renewed interest in the importance of property in Western countries (and especially its relationship to the emerging environmental agenda);
- the importance of land and property reform to the international economic agenda (beginning with the focus on economic liberalization and the so-called Washington consensus through to the Millennium Development Goals); and
- the need for social stability following the collapse of communism (which led to fast, innovative programmes for the restitution of land and compensation to the former owners).

The World Bank in particular played an especially important role in shaping this agenda, arguing for example in its 2003 Annual Report, that "increasing land rights for poor people is the key to reducing poverty and stimulating economic growth". This has been reflected through the World Bank funding loans of over US$1 billion to 40 land projects in 23 Europe and Central Asia countries in support of the land and property sector (Satana et al. 2014). This is the largest programme of land reform the world has ever seen and has included: land privatization, especially farm restructuring; business, housing and enterprise privatization; restitution; systematic registration and improved services. New programmes include land consolidation, NSDI, state land management, planning, property taxes and e-government.

Another important dimension in this third wave was the re-emergence of land administration as a significant field of endeavour. Early attempts at automation, especially in the field of land registration, were followed by the introduction of modern systems engineering concepts and the evolution of new models for the integration of the various components of land administration (including surveying, registration, valuation and so-forth). Beyond all of this, the concept of the land parcel as a fundamental window into the information world (introduced, for example, by McLaughlin and Palmer (1996) in his work from the multi-purpose cadastre through to the spatial data infrastructure concept and published in a series of seminal US National Academy of Sciences publications). Out of all of this came a series of major programme initiatives in Canada, Australia, Scandinavia, and elsewhere, led by such iconic programs as the Land Registration and Information Service (LRIS) in the Maritime Provinces of Canada. These influential initiatives provided the framework and thinking for citizen services and citizen engagement in land management under the emerging e-government agenda, and were forged with increasing partnerships with the private sector (U.S. National Research Council Committee 1980, 1993).

Building on this brave new world, and with the emergence of a new generation of creative and motivated land administration officials, focus shifted to the developing world. The Thailand Land Titling Project (an initiative of the Royal Thai Government, the World Bank and the Australian Agency for International Development) can serve as perhaps the iconic initiative of this chapter, receiving the World Bank Award for Excellence in 1997. The project stood out for its ambitions goals (including both institutional strengthening of the Thai Department of Lands and its commitment to delivering approximately 13 million titles to Thai landowners); it also became a major international reference site due to the extensive assessment of its progress by Gershon Feder and his colleagues (see, for example, Feder et al. 1988).

The intellectual foundations for this chapter built on a significant post-war literature, especially the richly documented case studies of organization such as FAO and the incredibly useful depositories provided by some very special libraries (the Office International du Cadastre et du Régime Foncier in the Netherlands and the library of the Land Tenure Center at the University of Wisconsin come immediately to mind). As well, there was a small, but immensely influential professional practice literature represented (in the English-language) by such works as S.R. Simpson's *Land Law and Registration* (1976), a seminal work at the time which contained such memorable pearls of wisdom at "land registration is only a means to an end. It is not an end in itself. Much time, money, and effort can be wasted if that elementary truth be forgotten". Indeed!

The third wave of property reform has subsequently benefited from a series of extensive programme reviews, which have explored (from a variety of perspectives) its significance to economic and social development under many different circumstances. For example, Feder and Nishio (1999) undertook a rigorous examination of the benefits of land registration and titling, concluding that "there is convincing evidence from around the world that land registration has led to better access to formal credit, higher land values, higher investments in land, and higher output/income." They went on to note, however, that "there are prerequisites for land registration to be economically viable, and social aspects which need to be considered when designing a land registration system".

Following the significant investments in countries in transition in Central and Eastern Europe through the early 1990's, the UN Economic Commission for Europe collated the experiences in a set of land administration guidelines (UN ECE 1996). This provided an important framework to guide investments in land administration in the region and influenced the significant reforms implemented in the Baltic countries, especially Lithuania.

Further east in Central Asia, the World Bank provided significant loans to countries of the former USSR for land administration and management programmes. The World Bank had learned that speed, innovation and Fit-For-Purpose were key characteristics of a new generation of land administration programmes. The World Bank has enabled the implementation of some very successful programmes in Kyrgyzstan, Russia and Georgia, for example. In Kyrgyzstan over five million parcels were registered in 3 years using para-surveyors and this resulted in the annual number of mortgages doubling between 2002 and 2007 and value increasing from US$85 million to $1 billion; this represented about 30% of GDP in 2007. Georgia is now the number one in the World Bank's 'doing-business' league table for registering a property. The experience from these projects is influencing approaches in the developing world. The best example is Rwanda where a nationwide systematic land registration started after piloting in 2009 and was completed in 2013 using para-surveyors. 10.4 million parcels were registered and 8.8 million of printed land lease certificates were issued. The unit costs were about 6 US$ per parcel. This is an example of a Fit-For-Purpose approach (FIG/World Bank, 2014) that is significantly influencing the fourth wave.

Closer to home, the Cadastre Modernization Project for Turkey, with major funding from the World Bank, provides a significant case study of the potential role of land administration reform to the broader e-government agenda (World Bank 2015).

More recently, a systematic review of the quantitative literature on the effects of tenure formalisation in developing countries funded by the U.K. Department for International Development (Lawry and Samili, 2014) concluded that formal registration of individual land rights increases investment, productivity, and household consumption (although this review also included the important caveat that productivity had not risen as much in Africa as in Asia and Latin America). These findings, coupled with a review of the literature on best practices and policy direction (the *Voluntary Guidelines on the Responsible Governance of Tenure* (FAO 2012) providing an especially interesting and important example) will be briefly addressed in our talk.

The Committee on World Food Security has formally endorsed these guidelines, which resulted from an unprecedented negotiation process chaired by the United States, and which featured broad consultation and participation by 96 national governments, more than 25 civil society organizations, the private sector, non-profits and farmers' associations over the course of almost 3 years. The new guidelines provide a set of principles and practices that can assist countries in establishing laws and policies that better govern land, fisheries and forests tenure rights, ultimately supporting food security and sustainable development.

Towards a Fourth Wave

The property reform story, and the crucial contributions being made by the land administration community, continues to evolve and feature prominently in the international development agenda. And while much of the professional practice literature continues to be based on paradigms developed in the West, there is a significant and growing contribution by academics and practitioners based elsewhere (the recent paper by Demir et al. (2015) being a good example). But at a deeper level, we are also witnessing the evolution of a new narrative: about the nature and importance of property, the institutional and administrative underpinnings required and the role of citizens and civic society for the successful and sustainable implementation of reform.

The beginnings of this new narrative date back decades. One is reminded, for example, of the pioneering work done by Solon Barraclough and his colleagues at FAO a half a century ago on the need to recognize the importance of communal land tenures, and the overarching commitment to providing more equitable access to agricultural land. These themes were often marginalized in the heady days of the neo-liberal agenda, but are very much back on the table today. Similiary, Hernando de Soto, a prominent Peruvian development economist has made a huge contribution to the way we think about property and its role in civil society through his framing of the narrative through the formal/informal lens. His work, from *The Other Path* (1989), through *The Mystery of Capital*, to his most recent documentary for Public Broadcasting in the US (*Unlikely Heroes of the Arab Spring)* have been especially successful in connecting with the most senior leaders in the political and business worlds.

Another major strand in advancing a fourth wave narrative relates to Deng Xiaoping and his rise to power following the Third Plenum of the Central Committee Congress of the Communist Party of China in December 1978. The household-responsibilty system and the famous experiments in Xiaogang village, Anhui, and subsequently in Sichuan and Anhui provinces, which led to dramatic increases in agricultural productivity and nationwide adoption since 1981 have fundamentally changed the world!

Furthermore, while there seemed to be a widespread perception in the West that China somehow created capitalism out of thin air, without the initial imperative of securing private-property rights and imposing limits on state power, in fact this view is wrong. As the economist Yasheng Huang from MIT has argued, institution in fact have mattered as much in China as elsewhere. While China doesn't have well-specified property rights security, in the early 1980s it moved very far and very fast toward establishing security of the proprietor. "One should never underestimate the incentive effect of not getting arrested" (Huang 2008). See also Caryl (2013) for an excellent review of this remarkable story.

Social-media is also reshaping how land administration services are being provided and how citizens and communities are engaging in the process. A movement of democratisation of land rights is emerging that will allow citizens to directly record their evidence of land rights and post it on a global platform. This is outside the formal land administration domain and is based on trust and information transparency. This has inherent risks that have to be managed effectively, but has the promise to be inclusive and scalable – something that hasn't been achieved in the past.

Future land administration services must also increasingly support solutions to the twenty-first century global challenges of climate change, critical food and fuels shortages, environmental degradation and natural disaster as today's world population of 6.8 billion continues to grow to an estimated nine billion by 2040 when over 60% will be urbanised. This is placing excessive pressure on the world's natural resources. This support will be reflected through the inclusion of global land indictors in the post-2015 Sustainable Development Goals currently being negotiated by the United Nations to replace the Millennium Development Goals. There will be no hiding for land sector community! As well, we are beginning to witness the development of new administrative and professional practice models which (although they still embed much of the thinking and experience from the West) reflect a very different set of priorities and realities. In this regard, the increasing importance of the urban agenda will inevitably lead to new land administration priorities and practices in local government (such as can be seen in the emergence of new land taxation strategies in China, where local government carry out over eighty percent of the country's public spending but receive less than half of the taxes). From a surveyor's perspective, the FIG/World Bank statement on Fit-For-Purpose is expected to be especially influential in shaping the professional practice model (Enemark et al. 2014).

We will conclude our paper with a few thoughts on this theme, arguing that increasingly the leadership in our profession is going to come from a new generation of land administration specialists largely based in the developing world. An excellent example is provided by an outstanding property specialist, Dr. Clarissa Augustinus (Chief of the Land and Tenure Section at UN-Habitat), and her colleagues in fashioning the Social Tenure Domain Model (Lemmen 2010) and coordinating the Global Land Tool Network that is delivering pro-poor solutions. This transition in leadership will be dependent on how effectively capacity is built in developing countries, especially at the management level, and how successfully new innovative approaches, driven by Fit-For-Purpose, are accepted and implemented. If successful, Africa in particular has the potential to become a land administration powerhouse.

To provide context for that discussion, we rather arbitrarily divide the property world into three components.

Different Worlds, Different Agendas

At one end of the spectrum lie those traditional economies ranked as low on the human development scale by the UNDP (including Nepal, Kenya, Nigeria, Yemen, Haiti, Sierra Leone and Congo). These societies are largely outside the formal economy, confounded by the issues of deep poverty, food security, lack of institutional integrity, and so forth. The importance of property reform to the economic and social development agenda in these societies is vitally important – as recognized for example in emerging concerns about land grabbing (Pearce 2012). In this regard, the importance attached to property reform in the Sustainable Development Goals (the successor to the Millennium Development Goals scheduled to be agreed to by world leaders at the UN General Assembly) will be of special significance. However, any success in tackling the subject will continue to be frustrated by the severe limitations of the institutional foundation. As Deininger and Feder (2009) have put it, the realization of the benefits from land administration reforms (they focus on registration) depend "on the broader socio-economic and governance environment and the nature of interventions. Bad governance and an ineffective or predatory state will hinder benefits from such interventions, or even cause negative outcomes."

Deep, sustainable reform is unlikely to come from government any time soon; rather we anticipate real change coming much more from bottom-up initiatives. Some of this new direction is reflected in the current interest in the potential role of behavioural economics by the development community (the most recent World Development Report providing an instructive example). Ultimately far more important, however, are the emerging voices in the market and in civil society captured in the new social media world (through platforms such as http://timbuktuchronicles.blogspot.ca/ and http://africanarguments.org/).

At the other end of the spectrum are those post-modern economies (the EU 15, the US, Canada, Japan, etc.). In these countries the role of property and its supporting administrative infrastructure is moving beyond its traditional role of supporting the real estate market to being viewed as an important component in re-imagining the role of civil society (including a new dialogue with indigenous peoples) and the increasing importance of the environmental agenda – see, for example Grinlinton and Taylor (2011). In our talk we will very briefly discuss this theme from an institutional geography perspective. But it is a third group of nations that are mostly likely to provide the leadership for the next chapter.

The New Leadership Agenda

The heart of this next chapter of property reform will be based in a group of modernizing nations, which may be thought of as primarily (albeit not exclusively) the approximately 50 nations ranked in UNDP Human Development Index as "High

Development Nations". This will include such countries as Uruguay, the Russian Federation, Malaysia, Turkey, Mexico, Peru, Thailand, Tunisia, China and Ecuador.

It is in these countries where we can expect the next generation of academic and professional leadership to evolve. For example, we anticipate that prominent academic departments within these countries (such as the hosts of our conference – Geomatics Engineering at ITU – and others such as the Department of Cadastre at Warsaw University of Technology) and at the intersection of the advanced and developing worlds (such as the Department of Land Surveying and Geo-Informatics at Hong Kong Polytechnic University) will play an increasingly important role in fashioning the new intellectual and professional narrative. Similarly, we are already witnessing the professional centre of gravity moving to practitioners in these countries.

The International Federation of Surveyors, especially during the tenure of its President, Professor Stig Enemark, has been especially pro-active in nurturing this new agenda. See, for example, Enemark et al. (2009).

Acknowledgement Several people have been very helpful in the preparation of this work. In particular, we would like to thank Dr. David Palmer, with FAO in Rome, for his particularly useful insights.

References

Caryl, Christian. 2013. *Strange rebels: 1979 and the birth of the 21st century*. New York: Basic Books.

Dasgupta, Partha. 2010. *Selected papers of Partha Dasgupta: Institutions, innovations and human values*. Oxford: Oxford University Press.

Deininger, Klaus, and Gershon Feder. 2009. Land registration, governance, and development: Evidence and implications for policy. *The World Bank Research Observer* 24(2): 233–266.

Demir, O., B. Uzun, and Y.E. Coruhlu. 2015. Progress of cost recovery on cadastre based on land management implementation in Turkey. *Survey Review* 47(340).

de Soto, Hernando. 1989. *The other path*. New York: Harper & Row.

Enemark, Stig, Robin McLaren, and Paul van der Molen . 2009. *Land governance in support of the millennium development goals*.In FIG/World Bank conference, Washington, DC, March 2009.

Enemark, Stig, Keith Bell, Christiaan Lemmen, and Robin McLaren. 2014. *Fit-for-purpose land administration*, FIG publication No. 60. Copenhagen: FIG and World Bank.

FAO. 2012. *Voluntary guidelines on the responsible governance of tenure of land fisheries and forests in the context of national food security*. Rome: Food and Agriculture Organization of the United Nations.

Feder, Gershon, and Akihiko Nishio. 1999. The benefits of land registration and titling: Economic and social perspectives. *Land Use Policy* 15(1): 25–43.

Feder, Gershon, et al. 1988. *Land policies and farm productivity in Thailand*. Baltimore: Johns Hopkins University Press.

Grinlinton, David, and Prue Taylor. 2011. *Property rights and sustainability*. Leiden: Brill Academic Publishers.

Huang, Yasheng. 2008. *Capitalism with Chinese characteristics*. Cambridge: Cambridge University Press.

King, Russell. 1977. *Land reform: A world survey*. London: G. Bell & Sons.

Lawry, Steven, and Cyrus Samili. 2014. *Report to the annual World Bank conference on land and poverty*. Washington, DC.

Lemmen, Christiaan. 2010. *The social tenure domain model: A pro-poor land tool*, FIG publication No. 52. Copenhagen: The International Federation of Surveyors (FIG).

McLaughlin, John. 2010. *The next chapter in property reform*. An invited presentation to the FIG International Congress. Sydney, Australia.

McLaughlin, John, and David Palmer. 1996. Land registration and development. *ITC Journal* 96(1): S10–S18.

Pearce, Fred. 2012. *The land grabbers: The new fight over who owns the Earth*. Boston: Beacon Press.

Satana, Suha, et al. 2014. *Economic impact of 20 years of ECA land registration projects*. In Annual World Bank conference on land and poverty. Washington, DC. March 2014.

UN Economic Commission for Europe (UN ECE). 1996. *Land administration guidelines*. New York: United Nations Publication.

U.S. National Research Council Committee. 1980. *Need for a multipurpose cadastre*. Washington, DC: National Academy Press.

———. 1993. *Toward a coordinated spatial data infrastructure for the nation*. Washington, DC: National Academy Press.

World Bank. 2015. http://www.worldbank.org/projects/P106284/turkey-land-registration-cadastre-modernization-project?lang=en.

Chapter 2
From a Traditional to a Comprehensive Cadastre

Jürg Kaufmann

Development of Cadastres

The first cadasters date back to roman times to recover state owned lands that had been appropriated by private individuals, and thereby recover income from such holdings. With the fall of Rome the use of cadastral maps effectively discontinued. Medieval practice used written descriptions of the extent of land rather than using more precise surveys. In the sixteenth and early seventeenth centuries did the use of cadastral maps resume, beginning in the Netherlands. Napoléon, after taking the power about 1800, commanded to survey the parcels and to install cadastral systems for the land taxation wherever he invaded. Since then the official cadastre systems were spreading over the world and they served for the documentation of land rights and for land taxation.

These purposes remained unchanged for a long time until the issues of overcrowding and environment protection became obvious mainly after World War II. Emission cadasters, pipeline cadasters and multi-purpose cadastre arose, in many cases as parallel facilities to the property cadastre (Fig. 2.1).

In view of the developments taking place in the field of cadastre, FIG Commission 7 launched in 1994 a working group with the following terms of reference:

> Study cadastral reform procedures as applied in developed countries, take into consideration automation of the cadastre and the role of the cadastre as part of a larger land information system, evaluate trends in this field and produce a vision of where cadastral systems will be in the next 20 years, show the means by which these changes will be achieved and describe the technology to be used in implementing these changes.'

The result of the work was published in 1998 under the title CADASTRE 2014 – A Vision for a Future Cadastre System by the leader Jürg Kaufmann and the

J. Kaufmann (✉)
Swiss Land Management Foundation, Kaufmann Consulting, Zurich, Switzerland
e-mail: jkcons@swissonline.ch

T. Yomralioglu, J. McLaughlin (eds.), *Cadastre: Geo-Information Innovations in Land Administration*, DOI 10.1007/978-3-319-51216-7_2

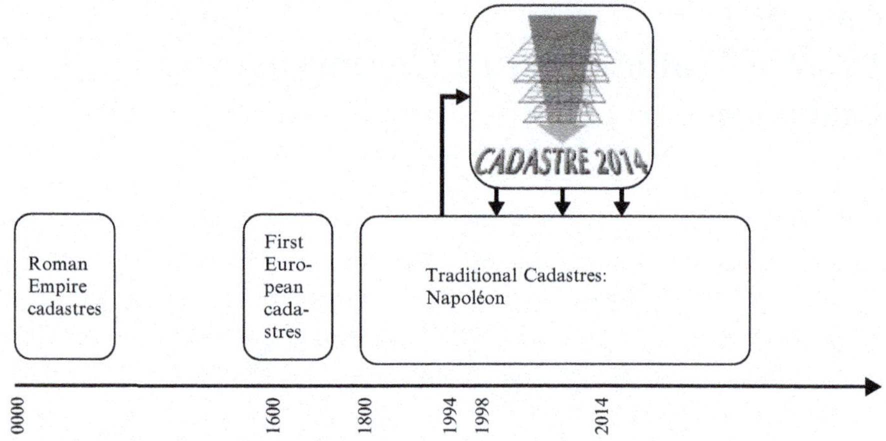

Fig. 2.1 Development of cadasters

Statement 1 -	Cadastre 2014 will show the complete legal situation in land, including public rights and restrictions.
Statement 2 -	The separation between maps and registers will be abolished.
Statement 3 -	Cadastre Mapping will be ded – long live modelling
Statement 4 -	Paper and Pencil Cadastre will have gone.
Statement 5 -	Cadastre 2014 will be highly privatized! Public and private sector working closely together.
Statement 6 -	Cadastre 2014 will be cost recovering.

Fig. 2.2 The six statements of Cadastre 2014

secretary Daniel Steudler with the Working Group 1 of FIG Commission 7. CADASTRE 2014 after the publication was translated in about 30 languages and influenced the thinking about cadastre systems.

The brochure Cadastre 2014 launched six statements showing the developments to be expected in the next 20 years (Fig. 2.2).

Statement 1 describes the idea of a Comprehensive Cadastre as a further development of the traditional cadastre to an infrastructure documenting not only the land property rights but also all the rights restriction and responsibilities imposed on land by official or traditional whether written or unwritten regulations.

The Comprehensive Cadastre must cover a wider field than the traditional cadastre has since its introduction. The circumstances of the resource land have changed significantly since its inception (Fig. 2.3). During the development of the legal systems the private laws were dominant. The constitutions of most countries defined the rights of the citizens, one of which is the guarantee to own property. Civil codes

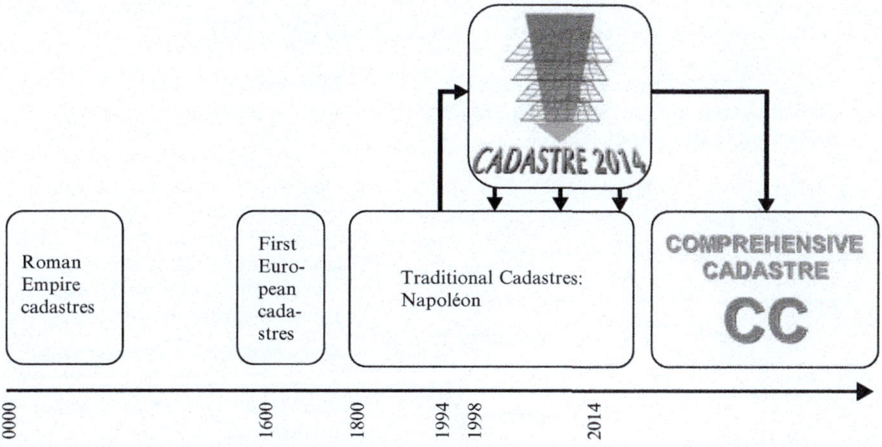

Fig. 2.3 The next development step: The comprehensive cadastre

have reinforced this guarantee and defined clear procedures and institutions to protect the rights of citizens against alienation.

The growing world population and the development of new technologies lead to an intensified use of natural resources including land. To protect the natural resources from being totally consumed, damaged or destroyed, restrictions of the absolute right to use the natural resources were defined in the name of the social necessity. Especially after World War II a growing number of new public laws were created. Land use planning, environment protection, noise protection, construction laws, protection against danger caused by natural phenomena, and so on, were regulated by public laws.

These definitions under public law can have an impact on the property right of the owner, but they are not part of the official register. The boundary definition process of the rights and restrictions defined under public law follows democratic legal rules. But there is no boundary verification, no title verification, and no registration in an official legal register (Fig. 2.4). Aside from land objects from private and public law, a third category of legal land objects occurs in several countries where traditional rights exist. In these cases, areas are defined where tribal land use rights exist. They can overlap other legal land objects, such as private property rights and public rights and restrictions, and concessions for the exploitation of natural resources. These traditional, customary rights are often not documented in a manner that creates the necessary legal security.

The Comprehensive Cadastre must correct this situation, which is becoming more and more precarious. It must document, in a safe manner, all legal aspects of land.

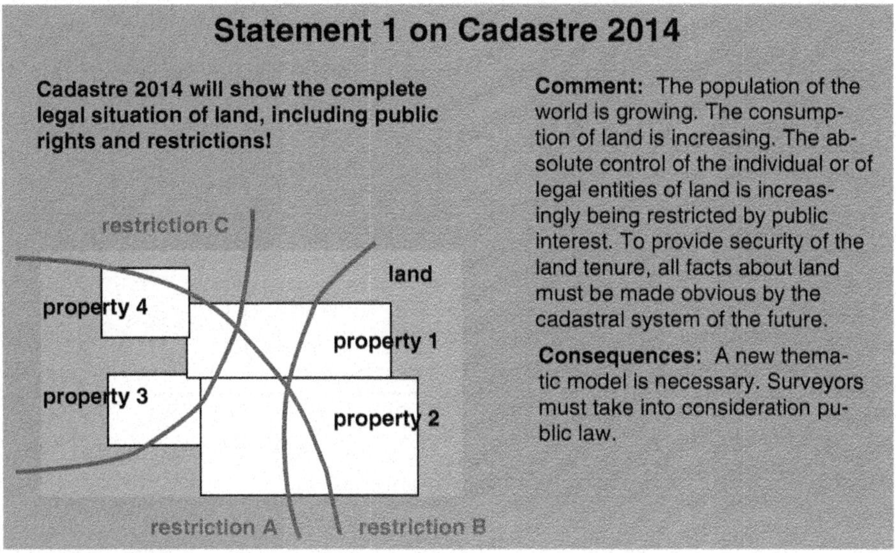

Fig. 2.4 Statement 1 of Cadastre 2014

Structure of the Comprehensive Cadastre

Cadastre 2014 has introduced some principles to be applied for the structure of the Comprehensive Cadastre. The structure of the Comprehensive Cadastre is to follow the principle of legal independence.

The principle stipulates that:

- *legal land objects, being subject to the same law and underlying a unique adjudication procedure, have to be arranged in one individual data layer; and*
- *for every adjudicative process defined by a certain law, a special data layer for the legal land objects underlying this process has to be created.*

The Comprehensive Cadastre is therefore based on a data model, organized according to the legislation for the different legal land objects in a particular country or district. While the traditional cadastre consists in general of one information layer representing the information about boundaries between different properties, in the Comprehensive Cadastre are added information layers representing the boundaries between land objects defined by different legal topics, which exist in a jurisdiction (Fig. 2.5).

Daniel Steudler and Abbas Rajabifard designate this principle in the Publication Spatially Enabled Society[1] as institutional independence. With this term, they indicate that this structure is suitable to assign the responsibility for the data layers to the authority charged with the enforcement of a certain Act.

[1] Spatially Enabled Society, FIG Publication No. 58.

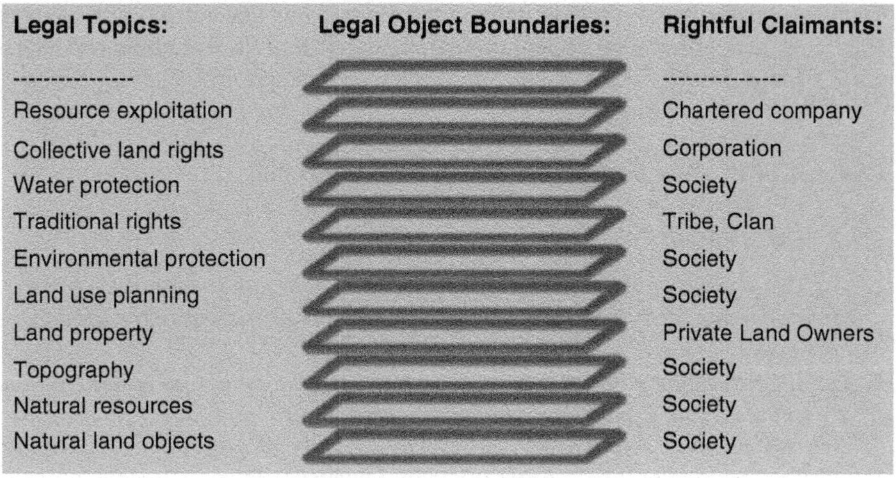

Fig. 2.5 The principle of legal independence

Precondition for the Comprehensive Cadastre

A further principle stipulated in Cadastre 2014:

> To make sure that legally independent organized land objects can be combined, compared, and brought into relation to each other, it is necessary that they will be localized in a common reference system. The combination and comparison of the thus located land objects can be realized by the method of polygon overlaying. This method was published already in 1973 by Kaufmann and Bigler.[2]

The Comprehensive Cadastre will only function in an efficient manner when the relations between land objects can be derived from their location. This avoids links between land objects in different information layers. According to experience in many cases, traditional and distorted maps are anyway to be replaced by data sets located in a common reference system in order to enable modern geographic information systems be able to render the expected services.

Steps to Implement a Comprehensive Cadastre Successfully

Introduce the Possibility for the Comprehensive Cadastre in Your Legal Framework

It is wise to fix the principle of a Comprehensive Cadastre before starting with the setup. Switzerland decided to introduce the cadastre of Public Law restrictions on Landownership, which can be considered as a first step of the Comprehensive

[2] Kaufmann and Bigler, 1973: New Techniques in Land Consolidation.

Cadastre. A short article was introduced in Switzerland's Federal Act of 5 October 2007 on Geoinformation (Geoinformation Act) http://www.admin.ch/ch/e/rs/c510_62.html:

Cadastre of Public-Law Restrictions on Landownership

Art. 16 Subject Matter and Form

1. *The Cadastre of public-law restrictions shall contain public-law restrictions on landownership rights which, in accordance with the provisions of the Civil Code are not part of the Land Register.*
2. *The Federal Council determines which official geodata under federal legislation are entered in the Cadastre of public-law restrictions.*
3. *The cantons may define additional official geodata of proprietary nature that must be recorded in the Cadastre of public-law restrictions.*
4. *The Cadastre of public-law restrictions shall be made available in electronic form either online or by any other method.*
5. *The Federal Council shall determine the minimum requirements with regard to the organization, management, data harmonization, methods and processes for the Cadastre of public-law restrictions.*

In the Principality of Liechtenstein the legal base was laid in the Law on the official surveying as follows:

Documentation of the Public-Law Restriction of the Landownership

Art. 57 Basic Principle

1. *The public-law restrictions with geometric characteristic as, in particular land use and development plans, protection zones or building lines, are represented in specific information layers.*
2. *The government determines the spheres, where information layers are defined.*

Develop a Short Enactment on the Comprehensive Cadastre

Because the rules for the Comprehensive Cadastre are the same as those for the traditional cadastre a regulation can be kept short. In Switzerland we developed an Ordinance on the Cadastre of Public-law Restrictions on Landownership (PLR-Cadastre) with 33 articles regulating the details (Fig. 2.6).

Fig. 2.6 The content of the Swiss Ordinance on the PLR-Cadastre

Section 1:	General provisions
Section 2:	Content and Information
Section 3:	Inclusion into the Cadastre
Section 4:	Forms of Access
Section 5:	Authentification
Section 6:	Function as official gazette
Section 7:	Organization
Section 8:	Financing
Section 9:	Participation
Section 10:	Final Provisions

Introduce Data and Representation Modelling as Mandatory

One important aspect for the successful implementation is the provision to use data modelling for the description of all data topics of the Comprehensive Cadastre and representation models to define how these data are to be represented on maps or other documents. Switzerland regulated this in the framework of the Federal Act of 5 October 2007 on Geoinformation. As modelling standard we use INTERLIS 2. Details may be found under www.interlis.ch.

Determine a Responsible Authority for the Comprehensive Cadastre

In every country a responsible authority must be designated to organize the Comprehensive Cadastre. To allocate this task to the authority already taking care of the traditional property cadaster seems to be appropriate and advantageous.

Scan Your Legal Framework Including Traditional Rules Authority

The first task of responsibility is the scanning of the existing legal framework and also all existing unwritten traditional legal arrangements. As soon as a law or a regulation contains arrangements concerning maps, sketches, schemes, boundaries, building lines, etc., it is to be supposed that the respective land objects are candidates for inclusion in the Comprehensive Cadastre.

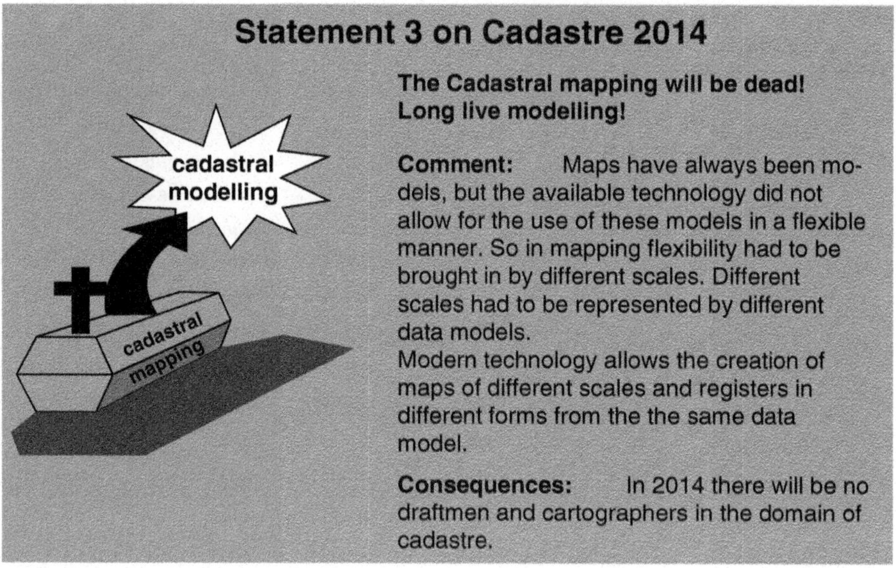

Fig. 2.7 Data modelling for the comprehensive cadastre

Identify the Stakeholders

A further result of this scan shows the institutions responsible for the enforcement of the law. These institutions are the stakeholders to be involved in the implementation of the Comprehensive Cadastre. Together with these stakeholders the further steps are to be executed.

Create Data Models for All Legal Topics Included into the Comprehensive Cadastre

It is absolutely important to describe all data of the Comprehensive Cadastre in a precise and easy to interpret manner in cooperation with the respective stakeholders. The modelling paradigm was launched by statement 3 of Cadastre 2014 (Fig. 2.7).

A tool for data modelling is determined in the ISO/TC211 – Geographic information/Geomatics Standards. The ISO 19152 standard deals with the Land Administration Domain Model (LADM) and was published in 2012. The standard describes the data model with Entity-Relationship-Diagrams but does not offer automatic model and data checking possibilities. Switzerland uses since 1993 the standardized data description language INTERLIS 2 which allows computer-assisted model and data checking. Recently the developers of the LADM from The Netherlands and Swiss data modelling specialists undertook an initiative to combine these modelling approaches by description of the LADM in INTERLIS 2 to profit from automatic checking facilities.

For all data topics to be included in the Swiss PLR-Cadastre data models have been developed. They are public and can be found on http://models.geo.admin.ch/.

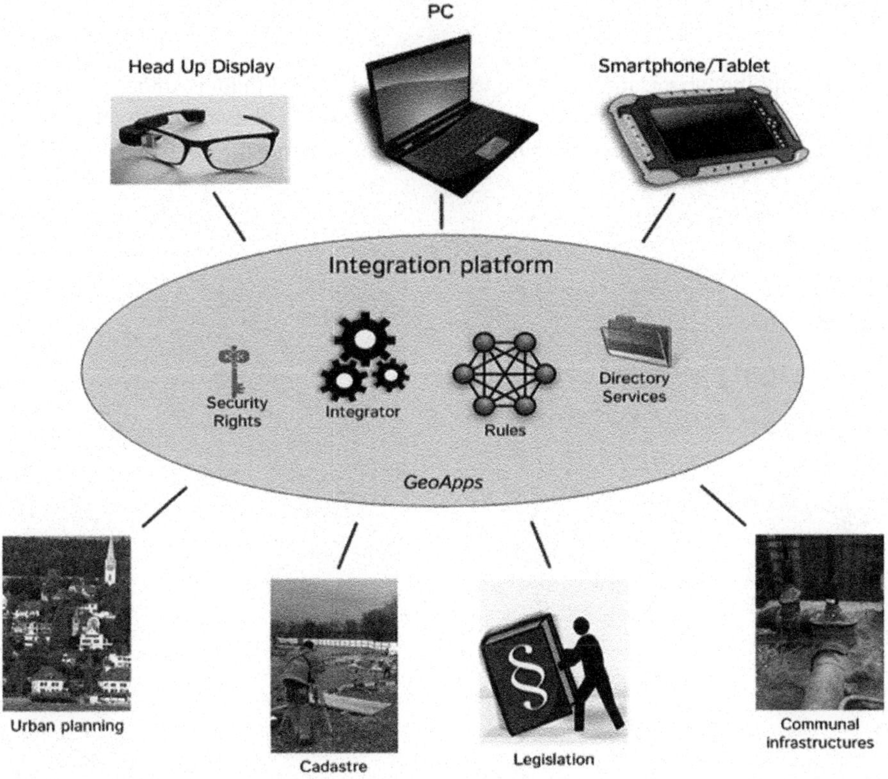

Fig. 2.8 Modern approach in Switzerland

Identify the Procedures for the Definition of Legal Arrangements

Similar to the traditional cadastre the effective procedures must be known and pursued by the Comprehensive Cadastre System to make sure it works correctly. Sometimes these procedures are complicated and in many cases not handled correctly. It is worthwhile to analyze the procedures carefully and to take the opportunity to simplify them if this is possible.

Develop a Feasible IT-Infrastructure

A Comprehensive Cadastre can only be realized with the help of IT. In a modern environment it makes sense to base the Comprehensive Cadastre on internet-technology. In Switzerland's PLR-Cadastre a modern solution GeoApp replacing WebGIS by Web-Application was chosen to realize an integration platform organizing the access to the different information systems of the stakeholders by governing the directories, controlling the access rights, integrate data from different sources, and managing the rules to be applied (Fig. 2.8).

Conclusion

The Comprehensive Cadastre is the tool to keep the land matters under control and to allow sustainable land management. Only a comprehensive land documentation makes land administration and land management possible.

Chapter 3
Cadastre or Land Administration: A Case Study of Turkey

Tahsin Yomralioglu and Mehmet Cete

Introduction

Cadastral systems have evolved over time primarily based on the changes in humankind to land relationship and technology. Land was regarded as a main symbol of wealth during the agricultural revolution and feudal system, and the cadastre recorded land ownership in this period. Cadastre became a tool to support land transfer and land markets during the Industrial Revolution when a process of strong physical ties to the land began. The post-World War II period with population boom generated awareness that land was a scarce resource. Countries preferred to address the scarcity with better planning in this period, and cadastre supported the planning process. Finally, in the 1980s, the focus was on wider issues of environmental degradation, sustainable development and social equity, and thus, land became a 'scarce community resource'. This forced the extension of cadastres into land administration systems (Ting and Williamson 1999; Williamson 2001a; Bogaerts et al. 2002; Steudler et al. 2004; Bennett et al. 2013; Cete and Yomralioglu 2013).

Land Administration System (LAS) is defined as "the processes of determining, recording and disseminating information about the tenure, value and use of land when implementing land management policies" (UNECE 1996). It means a LAS administers not only land tenure and ownership but also land value and land use data (Williamson 2001b; Enemark 2001; Bogaerts et al. 2002; Bandeira et al. 2010). This requires carrying out re-engineering processes in traditional cadastral systems which target to secure tenure and ownership to include land value and land use

T. Yomralioglu (✉)
Department of Geomatics Engineering, Istanbul Technical University, Maslak, 34469 Istanbul, Turkey
e-mail: tahsin@itu.edu.tr

M. Cete
Department of Geomatics Engineering, Izmir Katip Celebi University, Balatcik, 35620 Cigli/Izmir, Turkey

T. Yomralioglu, J. McLaughlin (eds.), *Cadastre: Geo-Information Innovations in Land Administration*, DOI 10.1007/978-3-319-51216-7_3

components, and their focus are needed to be evolved from market to an additional facilitative role for multipurpose spatial information infrastructures in order to support the implementation of sustainable development objectives (UN–FIG 1999; Enemark 2001; Williamson 2001a; Bogaerts et al. 2002; Wallace and Williamson 2006; Rajabifard et al. 2007; Bennett et al. 2008). In this context, the evaluation of national LASs has become more and more of an issue of concern over the last few decades worldwide (Williamson 2001b; Steudler et al. 2004; Robertson 2002; Wallace and Williamson 2006; Rajabifard et al. 2007; Mitchell et al. 2008). Turkey, having experience about 170 years in cadastre, is one of the countries carrying out reform projects to provide improvement in the system, and to address the current and future needs of cadastre (Cete and Yomralioglu 2013) However, there is a need for a more comprehensive reform in the country to upgrade the current system from cadastre to land administration.

This paper, firstly, provides an overview of the current cadastre, topographic mapping and real estate valuation systems of Turkey, and then, proposes an approach to upgrade the cadastre to land administration.

Current Land Administration System in Turkey

LASs determine, record and disseminate information about land tenure, land value and land use. Since Turkey does not have a unified system of land administration, this chapter evaluates land registration and cadastre, topographical mapping and real estate valuation systems in the country under the subtitles below.

Land Registration and Cadastre

Land Registration and Cadastre (LRC) is the core engine of spatially enabled land administration (Enemark 2012). Therefore, LRC data has a special importance in the LASs. Turkey is an experienced country in the LRC domain. The first cadastral organization was founded in 1847 in the country. The organization carried out land registration works until foundation of the Republic of Turkey. In 1924, firstly, the General Directorate of Land Registry was founded. Then, cadastre unit was attached to the General Directorate, and cadastral surveys were initiated. The current General Directorate of Land Registry and Cadastre (GDLRC) was established with a re-engineering process in LRC in 1936.

Main legislations regulating LRC services are the Land Registration Law and the Cadastre Law. The GDLRC and the District Directorates of LRC organize cadastral works throughout the country (Fig. 3.1). Directorates of Land Registry and Directorates of Cadastre are the responsible organizations from the services provided in the local level. In 2005, the Licensed Offices of Surveying and Cadastre (LOSC) were also introduced into the cadastre (Official Gazette 2005). During

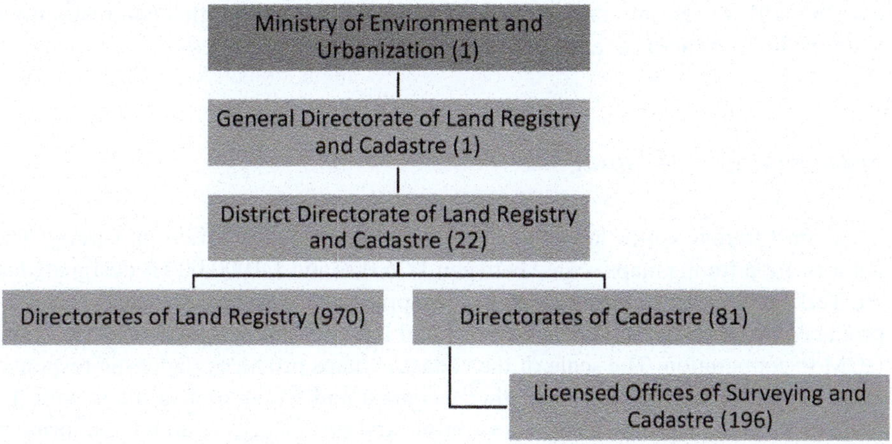

Fig. 3.1 Organizational structure of the Turkish land registration and cadastre system (TKGM 2015)

design of the LOSC, sub-districts of cadastre throughout the country were determined by taking workloads of the existing cadastre directorates into consideration. The LOSC have been authorized to carry out the cadastre works. Application of the cadastre maps into the field and showing boundaries of parcels in the relevant area are performed by the LOSC. These works are not subject to supervision by the cadastre administration. However, the LOSC works for use type change of a parcel, establishment and removal of easement rights and consolidation of parcels are supervised by the Province Directorates of Cadastre (Circular Letter 2010).

Except a few problematic units, establishment of land registration and cadastre has been almost completed in the country. However, cadastral surveys cover about 62% of the surface of the country. Active involvement of private surveyors into cadastral surveys after 2004 speeded up the cadastral works and played an important role in completion of the cadastre. Establishment of the Turkish Land Registry and Cadastre Information System (LRCIS) work which was initiated at the beginning of the 2000s is still in progress. Almost all land registry data is transferred into digital environment as a part of the project. However, transferring paper based cadastre maps into computer environment with high accuracies through digitization is not easy since most of the old cadastre maps have accuracy problems. These maps are transferred into digital environment with renovation projects and it takes time.

Turkey has a well-functioning LRC system but still there is a need for a reform project to increase data quality and to extend the content of the LRC. Sales prices of real estates are recorded in the land registry but those prices are not real market values. Owners of some real estates in the land registry are dead people since transfer of a real estate to heirs when a landowner dies is not compulsory (Cete et al. 2006). Use types of some parcels in the registry are not up-to-date. Roads and buildings are part of the cadastral maps but there is no dynamic or periodic process to update these data (Cete and Yomralioglu 2013). In addition, content of the Turkish cadastre still covers only data to secure the property. Public rights and restrictions and land use data are not represented in the cadastre.

Topographical Mapping

Maps in different scales are produced by different organizations in Turkey. The authorization for the maps scaled between 1:25,000 and 1:1000,000 is designated to the Turkish General Commandership of Mapping (GCM). 1:5000 scaled maps are produced by the General Directorate of Land Registry and Cadastre (GDLRC) and GCM in cooperation. The scale of the cadastral maps produced under the responsibility of the GDLRC is 1:1000. Other technical and topographical maps with the scale of 1:1000 are produced by many public and private organizations. Majority of the organizations producing 1:1000 scaled technical and topographical maps has caused some duplications in the production. The project called as "Information Bank of Maps" developed by the GDLRC in 2008 has reduced the duplications through providing a data bank about available maps. In the context of the project, institutions entered the metadata of the maps they produced into the web based system of the Information Bank of Maps. Currently, an organization can enter the system and query if a map for a specific area is available in another institution's hand or not. Nevertheless, there is still need for a national organization to organize production of maps and spatial information in all scales.

Real Estate Valuation

Turkey does not have a law on real estate valuation. Principles of the valuations are described in different laws and regulations like the Expropriation Law and the Taxation Law. Licensing procedures in real estate valuation are organized in the official notifications of the Capital Markets Board of Turkey (CMBT).

Turkey does not have a strong and well-functioning real estate valuation system in organizational means. Number of public institutions carrying out real estate valuations are more than twenty. Except for the CMBT, all the institutions work through real estate valuation commissions. A commission is made up of selected officials from the institution that needs real estate values for such purposes as taxation, expropriation, nationalization, etc. The officials do not have to have a license to take part in the commissions. Only in valuations for expropriation, a certificate is needed. This certificate is given by the relevant chambers attached to the Union of Chambers of Turkish Engineers and Architects. CMBT carries out valuations for capital market activities, and asks for a license from the appraisers. The only institution authorized to license real estate appraisers in the country is the CMBT (Cete 2008; Cete and Yomralioglu 2013). All faculty graduates can enter the license exams and become an expert on real estate valuation. It means, there is no professional restrictions to get the license in the country.

It is clear that real estate characteristics and sales prices data are two of the most crucial inputs in real estate valuation works. However, neither a systematic real estate characteristics nor sales prices databases are available in Turkey.

An Approach for Turkish Land Administration System

According to the Law on Organization and Duties of the General Directorate of Land Registry and Cadastre enacted in 1936, the main duty of the GDLRC is the determination, recording and sustaining of the legal and geometrical situations of real estates. Since then, the GDLRC has worked to fulfill this duty but the issues experienced in the data quality today in both land registry and cadastre shows that the General Directorate couldn't achieve this duty in the proper sense. Furthermore, modern trends of the cadastre urges countries to evolve their traditional cadastral systems towards land administration. This requires that content of the traditional cadastres be extended to include land value and land use data. In addition, cadastre should show the complete legal situation of land including public rights and restrictions. It is clear that accomplishment of all these duties with current legislation, organizational structure and technical tools of the GDLRC is almost impossible in Turkey. Therefore there is a need for re-engineering in the Turkish cadastral system.

The overall principle of re-engineering processes is that land policy drives legislative reform which in turn results in institutional reform and finally the implementation with all its technical requirements (Williamson 2001b). This study proposes an approach for re-engineering of the Turkish Land Administration System (LAS) by considering this principle. The vision is composed of three main components: (1) legal arrangements; (2) organizational structure; and (3) technical organization (Fig. 3.2).

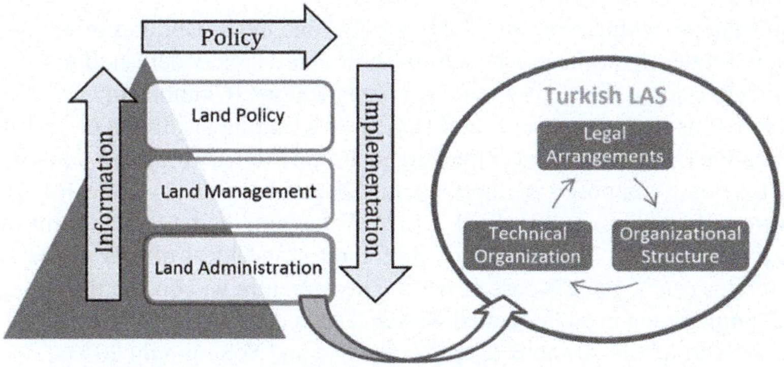

Fig. 3.2 Functionality of the proposed Turkish LAS approach (Cete 2008)

Legal Arrangements

Turkey needs for a 'Land Law' reorganizing existing scattered land related laws, eliminating duplications and gaps in the current laws, and providing an appropriate legal basis for a well-functioning LAS. The Land Law should mainly include regulations on land registration, cadastre, real estate appraisal, geographical information management, and land development. The law should be built on appropriate land policies and sustainable development objectives. Therefore, current land policies, and gaps in policies need to be revised first. Active participation of the relevant experts in the preparation of the law, which are generally disregarded aspect in Turkey, is other important point in this process (Cete 2008; Cete and Yomralioglu 2013). After preparation of the law, regulations and guiding documents also need to be prepared for each domain in the law, otherwise, implementation of the rules defined in the law appropriately and providing unities in the land-related works throughout the country would not be easy.

Organizational Structure

Turkey needs a 'leading institution' in the land administration to eliminate current gaps and duplications, and to provide effective coordination in the domain. Someone may think that the General Directorate of Land Registry and Cadastre may be a suitable administration to become the leading institution. Considering that the General Directorate has some difficulties in fulfilling the existing duties assigned to it and the modern cadastre requirements, it is difficult to make the General Directorate the leading institution responsible for the land administration in its current organizational structure. The leading institution will be responsible not only for land registry and cadastre but also for topographical mapping and real estate valuation. Therefore, the institution should be placed on a higher level than a general directorate in the Turkish administrative hierarchy. The leading institution is advised to be established as an undersecretariat of the prime ministry.

Carrying out all functions of a LAS goes beyond the capabilities of a single organization because requests in land administration are mostly delivered through business processes that run across multiple organizations (Chimhamhiwa et al. 2009). Therefore, this study proposes establishment of a leading institution of Turkish LAS named as the Undersecreteriat of the Turkish Prime Ministry for Land Administration (UPMLA) and composed of the General Directorates of (1) Land Registry and Cadastre; (2) Mapping; (3) Real Estate Valuation; and (4) Land Information Management (Fig. 3.3). The district directorates and local offices of the General Directorates can be built in case of need. This structure will ensure the operation of land administration in an integrated way, and each component will be carried out by its own expert administrations (Cete 2008; Cete and Yomralioglu 2013).

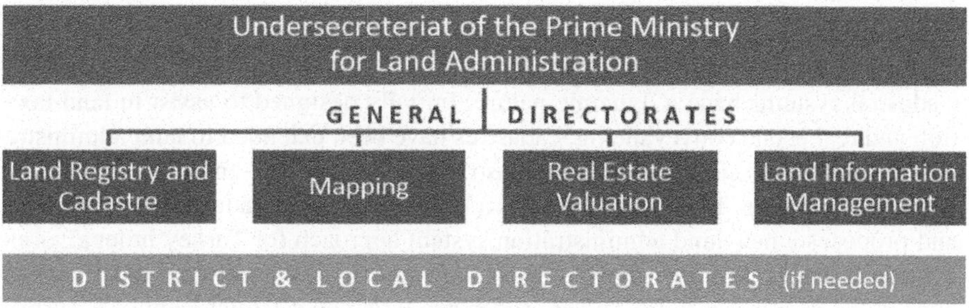

Fig. 3.3 Organizational structure of the proposed Turkish LAS

Technical Organization

LASs target to effectively handle land information through efficient and effective land information infrastructures (Thellufsen et al. 2009; Bennett et al. 2012). Therefore, LASs are increasingly evolving into a broader land information infrastructure which supports economic development, environmental management and social stability in both developed and developing countries (Williamson 2001b). Holistic treatment of land information is no longer arguable; it is essential (Bennett et al. 2008, 2012). However, the organizational framework that many public organizations are placed in often makes difficult the development of efficient and effective land information infrastructures. Due to historical reasons LASs typically consist of various governmental organizations located in separate ministries in many countries. This fragmentized structure leads to issues concerning inter-organizational collaboration, which are critical for the function of the systems (Thellufsen et al. 2009). The proposed Turkish LAS will provide an infrastructure for building up and sustaining an efficient Land Information System (LIS). The Undersecreteriat of the Turkish Prime Ministry for Land Administration (UPMLA) will make land related data available for governmental organizations and private corporations through the LIS. This will minimize duplication of data and provide efficiency. The LIS will organize not only the data produced by the UPMLA but also the land related data produced by other organizations. This study proposes that management of the LIS is carried out by the General Directorate of Land Information Management by taking international standards into consideration (Cete 2008; Cete and Yomralioglu 2013). During technical development of the proposed Turkish LAS, some emerging and important issues such as Land Administration Domain Model (Lemmen et al. 2015), 3D/4D cadastres (Van Oosterom et al. 2006a, 2006b; Döner et al. 2010) and registration of the Rights, Responsibilities and Restrictions (Kaufmann and Steudler 1998; Lemmen et al. 2010) should also be taken into consideration by the UPMLA.

Conclusion

Cadastral systems have a dynamic nature. Initially designed to assist in land taxation and real estate conveyancing, cadastres have been extended to land administration systems. This situation forces cadastral systems to be re-engineered over time to meet the change. This paper provides a brief overview of cadastral developments and proposes a new land administration system approach for Turkey in legal, organizational and technical means both to eliminate existing issues and to fill current gaps in the existing system. The approach proposes the establishment of a Turkish Land Law in a participatory way to bring together the existing scattered laws, to eliminate duplications and gaps in the current regulations, and to provide an appropriate legal basis for well-functioning land administration system. In organizational means, a leading institution named the Turkish Prime Ministry for Land Administration are advised to be established, and all land administration works are organized and supervised by this institution. A Land Information System managed by the proposed General Directorate of Land Information Management is the technical component of the approach. The approach provided in this paper is recommended to be implemented in incremental steps since implementation at once could lead to some disruption and malfunctions in services during the re-engineering and subsequent processes.

References

Bandeira, P., J.M. Sumpsi, and C. Falconi. 2010. Evaluating land administration systems: A comparative method with an application to Peru and Honduras. *Land Use Policy* 27(2): 351–363.

Bennett, R., J. Wallace, and I. Williamson. 2008. Organising land information for sustainable land administration. *Land Use Policy* 25: 126–138.

Bennett, R., A. Rajabifard, I. Williamson, and J. Wallace. 2012. On the need for national land administration infrastructures. *Land Use Policy* 29: 208–219.

Bennett, R., N. Tambuwala, A. Rajabifard, J. Wallace, and I. Williamson. 2013. On recognizing land administration as critical, public good infrastructure. *Land Use Policy* 30: 84–93.

Bogaerts, T., I.P. Williamson, and E.M. Fendel. 2002. The role of land administration in the accession of central European countries to the European Union. *Land Use Policy* 19: 29–46.

Cete, M. 2008. *An approach for Turkish land administration system*. PhD thesis, Graduate School of Natural and Applied Sciences, Karadeniz Technical University, Trabzon, Turkey.

Cete, M., and T. Yomralioglu. 2013. Re-engineering of Turkish land administration. *Survey Review* 45(330): 197–205.

Cete, M., H. Magel, and T. Yomralioglu. 2006. *The needs for improvement in Turkish land administration system: Lessons learned from German case*. Proceedings of the XXIII FIG Congress, 8–13 October 2006, Munich, Germany.

Chimhamhiwa, D., P. van der Molen, O. Mutanga, and D. Rugege. 2009. Towards a framework for measuring end-to-end performance of land administration business processes – A case study. *Computers, Environment and Urban Systems* 33: 293–301.

Circular Letter. 2010. *Circular letter about licensed surveyors and surveying offices*. Official Web site of the General Directorate of Land Registry and Cadastre. http://www.tkgm.gov.tr, 23.03.2015.

Döner, F., R. Thompson, J. Stoter, C. Lemmen, H. Ploeger, P. van Oosterom, and S. Zlatanova. 2010. 4D cadastres: First analysis of legal, organizational, and technical impact – With a case study on utility networks. *Land Use Policy* 27: 1068–1081.

Enemark, S. 2001. Land administration infrastructures for sustainable development. *Property Management* 9(5): 366–383.

———. 2012. *From cadastre to land governance*. FIG Working Week, 4–10 May 2012, Rome, Italy.

Kaufmann, J., and D. Steudler. 1998. *Cadastre 2014 – A vision for a future cadastral system*. Results 1994–98 of Working Group 1 of FIG Commission 7.Rüdlingen and Bern, Switzerland. http://www.fig.net/cadastre2014, 23.03.2015.

Lemmen, C., P. van Oosterom, C. Eisenhut, and H. Uitermark. 2010. *The modelling of rights, restrictions and responsibilities (RRR) in the land administration domain model (LADM)*. Proceedings of the FIG Congress, 11–16 April 2010. Sydney, NSW, Australia.

Lemmen, C., P. van Oosterom, and R. Bennett. 2015. The land administration domain model. *Land Use Policy* 49: 535. http://dx.doi.org/10.1016/j.landusepol.2015.01.014.

Mitchell, D., M. Clarke, and J. Baxter. 2008. Evaluating land administration projects in developing countries. *Land Use Policy* 25: 464–473.

Official Gazette. 2005. *The law about licensed surveyors and surveying offices*. Law No. 5368, The Turkish Official Gazette, Date: 29.06.2005, No: 25860.

Rajabifard, A., I. Williamson, D. Steudler, A. Binns, and M. King. 2007. Assessing the worldwide comparison of cadastral systems. *Land Use Policy* 24: 275–288.

Robertson, W.A. 2002. Anticipating the further development of cadastral systems. *Computers, Environment and Urban Systems* 26: 383–402.

Steudler, D., A. Rajabifard, and I.P. Williamson. 2004. Evaluation of land administration systems. *Land Use Policy* 21(4): 371–380.

Thellufsen, C., A. Rajabifard, S. Enemark, and I. Williamson. 2009. Awareness as a foundation for developing effective spatial data infrastructures. *Land Use Policy* 26: 254–261.

Ting, L., and I.P. Williamson. 1999. Cadastral trends: A synthesis. *The Australian Surveyor* 4(1): 46–54.

TKGM. 2015. Official website of the General Directorate of Land Registry and Cadastre. http://www.tkgm.gov.tr, 23.03.2015.

UNECE. 1996. Land administration guidelines. http://www.unece.org, 23.03.2015.

UN–FIG. 1999. The Bathurst declaration on land administration for sustainable development. http://www.fig.net, 23.03.2015.

Van Oosterom, P., H. Ploeger, J. Stoter, R. Thompson, and C. Lemmen. 2006a. *Aspects of a 4D cadastre: A first exploration*. Proceedings of the XXIII FIG Congress, 8–13 October 2006, Munich, Germany.

Van Oosterom, P., C. Lemmen, T. Ingvarsson, P. van der Molen, H. Ploeger, W. Quak, J. Stoter, and J. Zevenbergen. 2006b. The core cadastral domain model. *Computers, Environment and Urban Systems* 30: 627–660.

Wallace, J., and I. Williamson. 2006. Developing cadastres to service complex property markets. *Computers, Environment and Urban Systems* 30: 614–626.

Williamson, I.P. 2001a. Re-engineering land administration systems for sustainable development – From rhetoric to reality. *International Journal of Applied Earth Observation and Geoinformation* 3(3): 278–288.

———. 2001b. Land administration ‘best practice’ providing the infrastructure for land policy implementation. *Land Use Policy* 18: 297–307.

Chapter 4
A Performance Assessment Model for Cadastral Survey System Evaluation

Haodong Zhang and Conrad Tang

Introduction

The most essential function of a cadastral survey system is to provide reliable information and descriptions of land parcels. Land parcels are the basic spatial unit in the operation of a land administration system. A cadastral survey system targets societal requirements by implementing cadastral survey and mapping activities and then supplies spatial related information to users. Together with land registration, these core cadastral components (FIG 1995) become an important infrastructure to facilitate the implementation of land use policies (UN-FIG 1999).

As indicated by Dale (1979), the cadastral survey and mapping activities are regulated by its own legal and institutional arrangements. Nowadays, as land becomes scarce community resource (Ting and Williamson 1999), social and environmental interests are more seriously considered by land stakeholders for sustainable development (Bennett et al. 2008). Ruled by each jurisdiction's own background, no unique model can effectively fulfil its societal requirements world-wide. This notion cannot be regarded as a brand-new concept in the field of cadastre (see Dale 1976 and Williamson 1985). Still, unsatisfied land administration projects are continuously being reported during international conferences or in the publications. Merely complying with top-end technological solutions and rigid regulations for accuracy are summarized as the reason for most unsuccessful land administration projects (FIG 2014). In addition, Enemark (2013) and Enemark et al. (2014) highlight the influence of unfit-for-purpose solution in building spatial framework for a land administration system.

A cadastral survey system produces spatial related datasets in building and maintaining the spatial framework of a cadastral system or land administration system.

H. Zhang (✉) • C. Tang
Department of Land Surveying and Geo-Informatics, The Hong Kong Polytechnic University, Hung Hom, Kowloon, Hong Kong
e-mail: hd.zhang@connect.polyu.com

T. Yomralioglu, J. McLaughlin (eds.), *Cadastre: Geo-Information Innovations in Land Administration*, DOI 10.1007/978-3-319-51216-7_4

The appropriateness of a cadastral surveying system-design directly influences the performance of land registration in each jurisdiction. However, as an indispensable land administrative function, the end results of a cadastral survey system have rarely been evaluated. Most assessment projects in the field of cadastre and land administration are either focusing on a broad aspect of land matters (see Steudler et al. 1997; Williamson 2001; Mitchell et al. 2008) or using a specific cadastral activity to evaluate the cadastral survey system (see Chimhamhiwa et al. 2011). Furthermore, there are very few projects with special focus on cadastral survey system performance in developed land markets nowadays.

This on-going research project aims to build a self-assessment framework for any cadastral survey systems in both developed and developing land markets. The general successfulness or fitness of a cadastral survey system will be examined through a structured multi-criteria assessment model. With a specific focus on the technical, economic, legal and institutional arrangements, we propose four criteria termed: *Capability*, *Cost*, *Security* and *Service*. Under each criterion, performance indicators are selected and evaluated by assessors for gap analysis. Land stakeholders, especially cadastral surveyors who are the key operators of the system are expected to give their judgements on what the "purpose" (optimum societal requirements) is and how well the system "fits for" it.

This paper is structured as follows. First, an explanation on the structure of the established assessment model is introduced. Second, the assessment methodology and strategy are discussed with highlights of utilization of the adopted multi-criteria decision analysis methodology: Analytic Hierarchy Process (AHP). The initial results of a pilot study on the performance assessment of Hong Kong cadastral survey system will be followed. Finally, the paper discusses the initial findings of the case study of Hong Kong and suggests the directions for further development of the proposed assessment model.

Assessment Framework

A Performance Assessment Model

In general, each cadastral survey system has its unique characteristics. To thoroughly assess a cadastral survey system, the assessor needs to have extensive resources of the system design and deep understanding of its jurisdictional background. In addition, system users' satisfaction level should be considered, which is also resource demanding. Neely et al. (2005) commented that the performance of a system is more practicable to be assessed and it is easier to quantify the efficiency and effectiveness of the system. Indeed, the strategy of conducting performance assessment is widely applied in land administration systems evaluation projects.

In this project, we built an assessment model to measure the efficiency of each individual cadastral survey system by its performance. Furthermore, the established

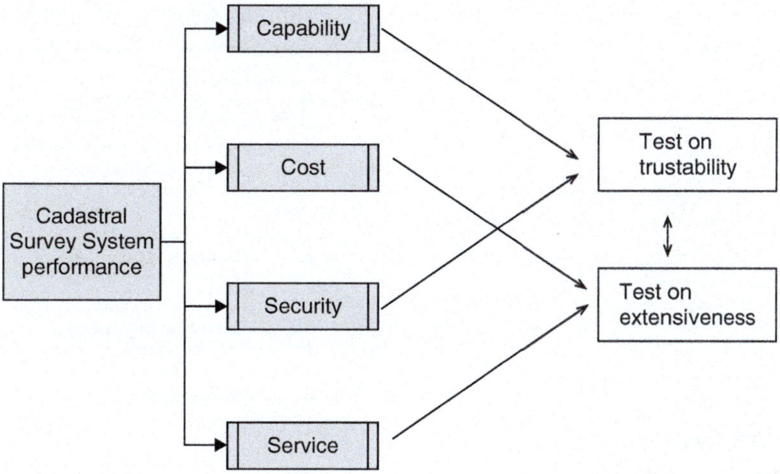

Fig. 4.1 Logic connections of the criteria set

assessment model will be applied to evaluate how well an individual cadastral survey system fulfils its society's requirements. Therefore, the performance gaps between the optimum societal requirements and the achieved performance can be identified and evaluated.

From an overall aspect, Williamson (2000) concludes a desired cadastral or land administration system performance should be decided by two key performance indicators: (1) whether the system was trusted by the general populace; and (2) whether it was extensively used by land stakeholders. Following these two key principles, for the assessment of cadastral survey system, we defined four key system performance aspects as the assessment criteria termed: *Capability*, *Cost*, *Security* and *Service*. The logic connections are represented in Fig. 4.1.

A Multi-criteria Assessment Model

To the best of our knowledge, very few assessment projects in the field of cadastre or land administration adopted single criterion to perform whole system evaluation process. In general, the term "multi-criteria" represents the utilization of Multiple Criteria Decision Analysis (MCDA) methodology. In this project, we selected Analytic Hierarchy Process (AHP) as the applied MCDA methodology for structured evaluation of the system performance. The detailed introduction of AHP will be addressed later. In this section, the selected assessment content is focused.

The structure of the adopted criteria and their sub-criteria are illustrated in Fig. 4.2. The overall performance of a cadastral survey system is divided by four criteria: *Capability*, *Cost*, *Security* and *Service*. These criteria set covers the technical,

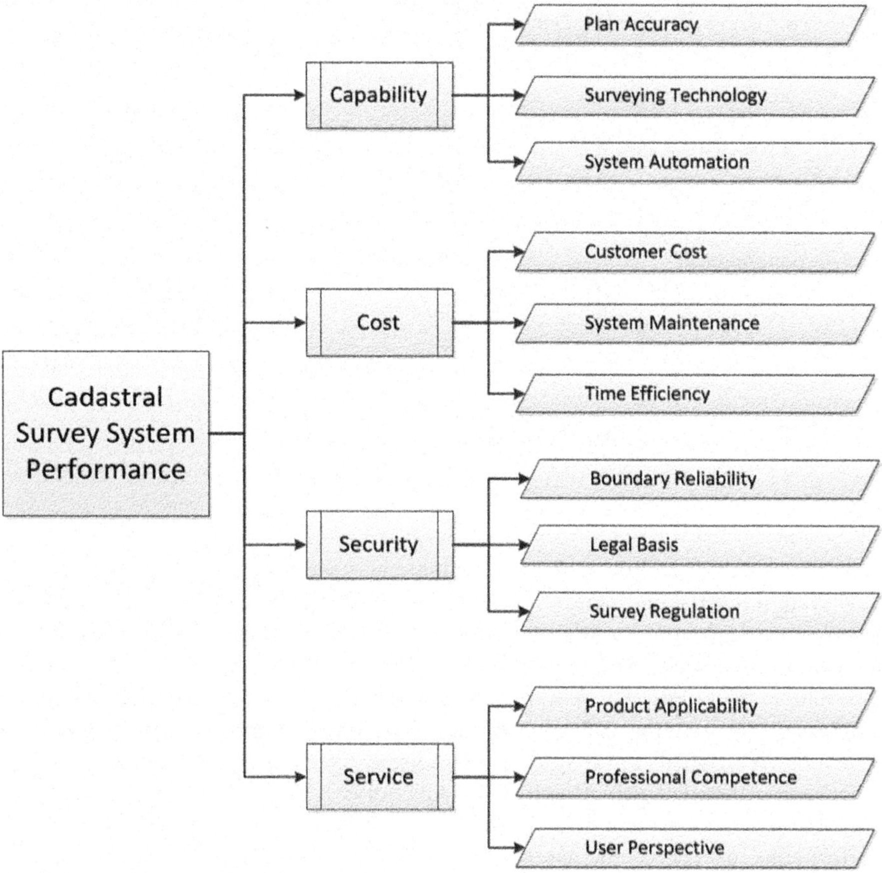

Fig. 4.2 The structure of adopted criteria set

economic, legal and institutional aspects of a system. Under each criterion, three sub-criteria are selected to measure the performance specifically.

Capability

Capability aims to evaluate the system performance with highlights on its technical dimension. In short, it asks the question "what can be done" by the current system. The sub-criteria set includes: (1) Plan Accuracy; (2) Surveying Technology; and (3) System Automation.

Plan Accuracy targets the horizontal accuracy of the currently produced land boundary plan. *Surveying Technology* exams the current adapted level of surveying technology to producing cadastral survey datasets. *System Automation* measures the level of system automation process with a focus on the database and data model approach.

Cost

Cost measures the performance of the system from the economic aspect. In short, it asks the question: "what is the cost" in providing cadastral survey and mapping activities to the society. Three sub-criteria are applied: (1) Customer Cost; (2) System Maintenance; and (3) Time Efficiency.

Customer Cost indicates the individual cost of using cadastral survey service. *System Maintenance* focuses on the system cost of maintaining the cadastral survey services. *Time Efficiency* considers the cost in the time dimension by measuring the time spent on using cadastral survey services.

Security

Security evaluates system performance from the legal aspect. It asks the question: how reliable is the service? Three sub-criteria are selected: (1) Boundary Reliability; (2) Legal Basis; and (3) Survey Regulation.

Boundary Reliability measures the potential boundary disputes of surveyed parcels and the efficiency of the surveyed boundary (e.g. would it be overridden easily by newly discovered evidence or other rights, such as adverse possession?). *Legal Basis* intends to exam the performance of the updated legislation for the operation of cadastral survey system and the authorization of legal boundary for surveying. *Survey Regulation* assesses the technical and administrative guidance for the cadastral survey industry.

Service

Service measures the development of a cadastral survey system as a service provider. It contains three sub-criteria: (1) Product Applicability; (2) Professional Competence; and (3) User Perspective.

Product Applicability measures the level of adopting cadastral survey products by land related professions and the involvement of current cadastral survey products for further system development (e.g. Spatial Data Infrastructures and Building Information Modelling). *Professional Competence* assesses the efficiency of professional service to fulfill client's requirements and the appropriateness of the licensing or practicing system. *User Perspective* checks the quality of the data and the overall satisfaction of general public.

Assessment Methodology

General Procedures

The established criteria set is closely connected to the fundamental functions of a cadastral survey system. It defines the assessment content in a flexible way. The assessor are required to give their judgements on what constitutes (which criterion contributes more to) a desired performance for his (including assessors of both genders) specific system. In this research project, the AHP pairwise comparisons are applied to determine the weight of each criterion. Figure 4.3 gives an example of AHP derived criteria weight distribution.

Theoretically, the criteria weight distribution reflects the constitution of a desired system performance. Further, it aims to give hints on what the "purpose" is for the cadastral survey system based on a summarization of different AHP evaluation results from relevant stakeholders.

The next fundamental question in the assessment model is how well the current system fits for the "purpose". Benchmarking the current optimum societal required performance (should-be performance), stakeholders as the assessors are required to rate on the currently achieved performance. Gap analysis will then be conducted to evaluate the relative strengths and weaknesses of the current system (Fig. 4.4). Furthermore, the overall weighted scores of current system rated by each individual assessor can be calculated based on his criteria weight distribution results. Thus, the normalized satisfaction level of the current system from different stakeholders can be assessed and compared.

It should be noted that the previously mentioned two types of judgements are both subjective individual opinions. It fits for the purpose of this assessment which is to provide a platform to show different understandings on the cadastral survey system performance. Sufficient feedbacks are needed to reflect the performance level of a cadastral survey system. There are two strategies adopted in the assessment to increase its reliability: one is to categorize different stakeholder types based

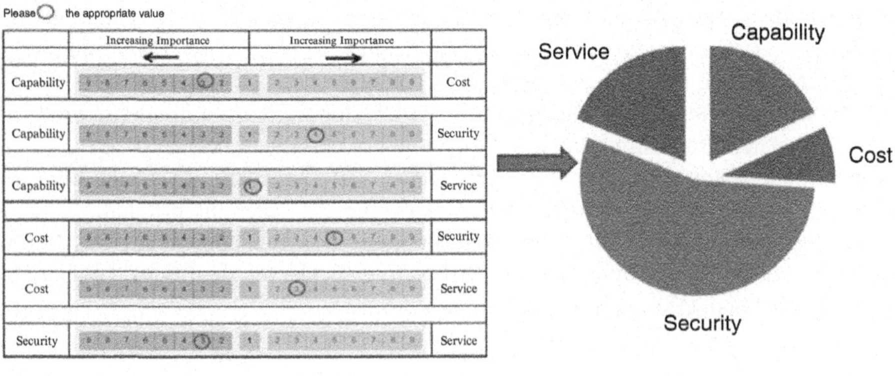

Fig. 4.3 A sample of AHP pairwise comparison methodology

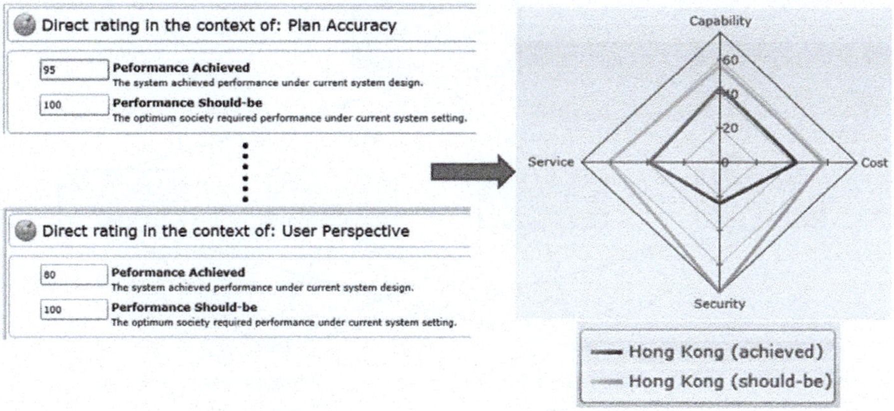

Fig. 4.4 Gap analysis of current system

on the assessors' professional backgrounds; the other is to invite informative answers from assessors on the specific performance datasets of the system. Correlation may be conducted between the given performance datasets and the previously defined performance satisfaction level. Therefore, with sufficient feedbacks, a robust multi-stakeholder assessment model containing actual performance of each cadastral survey system can be established.

Analytic Hierarchy Process

AHP served as the weight determination methodology in this assessment framework. The AHP method is a decision method for organizing and analyzing complex solutions. It was first introduced by Saaty in 1970s and widely used by researchers in different fields to transform qualitative and quantitative issues to the judgments about the data (Vaidya and Kumar 2006). As indicated by Macharis et al. (2004), the fundamental principles concerned with AHP are: hierarchy construction, priority setting and logical consistency.

In general, an AHP solution contains three layers: goal, criteria and alternatives (Fig. 4.5). In this project, the "goal" of the assessment project is an ideal/sound cadastral survey system performance that fits for the current optimum societal requirements. The detailed criteria set is introduced in Fig. 4.2. *Capability*, *Cost*, *Security* and *Service* are the four selected criteria. Under each criterion, three representative and measurable sub-criteria are defined. In this project, we aimed to build a self-assessment model to evaluate how well a system fit for the optimum societal requirements. Thus, only two alternatives for each system, termed as *Achieved Performance* and *Should-be Performance*, will be adopted.

In this model, priority settings of each criterion are derived from AHP pairwise comparisons. A matrix is used to calculate the priority values of those criteria with

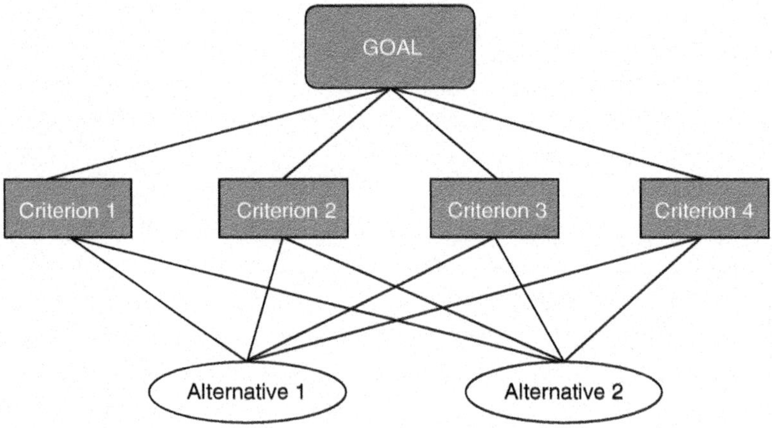

Fig. 4.5 Basic AHP structure

reference to the comparison attributes. Pairwise comparisons are provided to assessors to decide the relative importance of each pair of criteria in contributing to the goal. Here, the fundamental algorithm with the most common Saaty's 9-point pairwise comparison scale is applied. Table 4.1 lists the definition and explanations of each scale value.

The common steps to derive criteria weight can be generally illustrated by Fig. 4.6. In step 1, assessor is needed to pairwise compare the criteria set in the same layer. Step 2 shows the established comparison matrix based on assessor's judgements. In step 3, the weight of each criterion that is calculated by the AHP algorithm is listed. Also, inconsistency ratio of the assessor's judgment is presented. A thorough explanation of AHP algorithms will not be discussed in this paper, but can be found at Saaty (1980).

The inconsistency ratio reflects the logical stability of assessors' judgements, and further this value can be applied to weigh the influence of each assessor's judgements in summarizing stakeholders overall opinions.

Assessment Strategy

The established model intends to establish a platform to represent the development of cadastral survey systems world-wide. It does not aim to evaluate which system is better than the other. Under each jurisdiction, land stakeholders can express their ideas on the relative importance of different performance aspects of current system and their satisfaction level on each performance aspect. Correlating with achieved performance datasets, different stakeholders' opinions will show us a comprehensive performance level of the cadastral survey system in fulfilling its societal requirements.

Table 4.1 Saaty's 9-point pairwise comparison scale (Saaty 1980)

Intensity of importance	Definition	Explanation
1	Equal importance	Two elements (criteria/alternatives) contribute equally to the goal
3	Moderate importance	Judgment is slightly in favour of one element over another
5	Strong importance	Judgment is strongly in favour of one element over another
7	Very strong importance	One element is favoured very strongly over another
9	Extreme importance	There is evidence affirming that one element is extremely over another
2, 4, 6, 8	Intermediate values between above scale values	Should be the intermediate value adopted by the assessor

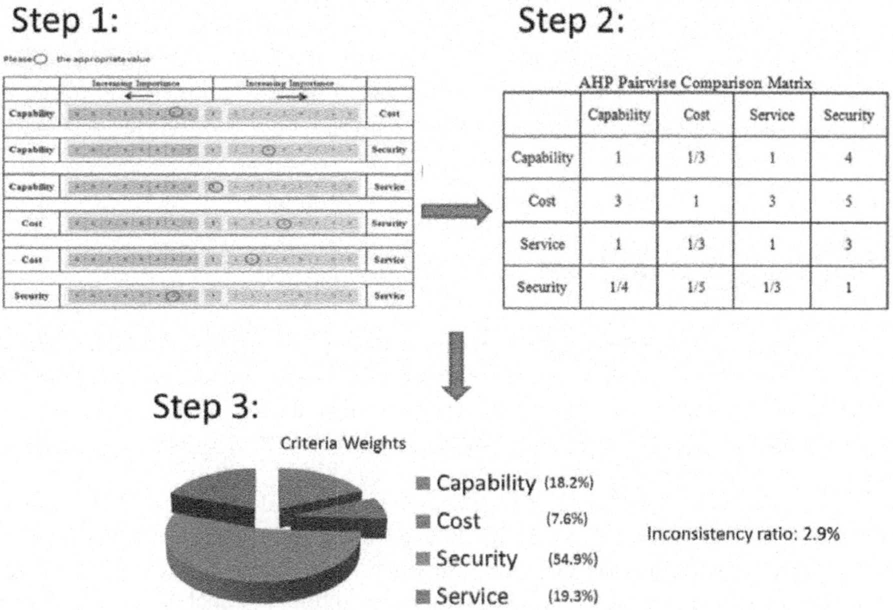

Fig. 4.6 Steps in AHP weight determination

The core task of implementing this model is to collect judgements and performance datasets from stakeholders. On one hand, international cooperation is sought. On the other hand, a pilot study of local cadastral survey system performance is currently conducted in Hong Kong under the coordination of the Land Surveying Division (LSD) of the Hong Kong Institute of Surveyors (HKIS). The strategy of implementing the established model in Hong Kong cadastral survey system assessment can be divided into three stages.

At stage 1, a consultancy panel is established. We supposed land surveyors are the type of stakeholders who know the system most. At this stage, as the key players of the system, thirteen land surveyors or surveying backgrounds members (four from public sector, three from private sector, three young surveyors and other three from academia) formed this consultancy panel under the coordination of HKIS. Through interview and questionnaire, opinions and comments are collected to calibrate the established assessment criteria and structured model.

At stage 2, an online questionnaire will be sent to all HKIS LSD members to collect their judgements on the performance level of our local cadastral survey system. In general, their professional backgrounds will be categorized into four types: public sector, private sector, academic and young surveyor.

At stage 3, this assessment model will be introduced to other relevant stakeholders through interviews or online questionnaire. Thus, comprehensive opinions can be collected to evaluate the actual performance of local cadastral survey system in fulfilling the requirements of the society.

With sufficient feedbacks, a robust platform can be established. To facilitate the process of opinions collection, a concise questionnaire was designed and utilized to collect assessors' judgements. Figure 4.7 shows the flowchart of the questionnaire. In general, an assessor can finish the questionnaire within 10 minutes by filling in all required questions. Information questions on the achieved performance datasets are optional.

Pilot Study in Hong Kong

The implementation of this assessment model is currently being conducted in Hong Kong cadastral survey industry. The formed consultancy panel has already been interviewed by our research team. Their opinions and judgements on the system performance are collected. Using the weight distribution of different assessment criteria as an example, summarized charts are listed in Fig. 4.8.

Characteristic weight distribution scheme can be found from these four summarized weight distribution results. Both of the group "Public Sector" and "Academic" have more concerns on the criterion "Security", and a reliable system is mostly expected by them. Comparatively, "Private Sector" prefers "Capability" and "Young Surveyor" considers the "Service" most. At this stage, we cannot conclude that the presented four charts can reflect the local cadastral survey industry opinions. But this initial results do provide us some clues on the requirements from different professions or stakeholders for the system.

Currently, an online platform has been established for relevant stakeholders. A concise online questionnaire will be sent to all HKIS LSD members soon. A set of more representative and comprehensive judgements on the system performance is expected.

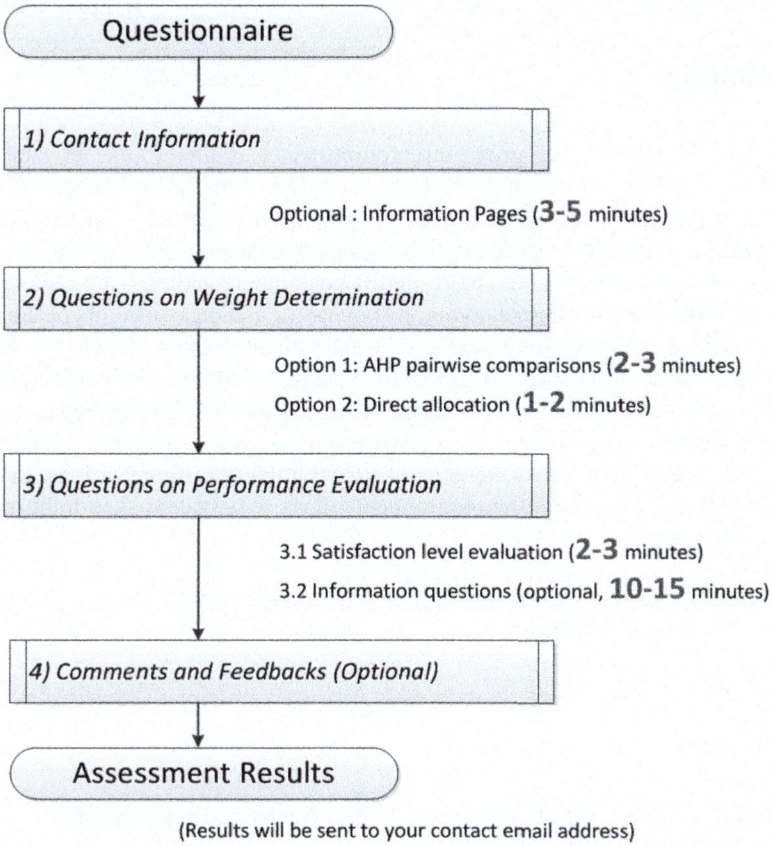

Fig. 4.7 Flowchart of the questionnaire

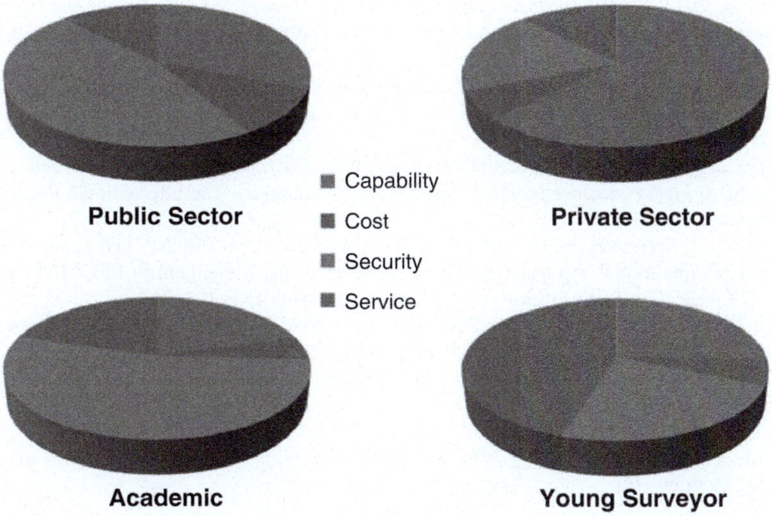

Fig. 4.8 Initial weight determination results in Hong Kong

Conclusions

This paper describes an on-going research project that aims to use an established model to answer questions on the cadastral survey system performance in fulfilling societal requirements. The established structural model settles the question of what to measure and how to measure through a set of criteria and performance indicators. Those model parameters intend to bring different understandings of a cadastral survey system performance into a common framework and measuring its performance by normalized yardsticks. Certainly this assessment framework cannot be well established without the involvement of relevant stakeholders. With sufficient feedbacks, a robust assessment results can be achieved and handily applied to measure the effectiveness and efficiency of a cadastral survey system. This research provides a scientific means to express the general successfulness or fitness of any cadastral survey systems in fulfilling the requirements of its society, and shed light on areas for improvement.

Acknowledgement This paper is supported by PolyU B-Q32N funding (RGC Ref No. 525712) and HKIS 2015 research funding.

References

Bennett, R., J. Wallace, and I.P. Williamson. 2008. A tool box for mapping and managing new interests over land. *Survey Review* 40(307): 43–53.

Chimhamhiwa, D., O. Mutanga, and P. van der Molen. 2011. Measuring quality performance of cadastral survey and deeds registration work processes. *Land Use Policy* 28(1): 38–46.

Dale, P.F. 1976. *Cadastral surveys within the Commonwealth*. London: HMSO.

———. 1979. A system view of the cadastre. *Survey Review* 25(191): 28–32.

Enemark, S. 2013. *Fit-for-purpose: Building spatial framework for sustainable and transparent land governance*. World Bank conference on land and poverty, 8–11 April, Washington, DC, USA.

Enemark, S., C. Lemmen, and R. McLaren. 2014. *Building fit-for-purpose land administration systems*. Proceedings of the XXV FIG international congress, 16–21 June, Kuala Lumpur Malaysia.

FIG. 1995. *Statement on the cadastre*. Copenhagen: The International Federation of Surveyors.

———. 2014. *Fit-for-purpose land administration*. Copenhagen: The International Federation of Surveyors.

Macharis, C., J. Springael, K. De Brucker, and A. Verbeke. 2004. PROMETHEE and AHP: The design of operational synergies in multicriteria analysis. Stengthening PROMETHEE with ideas of AHP. *European Journal of Operational Research* 153: 307–317.

Mitchell, D., M. Clarke, and J. Baxter. 2008. Evaluating land administration projects in developing countries. *Land Use Policy* 25(4): 464–473.

Neely, A., M. Gregory, and K. Platts. 2005. Performance measurement system design: A literature review and research agenda. *International Journal of Operations and Production Management* 25: 1228–1263.

Saaty, T.L. 1980. *The analytic hierarchy process: Planning, priority setting, resource allocation*. Texas: Mcgraw-Hill.

Steudler, D., I.P. Williamson, J. Kaufmann, and D.M. Grant. 1997. Benchmarking cadastral systems. *The Australian Surveyor* 42(3): 87–106.

Ting, L., and I.P. Williamson. 1999. Cadastral trends: A synthesis. *The Australian Surveyor* 4(1): 46–54.

UN-FIG. 1999. *The Bathurst declaration on land administration for sustainable development*. International conference on land tenure and cadastral infrastructures for sustainable development, 18–22 October, Bathurst Australia.

Vaidya, O.S., and S. Kumar. 2006. Analytic hierarchy process: An overview of applications. *European Journal of Operational Research* 169: 1–29.

Williamson, I.P. 1985. Cadastres and land information systems in common law jurisdictions. *Survey Review* 28(217): 115–129.

———. 2000. *Best practices for land administration systems in developing countries*. International conference on land policy reform, 25–27 July, Jakarta Indonesia.

———. 2001. Land administration "best practice" providing the infrastructure for land policy implementation. *Land Use Policy* 18(4): 297–307.

Chapter 5
Management of Natural Risks and Disasters in a River Basin Within the New Cadastre Concept

Rodolfo Salazar, David Miranda, and Urbano Fra

Introduction

Land administration plans and decision-making processes currently do not consider the analysis of multi-hazard and risk in land use planning or environmental protection integrated into a single system; consequently when an event occurs, government and citizens suffer human and economic losses which have not been foreseen or quantified.

ISO 19152, Land Administration Domain Model, LADM considers the elements of basic geoinformation (including water, land, and the elements above and under the earth´s surface) in a concept that contains: *parties* (persons and organizations); *rights*, responsibilities and restrictions; *legal objects* (parcels, constructions and network services); *spatial objects* (topography) and *spatial representations* (geometry and topology). It provides a terminology on land administration and allows combining information from different sources and themes like in the present case referred to multi-hazard and risk.

Some models were developed (Aubrecht et al. 2012 for a multi-level geoinformation inventory of people and their interests in land and risk, starting from global to local scales and urban development. These models allow including cadastre

R. Salazar (✉)
Departamento de Ciencias de la Tierra y de la Construcción, Universidad de las Fuerzas Armadas ESPE, Av. Gral. Rumiñahui s/n 171-5-231B, Sangolquí, Ecuador
e-mail: rjsalazar@espe.edu.ec

D. Miranda
Laboratorio del Territorio, Universidad de Santiago de Compostela Campus Universitario, 27002 Lugo, Spain

U. Fra
Escuela de Ciencias Sociales y Humanidades, Universidad de Extremadura, Campus Universitario, 10071 Cáceres, Spain

T. Yomralioglu, J. McLaughlin (eds.), *Cadastre: Geo-Information Innovations in Land Administration*, DOI 10.1007/978-3-319-51216-7_5

information in a land use planning system integrated with risk exposed zones at local level.

Ecuadorean Constitution of 2008 states that planning is a fundamental element of territorial development and it will be obligatory at all levels of decentralized autonomous governments, GAD. Jurisdiction of territorial planning and land administration must be coordinated within national, provincial, cantonal and parish levels. Art. 244 states that two or more provinces with territorial continuity can create autonomous regions; interregional balance, historical and cultural affinity, ecological complementariness and integrated basin management will be sought. At present, Ecuadorean administration is divided in nine Planning Zones, 24 Provinces and 221 municipalities (cantons).

Using CRED database information (Guha-Sapir et al. 2014), natural disasters that caused most damage in Ecuador (human and economic losses) are flooding (42%), earthquakes (41%), landslides (13%) and volcanic activity (4%). This study will focus only in flooding, and volcanic activity affecting land cover, roads and educational and health facilities. Afterwards, it will link this multi-hazard risk analysis of Esmeraldas river basin with cadastral parcels involved, according to ISO 19152 LADM. But, because of the lack of municipal cadastral parcel information, the block (manzana) level provided by the National Institute of Statistics and Census (INEC 2010) will be used in this case.

The standardization of municipal cadastral information systems allows having uniform and homogeneous land administration data regardless of the existing administrative boundary between two neighbouring jurisdictions. This unified model makes decision-making uniform, timely, consistent and efficient in all administrations managing the watershed.

Methodology

A review of the ISO 19152 LADM standard and the ability to couple it with the risk variable was performed by analyzing the legal basis that allows its modelling and implementation. In this context, the methodological process (Fig. 5.1) starts by analysing, evaluating and downloading geoinformation presented through the Geoportal of the National Information System, NIS that manages the National Ecuadorean Spatial Data Infrastructure (CONAGE 2010) to define the hazards to be analysed and the exposed areas. Then, the cadastral elements exposed (in this case the blocks because the lack of parcel data) to the volcanic activity and flooding hazard must be analysed in depth at local level and finally a conceptual model to integrate ISO 19152 LADM with the variable multi-hazard is proposed.

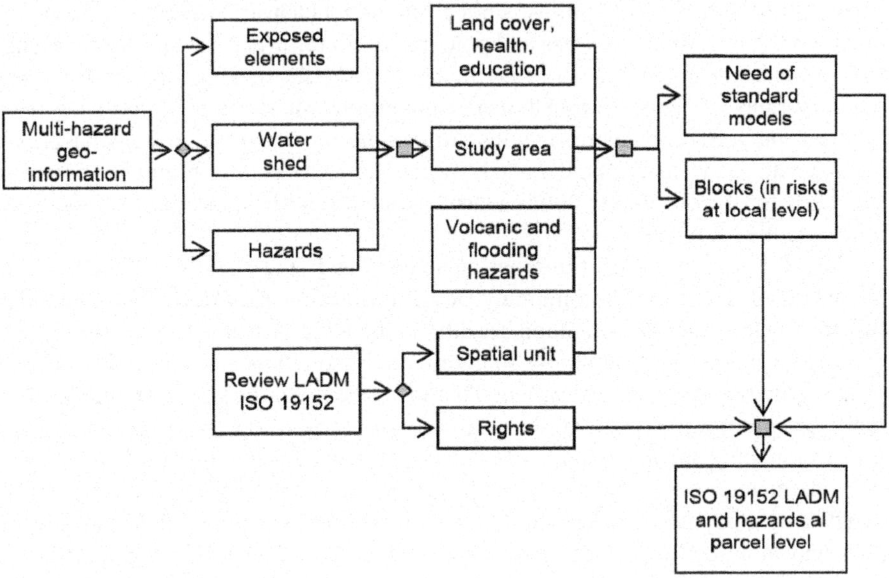

Fig. 5.1 Methodology to integrate multi-hazard analysis and ISO 19152 LADM

Table 5.1 Exposed elements in the Esmeraldas River Basin

Exposed elements	Volcanic hazard	Flooding hazard
Natural areas	33%	18%
Productive with human intervention areas	63%	79%
Urban areas	1%	1%
Other areas	3%	2%
Roads	26 sectors	28 sectors
Health centres	78	84
Educational centres	634	711

Analysis and Discussion

The study area is the Esmeraldas Basin Watershed located north-west of the Ecuadorian mainland, where exposed areas are grouped in three categories: (1) natural areas: páramo, wilderness, natural forest, shrub vegetation, natural grass; (2) productive with human intervention areas: pastures, harvested forest, oil palm, short-cycle crops, tropical tree crops, and (3) urban areas. Roads as linear elements; health and school units as point elements are also studied. The hazards analysed were only the volcanic and flooding, together with the exposed elements (Table 5.1). The official information was downloaded from the NSI.

When considering the Esmeraldas river basin as the unit of study and risk management, the joint intervention of municipal jurisdictions is essential. The need of standardization and compatibility among municipal cadastral systems in conurbations (continuous of urban areas and potential areas for urban development belonging to different municipalities) is the first problem. This is the case of the municipalities of Quito-Rumiñahui and increasing Mejia-Rumiñahui. They have different models of cadastre, making impossible an integrated risk management against natural events. Thus, the need to generate a model with minimum common standards for the development of cadastral systems and land use plans of the three municipalities is justified.

The Ecuadorean Constitutional mandate and the Código Orgánico de Organización Territorial Autonomía y Descentralización, COOTAD 2009, establish that the "mancomunidad/common wealth" is the legal instrument that allows integrated management of contiguous municipalities and in this case those that are part of the Esmeraldas River Watershed. This instrument can be used for standardization, geoinformation generation, analysis, research, modelling, training, joint decision-making and the creation of cadastral systems linked with multi-hazard data in a risk management system. The objective is to design a common model implementing the ISO 19152 LADM standard to unify the existent municipal cadastres to plan and monitor the use of land and resources at parcel level, considering the responsibilities and restrictions to the land under the existing risk, in addition to the rights and corresponding valuation.

A model for the relationship between the Infrastructure for Spatial Information in the European Community, INSPIRE Annex and ISO 19152 LADM has been proposed by Oosterom and compares the elements needed for land use planning. In Ecuador, a methodological guide has been prepared (Senplades 2011) to standardize plans of territorial ordinance which contains some similar elements than those of INSPIRE, but grouped in systems (environmental, economic, social, energy, mobility, connectivity and human). The concepts proposed in 4D research project (ITC 2013), and the Changes Project (European Community 2012) groups the same elements but in terms of climate change, land and socio economy, risk and vulnerability, and development scenarios. Integrating all this information in just one database will allow: (1) to select, validate, structure and standardize multi hazard information in a web geodatabase; (2) to associate land parcels to hazard, risk and main influence areas at local level; and (3) to model and simulate events for different hazards and risk types using ISO standards for geoinformation in order to integrate them in a Cadastral Data Infrastructure in terms of multi-hazard, multi-risk and multi-vulnerability.

The survey conducted by the Ministerio de Desarrollo Urbano y Vivienda, MIDUVI in (2012), highlights that only the 22% of 221 Ecuadorean municipalities have their cadastres based in one kind of GIS but still in two different systems (one for rural and one for urban areas). In an isolated system the Property Registry deals with tenure; however a new mandate states that it must be under the municipalities. ISO 19152 LADM proposes a system based in Cadastre 2014 concept integrating in one system three basic entities: subjects, land parcels and rights. The model to associate a land parcel as element of an integrated cadastre of urban and rural areas includes the interests as rights, restrictions and responsibilities. Then, risk within a

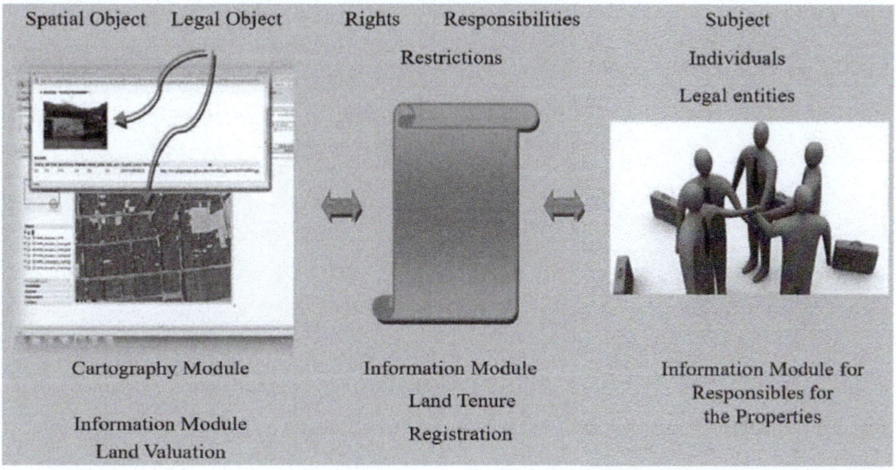

Fig. 5.2 Relationship object-rights-subject under Cadastre 2014 and ISO 19152 LADM concept

cadastral system provides rights, restrictions and responsibilities on the land parcel together with Property Register Agency (Fig. 5.2).

The *"legal object"* or parcel in traditional cadastral cartography is the geometric representation of the parcel and its buildings, boundaries and area, answering the question Where is it?. The entity is represented by spatial objects of the land parcel (spatial units) and its habitat by means of spatial objects and other representations like digital terrain model, hydrographic network, transportation network, geographic names, national and international boundaries, postal codes, services network, land cover, land use, buildings, natural (seismic, tectonic, volcanic, flooding, landslides) and anthropogenic hazards. This cadastral information is used to know the value, answering the question: How much does it cost?

The *"spatial object"* and the legal object with their components are considered part of the habitat and housing definitions. The *"rights"* or the legal parcel information contain rights, responsibilities, restrictions and tenure. Answers the question: What is the tenure?. Providing legal information of the relationship subject – parcel by rights (environmental and multi-hazard restrictions and responsibilities, among others). The *"subject"* is the entity that has the relationship with the legal object; this element has the attributes to describe it as name, ID number, address, etc., answering the question: Who is the subject related to the parcel?

For the concept based in object-rights-subject, work is necessary to integrate cadastral systems of urban and rural areas as well as records of the property that at present are separated, see Fig. 5.3a, in one unique municipal system. The ISO 19152 LADM standard in its modularity considers three elements – people, rights and properties – into a single model as shown in Fig. 5.3b and integrates through the parcel, all existing space objects in a jurisdiction including water, land and the elements above and below the earth's surface.

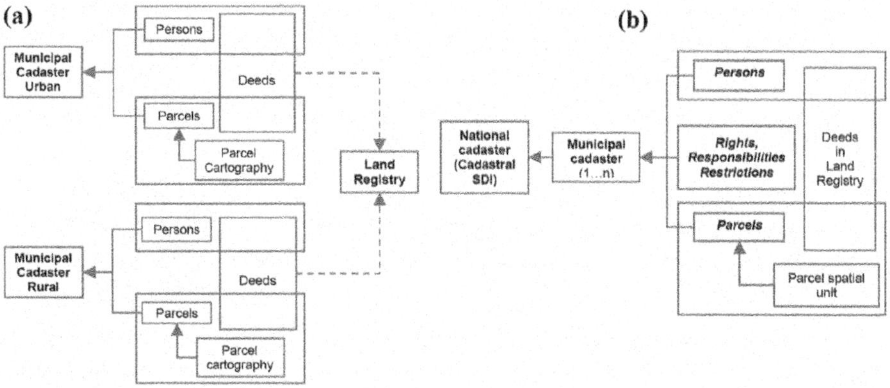

Fig. 5.3 Model of current cadastral systems (**a**), and model based on ISO 19152 LADM (**b**)

The subject module or Persons are natural and organizations to whom is assigned a portion of the earth's surface; module Rights, Responsibilities and Restrictions refers to all legal information in terms of a relationship with the person, not only from the point of view of ownership but with the potential use, land use, the multi-hazard, environment; module of legal objects (land, buildings and service networks) is made with the corresponding spatial objects (topography), spatial representations (property space) and the basic administrative units.

The information generated in this system is standardized and maintained in the respective databases, becoming the centre of the Municipal Cadastral System. In the case of a national system, this information easily through a distributed system allows the access to certain required municipal information distributed transparently for government decision-making and use of the citizens. The modernization of services and processes to improve the efficiency and effectiveness of the land administration model, requires minimal computational mechanisms.

The model to associate the premises as cadastral elements with risk elements is presented in Fig. 5.4 with a unified register for rural and urban areas including information of Land Registry. The relationship of risk is articulated in terms of tenure, restrictions and responsibilities. Any form of tenure may have beyond their own rights, responsibilities and restrictions under the optimum land use, risk or environmental protection. The cadastral model, under this concept, allows proper risk management and integrated planning as well as land use in a standardized Land Administration System in all municipalities that are part of the watershed.

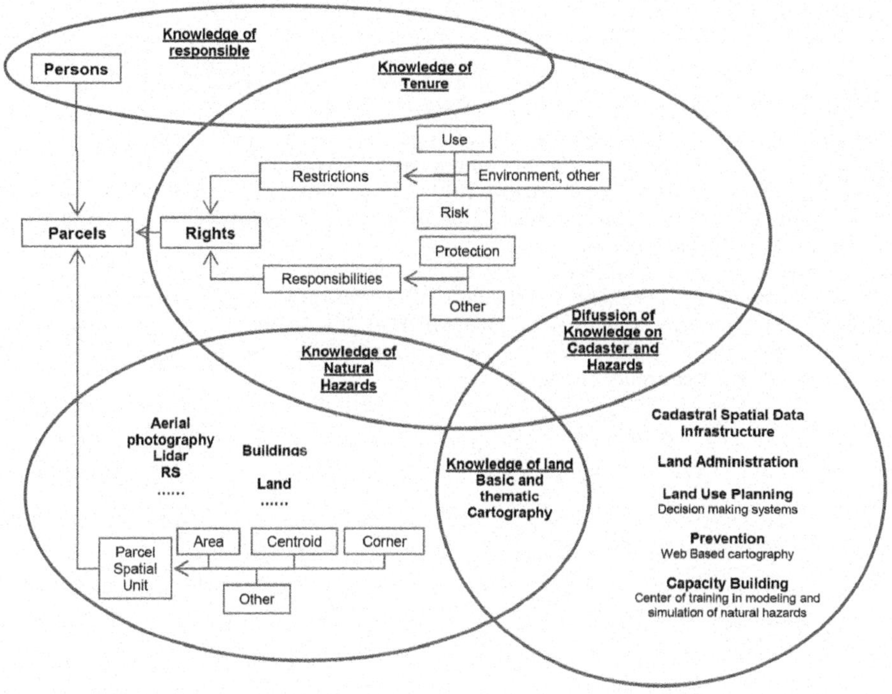

Fig. 5.4 Relationship of a risk management system integrated with a Land Administration System in ISO 19152, LADM

Conclusions and Further Remarks

The multi-risk analysis has no administrative boundaries. Its integrated management can generate a new proposal for administrative-political organization based on watersheds or encourage the use of legal mechanisms for joint management as in the case of the commonwealth. For this it is necessary to standardize a municipal cadastre model to incorporate the relevant variables to natural hazards in the basin.

Standardization, validation and definition of concepts related to integrated risk management in a system of territorial administration under ISO 19152 LADM are needed in:

- The type of tenure, land use, rights, responsibilities and restrictions in a parcel.
- The data model of integrated rural and urban cadastral system.
- The model of a system for spatial analysis and simulation that allows the training of those involved in risk management.
- Determining the economic valuation of potential losses of natural areas, disturbed areas of production and urban areas in the event of a disaster, with information on the proposed cadastral system.

References

Aubrecht, C. et al. 2012. Multi-level geospatial modeling of human exposure patterns and vulnerability. *Natural Hazards*, 147–163.

CONAGE. 2010. *Datos Geográficos Marco – Clasificación*. Quito: SENPLADES.

European Community. 2012. http://www.changes-itn.eu/. Accessed 17 Sept 2014.

Guha-Sapir, D., P.H. Hoyois and R. Below, 2014. *Annual Disaster Statistical Review 2013: The numbers and trends*. Brussels: Centre for Research on the Epidemiology of Disasters, University of Catholic Louvain.

INEC. 2010. *Censo de Población y Vivienda.*

ITC. 2013. http://www.itc.nl/4D-EARTH. Accessed 17 Sept 2014.

MIDUVI. 2012. *Decreto Ejecutivo 688*. Quito: MIDUVI.

SENPLADES. 2011. Registro Ofical, Ediccion Especial No.184, Secretaria Nacional De Planificación Y Desarrollo, Ecuador.

Chapter 6
Updating and Maintaining Land Parcel Types Through Crowd-Sourced Land Use/Cover Classification

Halil Ibrahim Inan, Abdurrahman Geymen, and Omer Faruk Inan

Introduction

Within the Turkish Land Administration (Land Registry and Cadastre) System (LAS), land parcel types are formally registered as attribute information. This information may include both the type of land use/cover and also natural and man-made features (trees, buildings, wells) (see Uzun and Inan 2007). This study focusses only on land use/cover information in this respect.

Updating registered land use/cover information within the Turkish LAS relies basically on related personal or governmental procedures (construction permit, subdivision, property tax, farmer support, land readjustment, land consolidation, etc.). Therefore, real time update of land use/cover information is impossible through such procedures. On the other hand, due to the fact that land/use cover type within a land parcel is not always homogeneous, determination of the exact amount of land use/cover is impossible. So, spatial representation of different land use/cover types within land parcels is the main focus of this study. This problem was previously introduced by Inan et al. (2010a, 2010b).

Incorporating land use/cover information as one of the most important components of modern LASs may be regarded as a basic requirement (see Enemark 2005). However, such an incorporation depends basically on the availability of land use/cover data. In addition, data quality, type of land use/cover classification (depending on the original production purpose), methodology of incorporation, and updating procedures may be listed as other related issues. This study basically aims at resolving or discussing these issues as part of a national scientific project no 112Y027

H.I. Inan (✉) • A. Geymen
Department of Geomatics, Erciyes University, 38039 Kayseri, Turkey
e-mail: hinan@erciyes.edu.tr

O.F. Inan
Tomarza Vocational High School, Erciyes University, 38900 Kayseri, Turkey

T. Yomralioglu, J. McLaughlin (eds.), *Cadastre: Geo-Information Innovations in Land Administration*, DOI 10.1007/978-3-319-51216-7_6

financially supported by the Scientific and Technological Research Council of Turkey.

Depending basically on the purpose and data quality, there may be several types of land use/cover classification systems. Internationally well known ones are FAO Land Cover Classification System (LCCS) (Di Gregorio and Jansen 1998), CORINE land cover (EC 1995) and INSPIRE land use/cover theme (INSPIRE D2.3 2007; INSPIRE D2.8.II.2 2013). Based on FAO LCCS, there has been an international standardization initiative (see ISO 19144-1 2009; ISO 19144-2 2012).

Inspired from agricultural policy implementation and related spatial data management issues in Europe (Kay 2002; Kay and Milenov 2006; Inan 2010; Inan et al. 2010a, b; Sagris et al. 2013) and Turkey (WB 2005; Goeman et al. 2007; Inan 2010), in this study, a new land use/cover classification system was developed (see Inan and Dursun 2014 for preliminary development stage) and incorporated with cadastral land parcels by defining robust geometry and topology rules (the so-called sub-parcel data model). Further, inspired from emerging crowd sourcing (McLaren 2011) techniques in land administration and related fields, in order to update related land parcel and spatially associated land use/cover information, work on multi-tier updating strategies with the contribution of related governmental institutions and also citizens has been continuing. Apart from its original development aim, the study will be presented, in the following sections, in terms of its contribution to the management of spatial land parcel types (incorporated land use/cover information).

Materials and Methodology

Pilot application of this study has been carried out in three districts (Elagoz, Karahoyuk and Vatan) in Kocasinan county of Kayseri province, Turkey.

For the production of land use/cover data, at the first stage, a tree level special classification system was developed. Number of designed classes is two, nine and eighteen in the first, second and third levels respectively (see next Section for classes). At the second stage, land use/cover classes were digitized using visual interpretation techniques on Very High Resolution (VHR) satellite imagery (WorldView-2). To assist in visual interpretation, VHR images of two different years (2010 and 2013) were used. Unchanged land use/cover information (fixed boundaries) were digitized and changes over years were also considered in accordance with the rules for digitization procedure. Sub-parcel data structure (see Inan et al. 2010a, b) was used as part of digitization procedure. To implement this data structure, cadastral land parcel boundaries data sets of three districts were used. As a result, using a total of 3640 land parcels (as spatial template) in three districts, a total of 6811 spatial sub divisions of land parcels were produced as spatial land parcel types (land use/cover classes).

Digitization process was done by trained operators. Later, at the data quality control stage, digitization and classification control processes were carried out by an expert. At the last stage of control procedure, topology control and accordingly semi-automated correction procedures were applied by a database specialist. In this stage, two topology rules namely "Must Not Overlap" and "Must Not Have Gaps" were utilized within an ESRI ArcGIS geodatabase, and as a result hundreds of topological errors were corrected.

The digitization process was carried out by using a static (un-changed) data structure. That is to say no updates to cadastral land parcels and similarly to sub-parcels have been applied. In fact, dynamic structures of these two data sets must be considered. To realize this fact, study on the integration of produced sub-parcel data set with an online cadastral land parcel data set by using WFS web service provided by the General Directorate of Land Registry and Cadastre (GDLRC) has been continuing. Similarly, to be able to update dynamic changes on sub parcels, study on serving this data set together with cadastral land parcels and background VHR images to related users via WMS and/or WFS has also been continuing. These users may be representatives of governmental institutions and also citizens who engaged in agricultural land use. Some of these users are planned to be data providers in different levels for dynamic updata procedures via crowd sourcing techniques.

Results and Discussion

Land Use/Cover Classes

Inspired from agricultural policy implementation and related spatial data management issues in Europe and Turkey, in this study, a new land use/cover classification system was developed (see Inan and Dursun 2014 for preliminary development stage).

In the development stage, possible use of internationally well known classification systems namely FAO Land Cover Classification System (LCCS) (Di Gregorio and Jansen 1998), CORINE land cover (EC 1995) and INSPIRE land use/cover theme (INSPIRE D2.3 2007; INSPIRE D2.8.II.2 2013) was considered and some application tests were carried out. However, it was experienced that visual interpretation on VHR is not a practical method for the classification of land outside agricultural areas. Therefore, a new hybrid classification system which classifies agricultural land in detail and classifies rest of the land only in generic classes is required. The developed classification system has classes in three levels. The first level only differentiates the main type of land which has or does not have agricultural potential. The second level includes eight major land use/cover types. The third level includes eighteen land use/cover types, yet focusses only on detailed classification of agricultural land (see Table 6.1).

Table 6.1 Hierarchic levels and classes of proposed land use/cover classification system

I level classes	II level classes	III level classes
Agricultural land	Cultivated agricultural	Fertile
		Protected
		Garden
	Planted/Perennial agriculture	Vineyard
		Orchard
		Olive grove
		Other planted
	Grassland/Pasture	Pasture
		Grassland
	Transition agriculture	Abandoned
		Uncultivated
		Immature
	Other agriculture	Mixed agriculture
		Uninterpretable
Non-agricultural land	Settlement	Settlement
	Built-up	Built-up
	Infertile	Infertile
	Other	Uninterpretable

Land Use/Cover Data Sets

In land use/cover production stage, by using a total of 3640 land parcels (as spatial template) in three districts, a total of 6811 spatial sub divisions of land parcels were produced as spatial land parcel types (land use/cover classes) (see Fig. 6.1).

In the production of land use/cover data sets, the main focus is the production of the most detailed Level 3 data. Data in lower levels are generalized versions of Level 3 data. Because instances of all classes designed in Level 3 (see Table 6.1) are absent in three application areas, the number of classes in Level 2 and 3 are not far different in Fig. 6.1.

Integrity of Cadastral Land Parcels and Land Use/Cover Data

Data Structure

Sub-parcel data structure (see Inan et al. 2010a, b) was used for the integrity between cadastral land parcels and land use/cover data. This data model requires the use of robust geometry and topology rules. Basic rules are as follows. A land use/cover unit (sub-parcel) must geometrically be a sub set of the specially coinciding cadastral land parcel. Similarly, data quality of cadastral land parcels must be used for shared boundaries although spatial data quality of cadastral land parcels and land

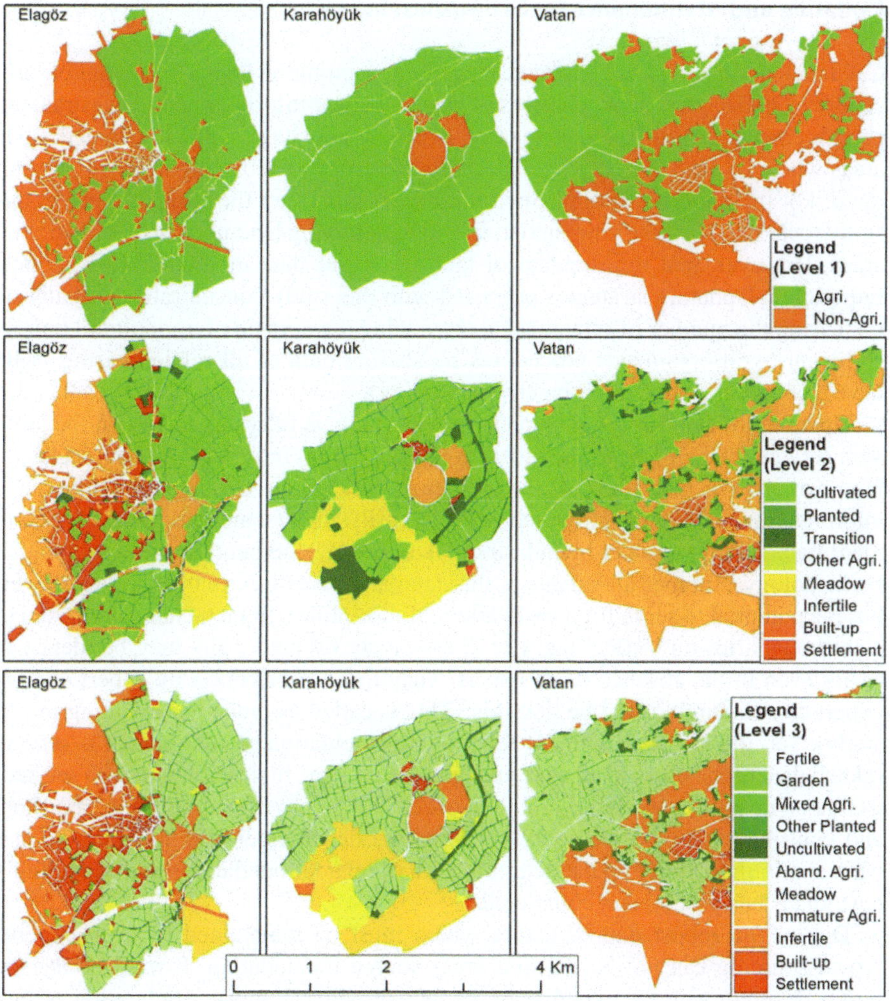

Fig. 6.1 Land use/cover data and three levels of classification in pilot application areas

use/cover units may be different. Land use/cover units within a cadastral land parcel must not overlap and may not have gaps only for exceptional cases. The produced data sets were checked and corrected against all these basic rules.

During implementation, these basic rules combined with updating and maintenance rules will always be implemented. The implementation strategy may differ depending on the type of process. Possible processes may be updating cadastral parcel boundaries and related land use/cover units, simply updating land use/cover units within a cadastral land parcel, a regular maintenance operation, a crowd source contribution with real geometry or only a spot indicating some sort of change etc. Study on these issues has been continuing.

Updating and Maintenance by Crowdsourcing

Cadastral land parcels live (updated) legally depending on land administration procedures. On the other hand, changes of land use cover information on cadastral land parcels cannot always be traced legally. This may only be possible in planned urban areas where all land use activity especially construction activities are legally monitored. For the majority of unplanned rural areas this is not the case. Therefore, this study focusses on classifying, updating and maintaining land use/cover data in rural areas. However, real time update of land use/cover data in such areas is almost impossible. Updating strategies proposed with this study concentrates basically on increasing the number of related legal (by adding the domain of agricultural policy) or illegal (well organized) procedures by sharing data of all related parties (who have a potential contribution by crowdsourcing).

Updating of cadastral land parcels, in this study, is fully dependent on the WFS web service which provide the most up-to-date digital information about cadastral land parcels. For land use/cover information updates, the classification and visual interpretation methodology which reflects easily understandable fixed or clearly identifiable land use/cover boundaries makes the development of a robust updating methodology. Another advantage is that land use/cover information in three levels has the potential to serve for a vast variety of disciplines. In fact, changes of fixed or clear boundaries on VHR images or alternatively by field visits may be identified easily even in the absence of an expert. Therefore, contribution by expert or non-expert crowdsourcing will be possible (This is called as multi-tier updating in this study). Yet, the key development for the implementation of such a task is serving related data to the users in an easy and convenient way. In this context, web services usable on desktop, tablet and mobile devices will be developed both for data share and contribution by crowd sourcing. Software development for this purpose has been continuing. Study on the determination of related foundations and accordingly custodians has partly been completed.

Due to the robust data structure which must be maintained by only experts, crowd sourcing can not be the first order source for updating. It must be at later orders depending on the expertise of the data provider. It must only be used as supporting evidence for updating procedures. For the management of such a complex, multi-tier updating strategy, management of metadata must have a crucial role.

Apart from crowd sourcing, some periodical update operations may be required. These may be done using VHR images and/or aerial photography. In the absence of up-to-date imagery or photography, field visits may also be another alternative. However, during visits, identification of changes may not be possible, and accordingly measurement of all (suspected) land use/cover boundaries may not be possible. Yet, a rapid field survey without boundary surveying may be useful for the determination of the need for imagery/photography update.

Conclusions

As part of modern land administration which serves not only for land market but also for other related disciplines (agricultural policy implementation in this study), management of up-to-date land parcel type information is of utmost importance. Similarly, instead of causing burden to related land administration foundations, using interoperable data structures/models (e.g. sub-parcel data structure) maintained in close collaboration with related foundations or even citizens is also required. However, management of conventional land parcel type information which is not capable of representing amount of land use/cover type spatially is not a convenient way. Instead, in terms of developing effective updating and maintenance procedures, using a well defined land use/cover classification system may be a robust solution. For the effectiveness of such multi-tier maintenance procedures, storing and managing related metadata about contributors, equipment used and also methodology of data acquisition are strictly required.

Acknowledgements This study is accomplished as a component of a national scientific project (no. 112Y027 and titled "A Spatial Data Model Design and its Application for the Management of Farmer and Agricultural Land") funded by the Scientific and Technological Research Council of Turkey.

References

Di Gregorio, A. and L.J.M. Jansen. 1998. Land Cover Classification System (LCCS): Classification Concepts and User Manual, Food and Agriculture Organization of the United Nations, Rome. http://www.birdlist.org/downloads/ecology/lccs_user_guide.pdf. Accessed 2 Mar 2015.

EC (European Commission). 1995. CORINE Land Cover, http://www.eea.europa.eu/publications/COR0-landcover. Accessed 2 Mar 2015.

Enemark, S. 2005. Understanding the land management paradigm. *FIG Com 7 Symposium on Innovative Technologies for Land Administration*, 19–25 June 2005, Madison, Wisconsin, USA. http://www.fig.net/council/enemark_papers/madison_2005.pdf. Accessed 2 March 2015.

Goeman, D., C. Kantor, V. Printzios, A. Zloty and E. Mercimek. 2007. *Final Report for Technical Assistance for the Ministry of Agriculture and Rural Affairs for the Design of a Functioning Integrated Administration and Control System (IACS) and a Land Parcel Identification System (LPIS) in Turkey*. The European Union's TR0402.08/002 Programme for Turkey, Ankara, 162 s.

Inan, H.I., 2010. *Arazi idare sisteminin tarim bileseni olarak konumsal veri modeli gelistirilmesi*. (PhD thesis in Turkish with English abstract. Institute of Applied and Natural Sciences, Karadeniz Technical University, Trabzon, Turkey.

Inan, H.I., A.C. Aydinoglu and T. Yomralioglu. 2010a. Spatial classification of land parcels in land administration systems. *Proceedings of International Conference on Spatial Data Infrastructures 2010* (September 15–17, Skopje, Macedonia).

Inan, H.I., V. Sagris, W. Devos, P. Milenov, P. van Oosterom, and J. Zevenbergen. 2010b. Data model for the collaboration between land administration systems and agricultural land parcel identification systems. *Journal of Environmental Management* 91: 2440–2454.

Inan, H.I. and I. Dursun. 2014. Methodology for the production and updating of agricultural land use/cover data set. *XXV FIG Congress*, June 16–21, Kuala Lumpur, Malaysia.

INSPIRE D2.3. 2007. *Drafting Team "Data Specifications" – deliverable D2.3: Definition of Annex Themes and Scope, version 2.0 2007-04-06*. http://www.ec-gis.org/inspire/reports/ImplementingRules/inspireDataspecD2_3v2.0.pdf. Accessed 03 Mar 2015.

INSPIRE D2.8.II.2. 2013. *Data Specification on Land cover – Draft Technical Guidelines*. http://inspire.jrc.ec.europa.eu/documents/Data_Specifications/INSPI RE_ DataSpecification_LC_v3.0rc3.pdf. Accessed 03 Mar 2015.

ISO 19144-1. 2009. *Geographic information – Classification systems – Part 1: Classification system structure*.

ISO 19144-2. 2012. *Geographic information – Classification systems – Part 2: Land Cover Meta Language (LCML)*.

Kay, S., 2002. *Monitoring and Evaluation of IACS implementation for the identification of agricultural parcels in Member States of the EU, Base document 2: Synthesis of technical alternatives (working draft)*. http://mars.jrc.ec.europa.eu/mars/Bulletins-Publications/Monitoring-and-Evaluation-of-IACS-implementation-for-the-identification-of-agricultural-parcels-in-Member-States-of-the-EU-archive-document-1994. Accessed 4 Mar 2015.

Kay, S. and P. Milenov. 2006. Status of the implementation of LPIS in the EU Member States. *12th MARS PAC Annual Conference*, November, Toulouse, France.

McLaren, R. 2011. *Crowdsourcing support of land administration – A partnership approach*. International Federation of Surveyors, Article of the Month – December 2011.

Sagris, V., P. Wojda, P. Milenov, and W. Devos. 2013. The harmonised data model for assessing Land Parcel Identification Systems compliance with requirements of direct aid and agri-environmental schemes of the CAP. *Journal of Environmental Management* 118: 40–48.

Uzun, B., and H.İ. Inan. 2007. Kadastral Verilerin CBS Ortamına Aktarılması ve Parsel–Mülkiyet Analizleri, TMMOB CBS Kongresi 2007, 30 Ekim – 02 Kasım. *Trabzon, Bildiriler Kitabı* 2: 518–526.

WB, 2005. *Republic of Turkey – Agricultural Reform Implementation Project (Loan 4631-TU) Proposed Amendment of the Loan Agreement*. http://www-wds.worldbank.org/servlet/WDSServlet?pcont=details&eid=000012009_20050225093742. Accessed 19 Mar 2015.

Chapter 7
Investigation of Availability of Remote Sensed Data in Cadastral Works

Selçuk Reis, A.T. Torun, and B.B. Bilgilioğlu

Introduction

Cadastre, which defines legal status and rights specifying the boundaries of the immovable property on the land and map, is very important in relation to property (Kadastro 1987). In countries like Turkey that require high precision, in cadastral survey data is used in the cadastral work to ensure that precision plays an important role. Less costly and more efficient studies should be used to improve suitability to the original on cadastral maps (NAP 1980). Mainly terrestrial methods in cadastral mapping studies, photogrammetric and remote sensing methods are also used. The uses of these methods appear in differences like necessary equipment, used techniques, accuracy requirements, staff and cost. Surveying of parcel boundaries and other details in cadastral works are performed using terrestrial measurement methods generally called as traditional method.

Traditional surveying concept has taken up new shape from disciplines such as geodesy, surveying, photogrammetry and cartography (Zahir 2012). Thanks to technological advances, more reliable, faster and less costly techniques emerged and other techniques were left useless (Silayo 2005). Nowadays, within the scope of multipurpose cadastre, geographical details (build, road, historical artifacts, art structures, water resources, energy resources etc.) which are outside of parcel boundaries are also required. Due to the dynamics of this type, data is needed updated at specific time intervals. Making with terrestrial, continuous measurement of these processes requires significant cost and labour. Therefore, availability of alternative measurement methods like photogrammetry and remote sensing in this kind of studies should be explored (Silayo 2005). Geographical data which generate with sub meter precision via photogrammetric and remote sensing methods will also help to gain time and speed in preparation of the cadastral maps (Yomralıoğlu 2000). According to

S. Reis (✉) • A.T. Torun • B.B. Bilgilioğlu
Geomatics Department, Aksaray University, Aksaray, Turkey
e-mail: sreis@aksaray.edu.tr

T. Yomralioglu, J. McLaughlin (eds.), *Cadastre: Geo-Information Innovations in Land Administration*, DOI 10.1007/978-3-319-51216-7_7

their studies, Mutluoğlu and Ceylan (2005) have revealed that use of orthophoto is 75% less costs than direct terrestrial survey. Alkan and Solal (2010), in their study, emphasize the importance of determining real property by using photogrammetric methods considering cost, labour and time. In this study, comparisons are made for obtaining, updating the present cadastral data and for better interpretation by using orthophoto data. Nowadays, using object and pixel based classifications which are commonly used in image processing methods, object extraction have been made from orthophoto. Also, hand digitizing has been made from orthophoto. The obtained results were compared with the existing cadastral data and interpreted availability of needed geographical data for cadastral works on orthophoto. Yalman area in the district of Aksaray/Gülağaç was determined as a pilot application area for the study.

Data Collection Techniques Used in Existing Cadastre

Existing cadastral measurement techniques can be classified as direct and indirect (Fig. 7.1). Direct techniques are the processes determining the points on land by measuring angles and distances with the help of classical measurement equipment. Coordinates and the area of each parcel in the field are calculated using mathematical methods. In indirect techniques, position information of the objects is obtained by using GPS, remote sensing and photogrammetry techniques. If we want to group the cadastral data collection techniques we can categorize as a simple; classical measurement techniques, GPS technique, photogrammetric techniques and remote sensing techniques. While terrestrial survey techniques are identified directly, other operations can be described as indirect techniques (Zahir et al. 2012).

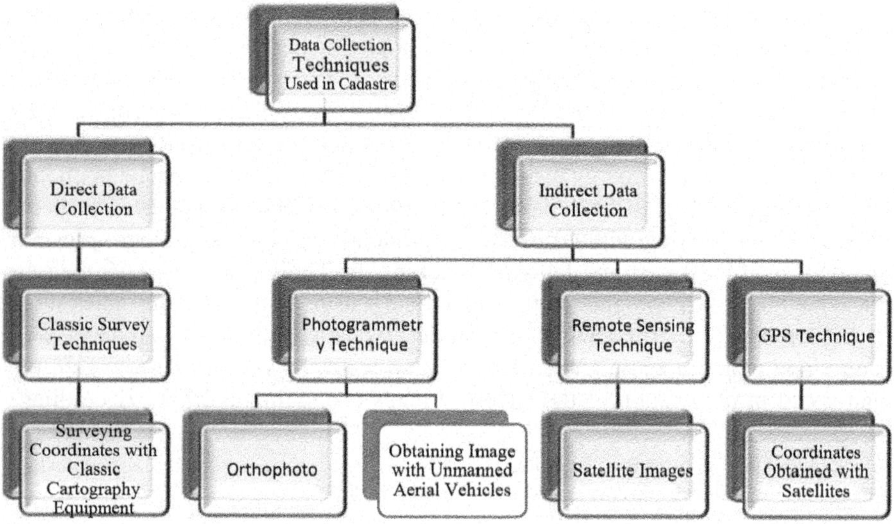

Fig. 7.1 Data collection techniques used in existing cadastre

After 1950, in our country, aerial photogrammetry was emphasized and was applied actively after 1955 in order to accelerate land cadastre works (Bıyık and Karataş 2002). Nowadays, orthophotos created by the elimination of geometric and projective errors is used to a significant extent in many disciplines. In many studies, orthophotos presented in precisions of submeter can minimize the level of labour and cost. In order to display and observe the Earth, the images taken from satellites sent to space were actively used in many areas (map, geology, forestry, agriculture etc.) due to improved technology. Nowadays, remote sensed data are used for many purposes. One of these applications is the use of very high resolution remotely sensed data in cadastre applications (Alkan and Marangoz 2009). Especially, satellite images, due to spatial resolution down to 40 cm, play an important role in for providing many spatial data (buildings, roads, historical, structures, water resources, energy resources, etc.) that is required in cadastre work (Zahir et al. 2012). Studies about the availability of remotely sensed data in the cadastral work has increased in recent years (Zahir et al. 2012; Zahir 2012; Mutluoğlu and Ceylan 2005; Yalçın and Erkek Yalcın and Erkek 2012. In addition, feasibility of the cadastre work with unmanned aerial vehicles was examined by Rijsdijk and van Heinsberg (Rijsdijk and van Heinsberg 2013).

Pilot Study Area

Aksaray which is selected as the pilot area has flat land structure in the central Anatolia region of Turkey (Fig. 7.2). Applications put in process at Gülağaç/Yalman district of Aksaray.

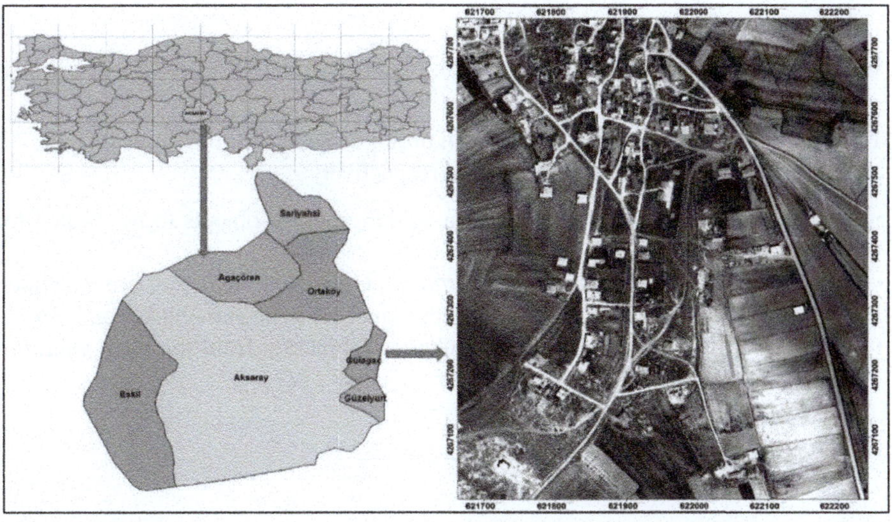

Fig. 7.2 Study area

Data Used in Application

In study, orthophotos of Aksaray was used as basic data which have 30 cm spatial resolution. Cadastral data of study area (parcel boundary, building, road, etc.) was used for the purpose of comparison. Application was benefited from: Erdas Imagine and e-Cognition Developer software for raster data and ArcGIS and NetCAD software for vector data.

Obtaining Geographic Data Using Orthophoto

A lot of object extraction method has come into use with remote sensing thanks to developing and innovating technology. The most preferred object extraction method is object based image classification. It was compared with other classification methods and it has been seen that object based classification is given the best reason every time (Kalkan and Maktav 2010; Oruç et al. 2007). Roads, buildings, rivers, farmlands and details can be easily determined with object extraction from remotely sensed data. In addition to that, extraction and detection of unregistered buildings by using orthophotos is predicted. Three methods were used for the purpose of extracted geographical objects from the orthophoto, including object-based classification, pixel-based classification and manual digitizing. With these methods, classification of buildings and roads has been applied by using basic cadastral data. Classification results were directly compared with cadastral data and accuracy was investigated.

Object Based Classification

The extraction of the object is made out of orthophotos in the pilot process (Figs. 7.3 and 7.4). Processing steps are as follows, respectively:

- *Segmentation Process:* Determination of the group of objects to be created for classification (Fig. 7.3a).
- *Determination Precision of Segmentation:* Generated segments are redefined according to the sensitivity of the object requested to be extracted (Fig. 7.3b).
- *Creating the Appropriate Function:* Generate function from the band values to distinguish details (Fig. 7.3c).

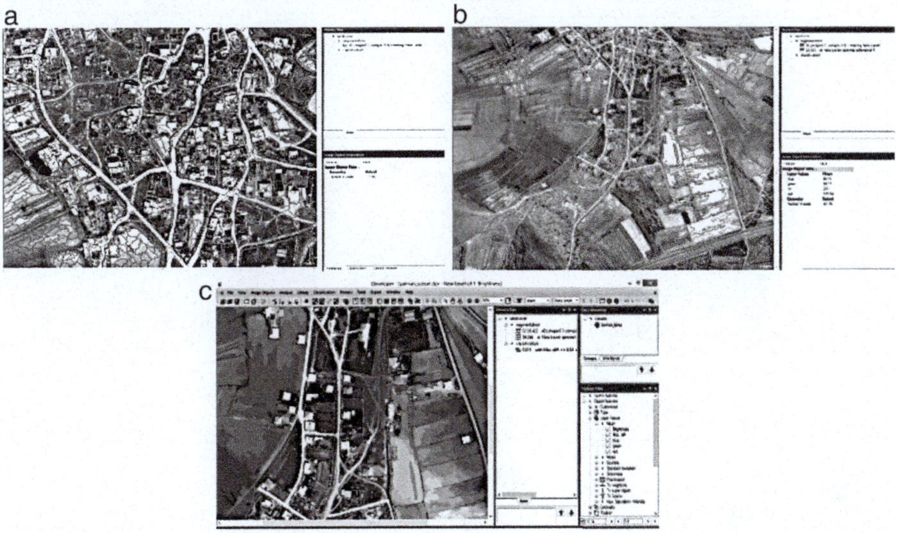

Fig. 7.3 Segmentation process (**a**), Detect precision of segmentation (**b**), and Generate function (**c**)

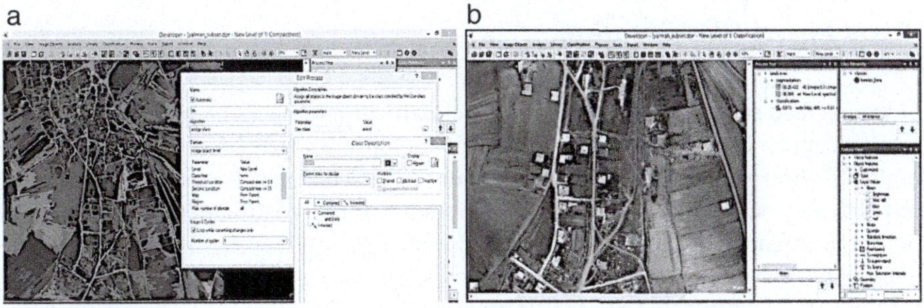

Fig. 7.4 Classification (**a**) and Detail control (**b**)

- *Classification:* Classification process according to the specified function (Fig. 7.4a).
- *Detail Controls:* Checking out the details of the classification result (Fig. 7.4b).

Proper method for classification process has been determined; function (rules) has been created and objects based classification process has been made. While making classification process, two main classes are determined. These are respectively: roads and buildings. The result data obtained is based from the specified classes, in order to be used in ArcGIS software arranged in the vector data format. The resultant vector data is compared with the vector data obtained from the cadastral terrestrial survey.

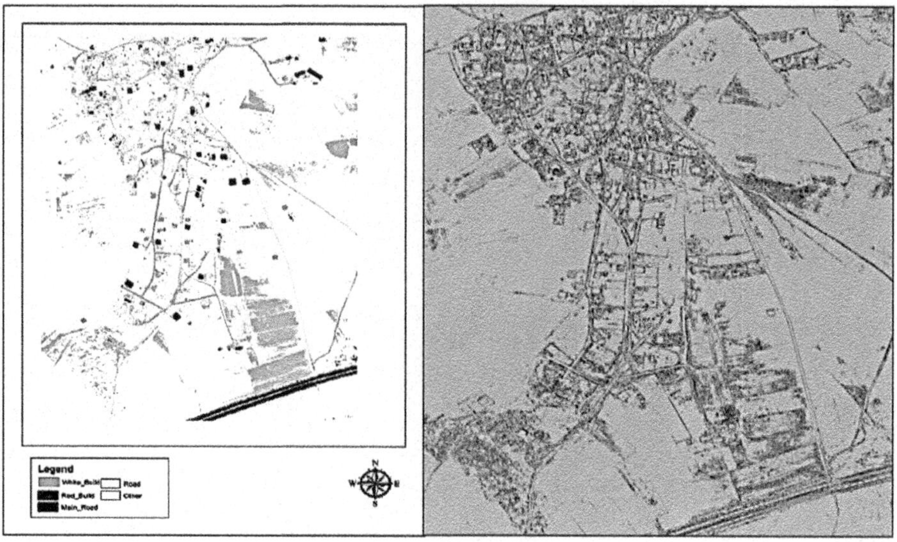

Fig. 7.5 Pixel based classification and Raster-vector conversion

Pixel Based Classification

Training data selected from object based classification was also used for pixel based classification. During application, classification process has been made with maximum likelihood algorithm which is commonly used. Land cover of classification process is generated from three classes including build, roads and other. Because of the performance on the spectral brightness of pixel values, classified or misclassified pixels may occur. Raster data obtained from pixel based classification has been converted to vector data for use to comparison (Fig. 7.5).

Digitizing and Present Cadastral Data

Orthophoto map is digitized by hand for comparison with cadastral data and the other obtained data. Manual digitized map is shown in Fig. 7.6. Present cadastral data belonging to Aksaray province is used as main comparison material. In Fig. 7.7, screenshot of the cadastral data is displayed. There are all objects that are subject to real property in the cadastral data.

Fig. 7.6 Manual digitizing

Fig. 7.7 Existing cadastral data

Comparison of Cadastre and Produced Data from Orthophoto

Some selected objects produced from orthophoto are compared with present cadastral data. These are five buildings, three dirt roads and an asphalt road. Details and geometric comparisons of five different buildings produced with terrestrial data (a), manual digitizing (b), object based classification (c) and pixel based classification

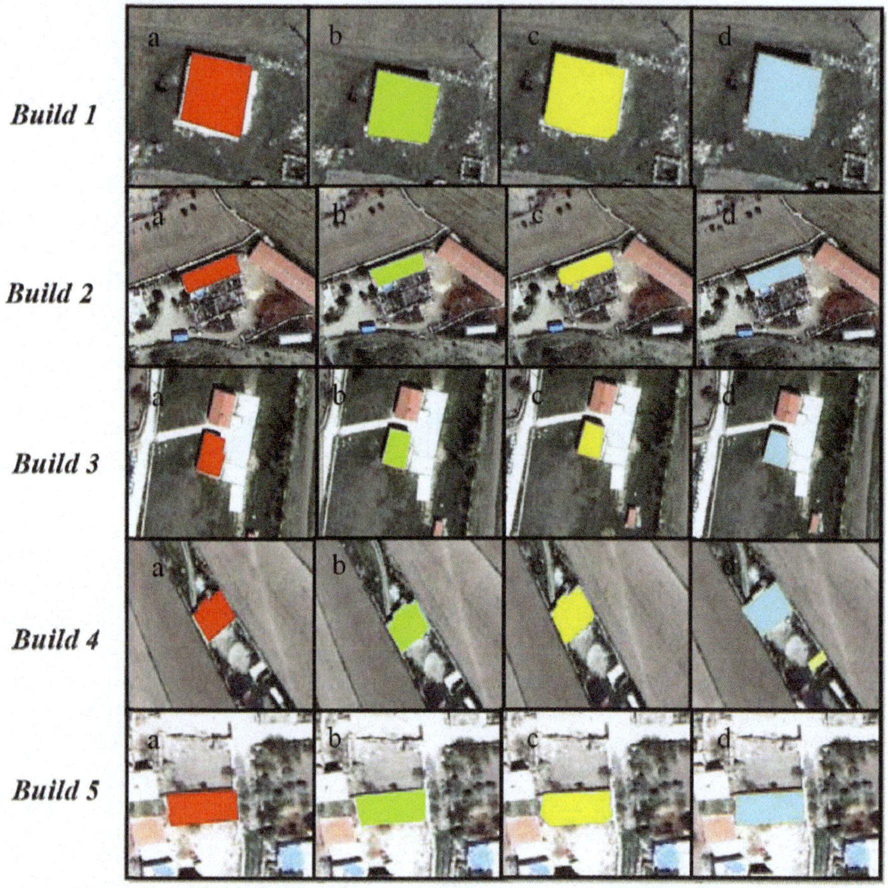

Fig. 7.8 Comparison of buildings: (**a**) Cadastral data, (**b**) Manual digitizing, (**c**) Object based classification and (**d**) Pixel based classification

Table 7.1 Comparison of building areas by using different methods

	Cadastre (Reference)	Digitizing		Object-based		Pixel-based	
Detail	m^2	m^2	%	m^2	%	m^2	%
Build 1	157	146	92,99	142	90,44	139	88,53
Build 2	122	125	97,60	121	99,18	113	92,62
Build 3	156	163	95,70	167	93,41	160	97,50
Build 4	163	174	96,93	157	97,54	171	95,32
Build 5	234	226	96,58	221	94,44	253	92,49

(d) is shown in Fig. 7.8. Table 7.1 shows the area of the buildings in different methods. Samples of asphalt road and dirt roads and comparison results are seen in Figs. 7.9 and 7.10. The area values of the asphalt and dirt road are shown in Tables 7.2 and 7.3.

Dirt Road 1

Dirt Road 2

Dirt Road 3

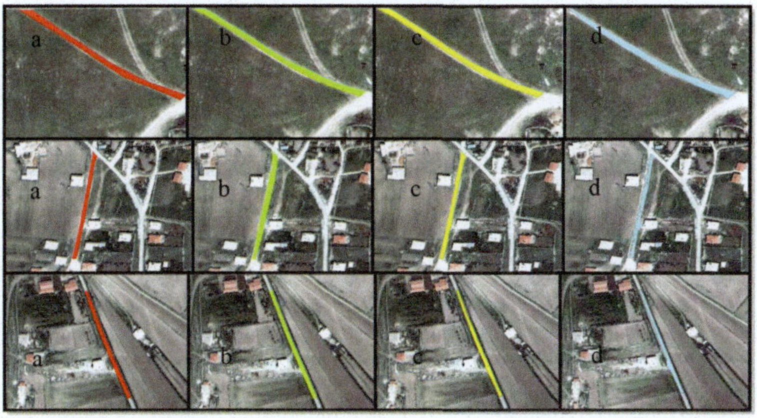

Fig. 7.9 Comparison of dirt roads results: (**a**) Cadastral data, (**b**) Manual digitizing, (**c**) Object based classification and (**d**) Pixel based classification

Asphalt Road 1

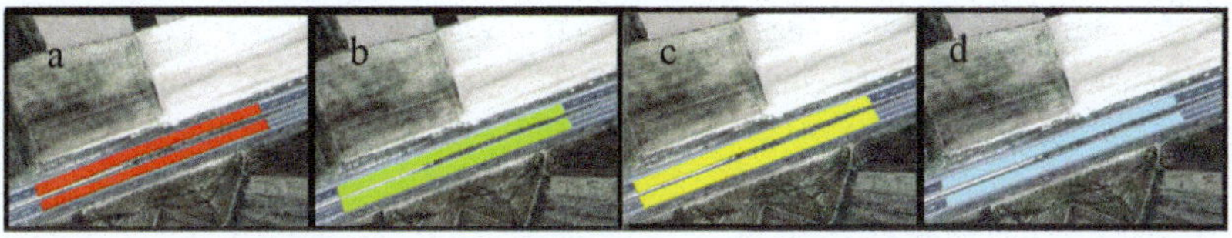

Fig. 7.10 Comparison of asphalt road: (**a**) Cadastral data, (**b**) Manual digitizing, (**c**) Object based classification and (**d**) Pixel based classification

Table 7.2 Comparison of dirty roads areas by using different methods

	Cadastre (Reference)	Digitizing		Object-based		Pixel-based	
Detail	m^2	m^2	%	m^2	%	m^2	%
Dirt Road 1	215	225	95,55	207	96,27	185	86,04
Dirt Road 2	736	713	96,87	721	97,96	647	87,90
Dirt Road 3	946	908	95,98	912	96,40	851	89,95

Table 7.3 Comparison of Asphalt Road area by using different methods

	Cadastre (Reference)	Digitizing		Object-based		Pixel-based	
Detail	m^2	m^2	%	m^2	%	m^2	%
Asphalt road	3001	2918	97,23	2892	96,36	2656	88,50

Statistical Analysis for Results

The coordinate values of each object (building and roads) are statistical analyzed by cadastral data as reference. Standard deviations Eq. (7.2) obtained using divisions Eq. (7.1) and means of the coordinate values. Shown in Eq. (7.1) is presented arithmetic means values of coordinates Eq. (7.3), N is number of point ($I = 1, 2, 3 \ldots N$) and x_i is coordinates of objects data. With presented s, standard deviation is obtained with benefit from differences and mean of coordinate values.

$$\overline{x} = \frac{1}{N}\sum_{i=1}^{N} x_i = \frac{x_1 + x_2 + \ldots + x_N}{N} \tag{7.1}$$

$$s = \sqrt{\frac{1}{N-1}\sum_{i=1}^{N}\left(x_i - \overline{x}\right)^2} \tag{7.2}$$

$$s_{\text{mean}} = \frac{s}{\sqrt{N}} \tag{7.3}$$

According to the t-distribution table, surveyed coordinates (cadastre survey) with mean values obtained from methods (pixel based class, object based class and digitizing) are expected to be equal. Therefore, foreseen that situation of between statistically expected difference values and calculated difference values equal to zero H_0: hypothesis. H_0 hypothesis is given in Eq. (7.4). Also Eq. (7.4) shows means of cadastre coordinates (x_{mean1}) and coordinates calculated from other methods (x_{mean2}).

$$\begin{aligned} &H_0: d = 0 \\ &d = x_{\text{mean1}} - x_{\text{mean2}} \end{aligned} \tag{7.4}$$

Whether there is statistically significant difference values are determined through as shown in Eq. (7.5) t-distribution confidence limit of test value $t_{\infty,\,1\text{-}\alpha/2}$.

$$T_i = \frac{|di|}{s_{di}} \sim t(f) \tag{7.5}$$

where s_{di} is standard deviation of the difference d_i. In case of $T > t_{\infty,1-\alpha/2}$ coordinates calculated for each object from the mean values of each method (object based class, pixel based class and manual digitizing), the deviation from the mean of the cadastral measurements are expressed as statistically significant. Otherwise ($T < t_{\infty,1-\alpha/2}$) change is statistically insignificant.

In Fig. 7.11, accuracy of methods between each other were determined using mean standard deviation of buildings and roads coordinates. Cadastral surveying results which are taken as a reference gave the best result with minimum standard deviation as expected. Then accuracies follow, respectively, digitizing, object-based and pixel-based methods. This result is similar to all buildings in the study field. This is an expected situation. There are some deviations originating from sensitivities of humans hand and eyes in manual digitizing process. Deviations in object-based classification process are related to deficiency of infrared-band and usage of band ratio. In pixel-based classification process the reason for the highest deviation ratio is algorithms which is used in classification process and another reason for these deviations is the unclassified pixels while performing the pixel-based classification process.

Each method was compared with cadastral data whose accuracy is accepted as higher than others (Fig. 7.12). Mean values of three methods are subject to cadastral

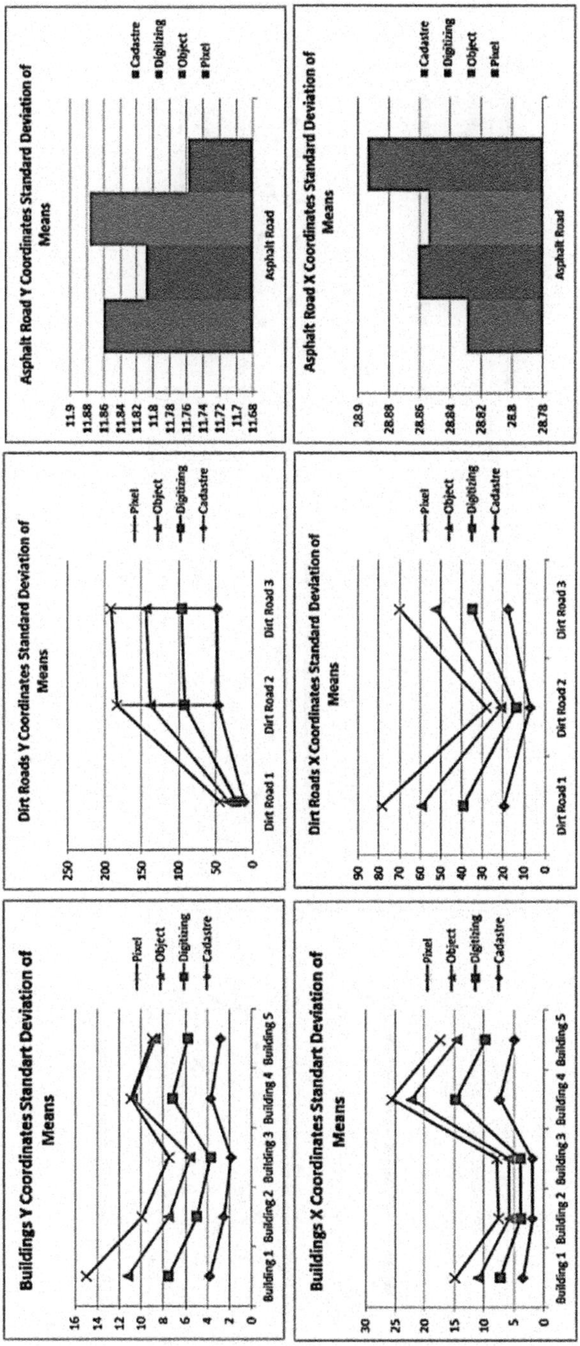

Fig. 7.11 Self accuracy determination of methods

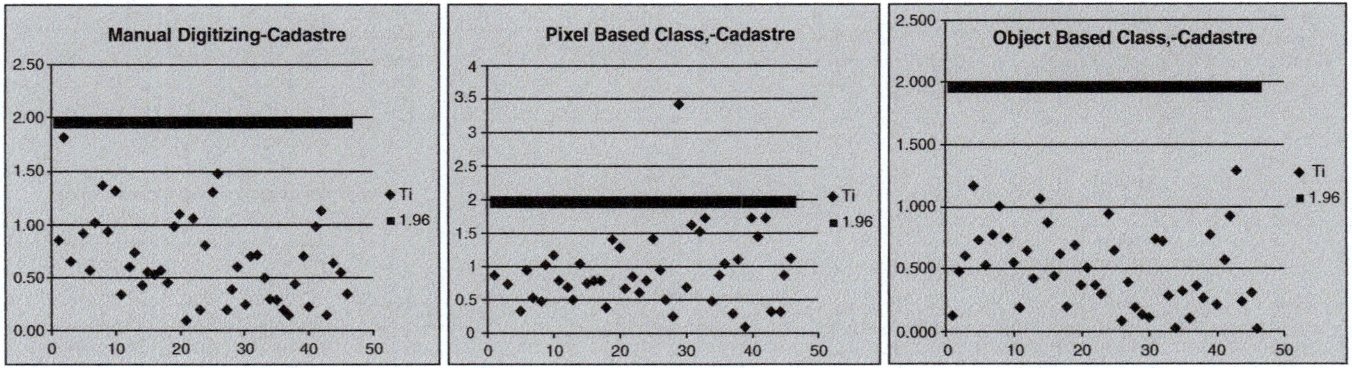

Fig. 7.12 According to *t*-distribution, results obtained from pixel, object and digitizing methods in comparison with cadastral surveys

mean values for t-distribution confidence interval. This result shows that these are statistically insignificant according to t-distribution ($t_{\infty,1-\alpha/2} = 1.96$) 95% confidence interval. Test values of dirt road results are statistically significant in pixel based classification method. For this reason, it can be said that there is unclassified pixels on some objects.

Pixel based classification distributed heterogeneous under the threshold. This shows that the coordinates have low accuracy of their own. The reason for this, unlike object based classification, is each pixel evaluates individual in pixel based classification. Besides, because digitizing was made by manual, it encountered some errors in coordinates. These errors originated from digitizer eye error and shadows of builds. As a result of statistics it is seen that coordinates didn't have homogenous distribution at manual digitizing technique. According to the results it can be seen that the optimal distribution of standard deviation of coordinates in object-based classification technique is meaningful. Methods compared statistically with each other show that object based classification technique gave higher accuracy results than pixel based classification and manual digitizing. These results supported by statistical analysis showed that object based classification method is available for object extraction such as buildings and roads from orthophotos in cadastral works.

Results and Suggestions

This study focused on using orthophoto for extraction of subject of real property and investigated availability of orthophotos in cadastral works. Different types of data can be collected from using orthophoto images. In this study, buildings and roads were selected and extracted from the orthophotos data. Objects such as buildings and roads can be extracted from orthophotos using image processing techniques. In this study, buildings and roads produced from orthophoto compared with cadastral surveyed data. The results show, firstly, that best accuracy result is got from object based classification when methods coordinates compared with cadastral coordinates. Secondly, the study shows, the digitizing method give the best results when compared with the methods of their own. Results have shown that producing data from orthophotos can provide high accuracy results depending on resolution of aerial photos. According to results of accuracy in this study, it could be said that digitizing and object based classification methods are useful in studies which needed precision similar with orthophoto resolution. For cadastral works, remote sensed data can be used to extract spatial data for second cadastre, cadastre renovation, cadastral update and multipurpose cadastre. In terms of labour, cost and accuracy, orthophotos can be used effectively in second cadastral works for countries needing high precision data such as Turkey. Besides, it can be said that unregistered or not yet registered structures/buildings which are kept rising day by day, easily detect on orthophotos. When spatial resolution of remote sensed data improves in the future, object extraction methods from orthophotos/satellite images can provide facilities for cadastral works.

References

Alkan, M., and M.A. Marangoz. 2009. Creating Cadastral maps in rural and urban areas of using high resolution satellite imagery. *Applied Geoinformatics for Society and Environment*, Stuttgart University of Applied Sciences.

Alkan, M., and Y. Solal. 2010. An investigation of 1:5000 scale photogrammetric data for Cadastral mapping uses: A case study of Kastamonu-Taskopru. *African Journal of Agricultural Research* 5(18): 2576–2588.

Bıyık, C., and K. Karataş. 2002. Yüzyılımızda Kadastroda İçerik Ve Kapsam, Selçuk Üniversitesi Jeodezi ve Fotogrametri Mühendisliği Öğretiminde 30. Yıl Sempozyumu,16-18 Ekim 2002, Konya.

Kadastro Kanunu. 1987. Numara 3402, 9/7/1987 tarihli Resmi Gazete.

Kalkan, K., and D. Maktav. 2010. Nesne Tabanli Ve Piksel Tabanli Siniflandirma Yöntemlerinin Karşilaştirilmasi (Ikonos Örneği), III. Uzaktan Algılama ve Coğrafi Bilgi Sistemleri Sempozyumu, Kocaeli, 11.10.2010–13.10.2010.

Mutluoğlu, Ö., and A. Ceylan. 2005. Dijital Ortofoto Haritalarda Konum Doğruluğu Ve Maliyet Karşılaştırması, S.Ü. Müh.-Mim. Fak. Derg., c.20, s. 1.

NAP. 1980. Need for a multıpurpose Cadastre. *Panel on a Multipurpose Cadastre Committee on Geodesy Assembly of Mathematical and Physical Sciences*. Washington, DC: National Research Council, Natıonal Academy Press.

Oruç, M., A.M. Marangoz, and S. Karakış. 2007. Pan-Sharp Landsat 7 Etm+ Görüntüsü Kullanılarak Piksel-Tabanlı Ve Nesne-Tabanlı Sınıflandırma Yaklaşımlarinin Karşılaştırılması, 11. Türkiye Harita Bilimsel ve Teknik Kurultayı, ODTÜ, Ankara, 02-06 Nisan.

Rijsdijk, M., and W.H.M. van Heinsberg. 2013. Unmanned aerial systems in the process of juridical verification of Cadastral border. *International Archives of the Photogrammetry, Remote Sensing and Spatial Information Sciences*, Volume XL-1/W2, 2013 UAV-g2013, 4–6.

Silayo, H.E. 2005. Searching for an affordable and acceptable Cadastral Survey method. Pharaohs to Geoinformatics FIG Working Week 2005 and GSDI-8 Cairo, Egypt.

Yalcın, G. and B. Erkek. 2012. Digital Photogrammetric Maps and Orthoimages for Cadastre Renovatıon Studies. TKGM.

Yomralıoğlu, T. 2000. *Geographical Information Systems: Basic Concepts and Implementations*. ISBN 975-97369-0-X, İstanbul.

Zahir, A. 2012. Assessing Usefulness of High-Resolution Satellite Imagery (HRSI) in GIS-based Cadastral Land Information System. *Journal of Settlements and Spatial Planning* 3(2).

Zahir, A., A. Tuladhar, and J. Zevenbergen. 2012. An integrated approach for updating cadastral maps in Pakistan using satellite remote sensing data. *International Journal of Applied Earth Observation ad Geoinformation* 18: 386–398.

Chapter 8
Usability of GNSS Technique for Cadastral Surveying

Reha Metin Alkan, I. Murat Ozulu, Veli İlçi, F. Engin Tombuş, and Murat Şahin

Introduction

The population of the world, which increases rapidly at an average rate of 1.2% annually, has exceeded seven billion. However, despite the increase in population, 150,000,000 km^2 terrestrial area, which composes the 29.2% of the surface of the earth, has not changed significantly. When this is taken into account, it is clear that the leading precious fact is land. Considering that not all of these areas are suitable for the life and use of human being, it is obvious that land and soil should be used with care and managed in accordance with professional policies.

What is undoubtedly essential to do in this regard is to identify the overall aspects of this value and keep records with the use of information systems. In this context, information about the ownership and boundary of a real property is specified with cadastral surveying. When the fact is not only the civilian but also boundary conflicts of many countries, resulting serious legal problems and even wars among countries, the significance of the issue can easily be understood.

In Turkey, cadastral surveys are conducted by General Directorate of Land Registry and Cadastre of Turkey (GDLRC). When Turkey Republic was established, cadastral studies were enacted in accordance with the 658 numbered Cadastre Law in 1925. Cadastral studies are still conducted with regard to 3402 numbered law in Turkey (İnam et al. 2011). The current status of the Turkey Cadastre is given in Table 8.1 (URL-1).

R.M. Alkan (✉)
Hitit University, North Campus, 19030 Çorum, Turkey

Istanbul Technical University, Maslak, 34469 Istanbul, Turkey
e-mail: alkan@hitit.edu.tr

I. Murat Ozulu • V. İlçi • F. Engin Tombuş • M. Şahin
Hitit University, North Campus, 19030 Çorum, Turkey

T. Yomralioglu, J. McLaughlin (eds.), *Cadastre: Geo-Information Innovations in Land Administration*, DOI 10.1007/978-3-319-51216-7_8

Table 8.1 Status of Turkey cadastre realization

	Realization			
Total units (district and village)	Before 2003	Between 2003 and 2015	On-going	Problematic units[a]
52,054	38,803	12,758	193	300

[a]Forest, Border disputes etc.

Regarding this data, it can be concluded that 99% of the cadastral studies have been completed. Although the first facility cadastre is about to be completed in Turkey, Cadastre Renovating/Updating studies are in progress because of the following main reasons (Çağla et al. 2011; Meha et al. 2013):

- Cadastral reconstruction,
- Technical inadequateness,
- Loss of application quality,
- Lack of information,
- Showing the boundaries of the ground as its original,
- Land consolidation in urban and rural areas,
- Regular maintenance of cadastre.

Turkish government aims at improving the cadastral system to maintain it in a more appropriate, effective and contemporary manner (Çete 2010). According to a study carried out by the GDLRC of Turkey in 2008, throughout the country under the Cadastre Law Article 22-a (Renovation Cadastre) that needed to renew approximately 8,100,000 parcels has been determined. A total of 5,642,255 parcels were auctioned by the World Bank and the GDLRC's their own budget between 2009 and 2014, and studies were completed in 3,600,000 parcels (URL-2). This means that about 44% of the parcels have been completed, and it is evident that a lot of studies would be done in this scope. It is generally said that there are plenty of studies in such fields as follows:

- Cadastre renovating/updating studies,
- Setting marks of parcels,
- The works carried out within the scope of land use conversion and amalgamation (measuring the buildings and other details on the parcel),
- Controlling maps and plans registered,
- Forest cadastre,
- Boundary dispute and novation,
- 3D cadastral mapping.

Historical background, current situation and emerging main problems of cadastral studies in Turkey can be found in detail in Çete (2010); İnam et al. (2011) and URL-2.

All these studies have been carried out either by related Official Cadastral Units or Licensed Surveyors. In those studies, cadastral surveys have been conducted with the use of conventional terrestrial methods/equipment; i.e. steel band, EDM, theodolite and for a couple of decades total stations and etc. Nevertheless, Global Navigation Satellite System (GNSS) (especially GPS and GLONASS), which have many advantages over conventional methods, have begun to be widely used all around the world. GNSS systems are competing with conventional methods in almost all fields of surveying applications including cadastral surveying. Although the GNSS systems have considerably facilitated the measurements, there are still some limitations in their usability in some cases like in densely urban areas, mountains, heavy tree cover, ravines and similar places. Even though the combination of GPS and secondary satellite system like GLONASS observations can overcome the abovementioned problems, conventional surveying techniques are still required in some instances.

A Review of Global Navigation Satellite System

The term GNSS is now used to describe a many different satellite positioning systems operated by different countries. Among them, GPS (Global Positioning System) is the very well-known and widely used one while the others, i.e. GLONASS, Galileo etc. have started to be used or about to be used in the near future. Twenty years ago, the usage of the GNSS by civilian was mainly limited to merchant ship crews and surveyors, but today there are millions of receivers which are used to achieve different tasks. Moreover, GNSS technology is becoming the most effective positioning methods for all types of engineering projects (Bonnor 2012). GNSS is competing with traditional surveying techniques and is widely used for marine, air and land applications for achieving navigation, surveying and scientific purposes because of the advantages such as being almost independent of weather conditions, not requiring surveying points seeing each other, being able to manage surveying during the day and night, and etc. The systems that constitute the GNSS are briefly given in the following sections.

Global Positioning System (GPS)

GPS is a satellite-based radio navigation system which was maintained by the United States Department of Defense. Although it was originally designed as a military system, it has become a global utility primarily used as a navigation system and its civil applications have grown much faster. The system provides position, velocity and time anywhere in the world. Since the launch of the first operational GPS satellite in 1978, as of today, the GPS has been widely used in marine, air and land applications. Positioning with GPS can be realized in two main ways: (i) absolute positioning and (ii) relative positioning. A general classification of the GPS positioning methods are shown in Fig. 8.1 (Kahveci and Yıldız 2012).

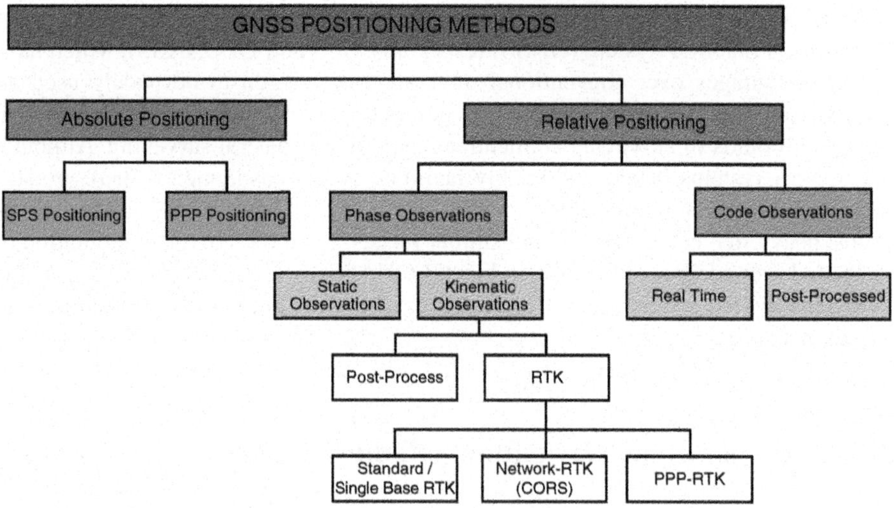

Fig. 8.1 General classification of GNSS positioning methods

Table 8.2 GPS constellation status (as of March 13, 2015) (URL-3)

	GPS
Total satellites in constellation	31 SC
Operational	30 SC
In commissioning phase	–
In maintenance	1 SC

The GPS was declared as Fully Operationally Capable (FOC) in July 1995 with 24 satellites. At present, there are 31 satellites 30 of which are operational (Table 8.2).

Network RTK/TUSAGA-Aktif (CORS-TR)

In recent years, Network-RTK (or commonly known as Continuously Operating Reference Station-CORS) has been used extensively while producing economical and rapid solutions with only single GNSS receiver. CORS networks have been established in many countries and regions all over the world. For instance, Germany has approximately 270 reference stations with an average spacing of about 40 km, Switzerland has 31 stations with an average spacing of 35–50 km, Sweden has 170 stations with an average spacing of 60–70 km, and Indonesia has more than 230 CORS stations (Reddy 2010).

Similar to the world samples, a network called as TUSAGA-Aktif has been established in Turkey by *Istanbul Kultur University* in association with the *General*

Directorate of Land Registry and Cadastre of Turkey (GDLRC) and *the General Command of Mapping of Turkey* and sponsored by *the Turkish Scientific and Technical Research Agency (TUBITAK)* in 2009 (Bakıcı and Mekik 2014). TUSAGA-Aktif system has 146 reference stations with an average spacing of 70–100 km covering entire Turkey and Turkish Republic of Northern Cyprus (Fig. 8.2).

The system aims:

- to determine positions fast, economically and reliably with cm accuracy within minutes, even seconds, 24/7 in real-time,
- to model the atmosphere (troposphere and ionosphere),
- to predict weather,
- to monitor plate tectonics,
- to determine datum transformation parameters between the old system and ITRF*yy* (Bakıcı and Mekik 2014).

In this system, the corrections are computed and transmitted by different methods such as Virtual Reference Station (VRS), Flachen-Korrektur Parameter (FKP), Master Auxiliary Concept (MAC), and Differential GPS (DGPS) techniques. RTCM 3.0 and higher protocols are used for communication between users and the control centre and thus GSM, NTRIP over GPRS/Edge and radio links are utilized (Mekik et al. 2011). The coordinates can be realized in ITRF96 datum and 2005.0 measurement epoch. In addition to the RTK correction, all the reference stations can also log the GNSS data for later use. Therefore, users can download 1 second-interval data with a limited fee and 30 seconds-interval data freely via web-site of the service. The user of the TUSAGA-Aktif has been increasing day by day and the number has reached to 6,128 as of March 15, 2015 (Bakıcı 2015).

The usage of the TUSAGA-Aktif is permitted by Large Scale Map and Map Information Production Regulation (LSMMIPR) in cadastral surveying. Usability of the TUSAGA-Aktif for cadastral surveying not only provides greater speed but also requires less time, less money and fewer personnel.

GLONASS

The Russian Federation operates their own satellite-based system called as GLObal'naya NAvigatsionnaya Sputnikovaya Sistema (GLONASS) and it was developed primarily for the use of the Soviet military. Development of the system began in 1976, but the first satellite was launched in October 1982 and 10 more satellites were launched between 1982 and 1985 (Bonnor 2012). Due to the financial recession in Russian Federation, GLONASS system continued to work with too few satellites in 1980s and 1990s. In 2000, after the economic recession period, Russian government made a considerable investment to reconstruct the GLONASS satellite system. The system was made fully available for civilian in 2007. In 2010, GLONASS achieved 100% coverage of Russia's territory. Full Operational

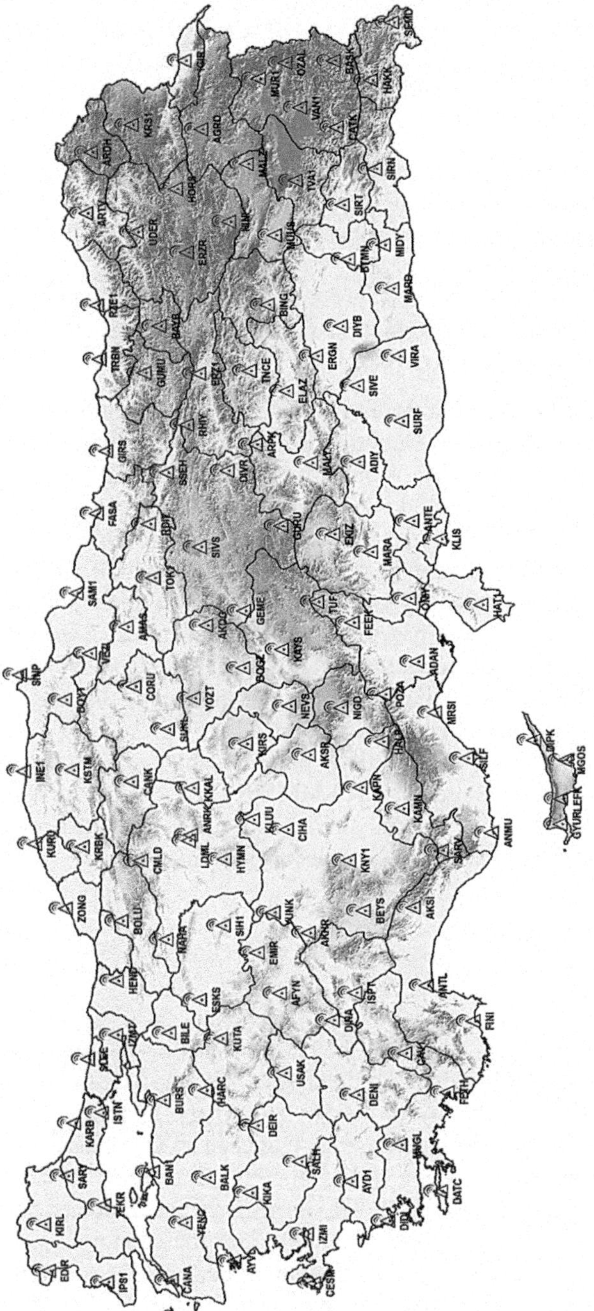

Fig. 8.2 Location of TUSAGA-Aktif reference stations (Bakıcı 2015)

Table 8.3 GLONASS constellation status (as of March 13, 2015) (URL-4)

	GLONASS
Total satellites in constellation	28 SC
Operational	24 SC
In commissioning phase	–
In maintenance	–

Capability (FOC) and thus full global coverage for GLONASS were attained with 24 satellites in 2011. As of today, GLONASS system has 28 satellites, 24 of which are operational (Table 8.3).

Galileo

European Commission decided to develop the Europe's own global navigation satellite system named as Galileo. Unlike GPS and GLONASS, construction and operation of Galileo will be under the control of civilians. Galileo was designed to cooperate with GPS and GLONASS. The full operational capability constellation of Galileo will consist of 30 satellites (27 operational + 3 active spares). These 30 satellites will revolve around the world in three circular Medium Earth Orbit (MEO) planes and at an inclination of the orbital planes of 56 degrees to the equator. On 21 October 2011 a Soyuz rocket from French Guiana launched two satellites, with two more following on 12 October 2012. The Galileo In-Orbit Validation (IOV) test campaign was conducted in 2013. The test shown that deployed four Galileo satellites and ground infrastructure work very well. They will be followed by several launches with Ariane-5 or Soyuz from the Europe's Space Port in French Guiana (URL-5; URL-6).

BeiDou (BDS)

BeiDou Navigation Satellite System (BDS) is the global navigation satellite system which has been established and operated independently by China. The system was planned to establish as three steps: demonstrational system, regional system and global system. The regional system is in full operation comprising five Geostationary Earth Orbit (GEO) satellites, five Inclined Geosynchronous Earth Orbit (IGSO) satellites and four Medium Earth Orbit (MEO) satellites. The system gives two types of service to their users as standard positioning service which is free for all civilian users and precise positioning service and system-embedded wide area differential positioning are only for authorized users (Xu et al. 2014). In 2020, the BeiDou will provide high accuracy and high reliability positioning to GNSS users in any time and all-weather conditions in global coverage (Li et al. 2014).

Japan Quasi-Zenith Satellite System (QZSS)

Japan is planning a Quasi-Zenith Satellite System (QZSS) using three satellites. In March 2013, Japan's Cabinet Office announced the expansion of the QZSS from three satellites to four. The plan was to have four satellites constellation in 2018. The system will include three satellites in the Inclined Geosynchronous Earth Orbit (IGSO) and one satellite in the Geostationary Earth Orbit (GEO) (Montenbruck et al. 2014). The first satellite (named as Michibiki) was launched in September 2010. At present, only one QZSS satellite is in operation. High elliptical orbits (more than 70 degree) increase the reliability of the satellites in service area more than 12 h a day (Bonnor 2012). Therefore, users can receive the signals with high elevation angle from at least one of the satellites of QZSS. Moreover, Japan and USA have agreed that GPS and QZSS have complete compatibility and interoperability of the systems. With small modifications of the GPS receiver, users can receive QZSS signal and calculate their position combining the GPS signals (Kogure et al. 2006).

GAGAN and IRNSS

India's satellite system is named GPS Aided GEO Augmented Navigation (GAGAN). The first GAGAN communication satellite was launched on 21 May 2011. The aim of the system is not only navigation but also communication and some other purposes. The Indian Regional Navigational Satellite System (IRNSS), which is an autonomous regional satellite navigation system, was approved by the Indian government in May 2006 to provide accurate position information in Indian region.

The first IRNSS satellite was launched in July 2013. Four Inclined Geosynchronous Earth Orbit (IGSO) and three Geostationary Earth Orbit (GEO) satellites were planned to be launched so that all satellites were visible in Indian Region (Majithiya et al. 2011). The entire IRNSS constellation of seven satellites is planned to be completed by 2015. IRNSS will provide two types of services; one is the Standard Positioning Services (SPS) available for all users and the other is Restricted Services (RS), available for authorized users.

Deployment status of global and regional satellite positioning systems are summarized in Table 8.4 (Montenbruck et al. 2014).

A total of six satellite-based positioning systems operated by different countries, i.e. GPS, GLONASS, Galileo, BeiDou (BDS), QZSS, and GAGAN&IRNSS, have become available for different type of users all over the world. The abovementioned systems will work together and will be compatible, interoperable and interchangeable among each other. This obviously increases the availability and the coverage while providing more robust and reliable services. The combined system provides more accurate positioning in some locations that have marginal sky visibility like

Table 8.4 Current status of space-based positioning systems

System	Blocks	Sats
GPS	IIA	8
	IIR-A/B	12
	IIR-M	7
	IIF	4
GLONASS	M	24
	K	1[a]
Galileo	IOV	4[a]
BeiDou	GEO	5
	IGSO	5
	MEO	4
QZSS	N/A	1
IRNSS	N/A	1[a]

[a]Have not been declared as operational

urban canyons, mountains, heavy tree cover, ravines, open-pit mines and extreme marine environments where satellite signals may either be blocked or strongly degraded by obstacles (Tamrakar 2013).

Combined Services Performances

By combining the multi-GNSS systems, improved performance in the following domains can be expected as:

- ***Availability:*** Using GLONASS in combination with GPS as an example, the number of available satellites will increase twice or more when compared with the GPS alone. With the addition of more GNSS system integration, more satellites will be available all around the earth. This is especially important in partially obstructed areas like urban canyon environments, heavy tree cover, ravines, mountains, open-pit mines.
- ***Position accuracy:*** Increasing availability of more satellites generally leads to better satellite geometry and enhanced better positioning performance.
- ***Integrity:*** Combined satellite systems will enhance the provision of integrity information.
- ***Redundancy:*** The combination of independent systems will lead to the required level of redundancy (Awange 2012).

The most important constraint that restricts the use of the systems can be the environments without clear view of the sky. Although the use of the combined systems provides a solution, even this cannot be sufficient in some cases. In that case, at least two or three control points should be established in the closest clear location of the study area. After determining the coordinates of those points with GNSS, measurements with conventional optical methods will be better than feasible.

Advantages of GNSS Usability

GNSS measurements, widely used in many areas, are prevalently used in cadastral measurements as well. The reason is the system provides important advantages to the users. Some of the advantages of GNSS surveying with respect to conventional methods in cadastral survey is given below (Tamrakar 2013).

- Inter-visibility between consequent stations are not required.
- Traverse stages in the field for providing control points for classic surveys are not needed.
- Establishment of the control points is more accurate, easier with less cost,
- The measurements can be made in day time and at night, under almost all weather conditions.
- Simple field operation.
- Continuous 3D positioning.
- All coordinates can be estimated in a global datum, i.e. ITRFyy or so on.

The abovementioned features make GNSS quite attractive for survey works.

Some Recommendations for GNSS Usability

Some recommendations as a minimum standard for the surveyors using GNSS is given below (Tamrakar 2013; URL-7; URL-8):

- GNSS surveyors should be trained before using this technique in cadastral surveys.
- As GNSS signals and fixed solutions can be affected from cell tower, radio station, and high-voltage transmission line, to establish geodetic control points must not be selected in the areas where they degrade GNSS signal quality.
- Reflecting surface like metal roofing, big vehicles and large water surface in surveying area may cause multipath error. Thus, it is suggested not to choose these kind of places.
- The location of the points must be well located in terms of ground stability, accessibility, clear sky views, protection from vandalism and disturbance.
- Although it is not necessary for the control points, the points should be intervisible for ease of subsequent use by conventional techniques.
- A proper observation log must be filled up.
- GNSS station obstruction diagram should be drawn carefully (obstructions and their elevation angle is shown).
- GNSS results can be affected by meteorological changes notably due to longer distance. thus instantly changing weather conditions must be written down in observation logs.
- GNSS receiver and antenna used in the survey should be of geodetic or survey grade equipment.

- GNSS measurement equipment should be calibrated and standardized periodically before being used in the field.
- The measurements should be carried out by considering the accuracy criteria accepted/recommended by the official authority/related regulation(s).
- The appropriate equipment and methods should be the choice to be able to meet the required accuracy.
- Zero baseline test should be conducted for a pair of GNSS receivers to verify the precision of the receiver, to prove that the receiver is operating correctly and to validate the data processing software.
- GNSS antenna height must be measured once before and after each survey.
- If possible, the GNSS surveys are conducted during the periods when the maximum number of satellites is in view, and the value of PDOP is at its minimum.
- The GNSS surveys should be observed with enough time to remove large outliers and to fix the reliable solutions.
- At least two independent observations (two baselines) should be carried out.
- Independent observations should be carried out after 30 minutes elapses (this allows the satellite constellation to change) by changing the height of the antenna (with cold start).
- Minimum 2 minutes observation is recommended for RTK and CORS RTK users.
- CORS RTK surveyors should use the nearest CORS base station for single-base users.

Whether conventional or GNSS system is used to get the required accuracy of the survey, the surveyor should understand and be aware of the following issues (URL-9):

- The limitation of the equipment to be used.
- The observation procedures.
- Processing techniques.
- Statistical analysis.

Usability of GNSS Technique for Cadastral Surveying in Turkey

Cadastral surveys have been carried out with different methods using various type of instruments with a compass and steel tape and then prism, tachometer, reduction tachometer, electromagnetic measurement instruments and more recently total station (Çağla et al. 2011). The rapid change in the world of science and technology has closely affected our profession and has started to use satellite-based positioning systems, especially GPS and GLONASS.

All the standards must be adapted to the government agency in Turkey with regard to map and map information production. Those standards were enacted in

Table 8.5 Some accuracy criteria for control and detail points measurements in Turkey

Point order	C1 (1st Order)	C2 (2nd Order)	C3 (3rd Order)	C4 (4th Order)	Detail
Properties	The points that are based on the high degree of networks and have 15-20 km distance length between them.	The points that are based on the high degree of networks and have 5 km average distance length between them.	The points that are based on the high degree of networks and have 3 km average distance length between them.	The points that are based on the high degree of closed traverse networks and photog. points.	Surveying point (parcel corner points, parcel border points, etc.)
Observation Mode	S	S	S/RS	S/RS/K/RTK	RTK (TUSAGA-Aktif/CORS)
S: Static					
RS: Rapid Static					
K: Kinematic					
RTK: Real-Time Kinematic					
Receiver frequency	Dual	Dual	Dual	Dual	Dual
Satellite Number (min.)	4	4	4	4	5
Elevation cut-off angle (min.)	15°	15°	10°	10°	10°
Measurement interval (max.)	15 s	15 s	15 s	10 s(S) 15 s(RTK)	5 s
Occupation time	2 h	45 min.	20 min. (per km +3 min. if > 5 km)	7 min. (S/RS) 5 epoch (RTK)	3 epoch (RTK)
Allowed accuracy values (in cm) – Horizontal (σ_φ ; σ_λ)	±3	±3	±5	±8	± 7
Allowed accuracy values (in cm) – Vertical (σ_h)	± 5	± 5	± 5	± 8	± 7
LSMMIPR article numbers	14–15	19–20	22	26 (43 for RTK)	45–46

Large Scale Map and Map Information Production Regulation (LSMMIPR) accepted in 2005. According to these regulations, cadastral measurements can be conducted with GNNS systems (in some situation conjunction with conventional terrestrial methods). In case of using the GNSS methods, the measurements should be conducted by considering the criteria given in the Regulation. Some measurement and accuracy criteria recommended by the LSMMIPR is given in Table 8.5 (URL-10).

Conclusions

In this study after explaining cadastre's current situation in Turkey briefly, it has been explained that areas of study and issues of study should be carried out in Turkey. As explained in the study, although 99% of cadastral surveys were completed in Turkey, there are still a lot of studies that will be carried out (like especially renovation, regular maintenance of cadastre, etc.). Even if optical methods can be still used, GNSS systems have become essential tools for cadastral surveyors thanks to their many advantages. Thus, indeed, it is widely used by providing flexibility and offering cost effective solutions around the world. By utilizing the type of CORS networks, positioning with high accuracy and in short time routinely is conducted in Turkey and across the world.

In summary if to say:

- The cadastral maps which constitute the basis for property right as one of the most fundamental human rights should be meticulously conducted by expert surveyors. Surveys carried out by unqualified technical staff will cause difficult and even irremediable problems.
- Training the users of this technology is extremely important. Those who carry out such measurements should be aware that they also produce important information. Thus, cadastral practitioners should be well trained and educated in the use of GNSS systems.
- Some control mechanism should be established and conducted in the field in addition to the ones that are already carried out in conventional methods.
- It is mostly possible to get a result even under the quite large DOP values, in poor satellite configuration and in multipath environment. Coordinate information by looking at the results cannot solve a problem without taking these factors into account. Furthermore, it could bring about new problems.

More information about the survey criteria can be accessed in URL-10.

References

Awange, J. 2012. *Environmental Monitoring Using GNSS: Global Navigation Satellite Systems*. Berlin/Heidelberg: Springer.

Bakıcı, S., 2015. Geoinformatik Teknolojilerinin Tapu ve Kadastro Genel Müdürlüğünde Kullanımı, 1. Konum Belirleme ve Uydu Navigasyonu Zirvesi-2015, 20 Mart, İstanbul (In Turkish).

Bakıcı, S. and Ç. Mekik, 2014. TUSAGA-Aktif: Delivering Benefits to Turkey. *Coordinates*, X(02), 19–26.

Bonnor, N. 2012. A brief history of global navigation satellite systems. *The Journal of Navigation* 65(1): 1–14.

Çağla, H., S. Akkus, E. Civcik and İ. Ozugur. 2011. Konya Example of Cadastre Renovation Work in Turkey. FIG Working Week, 18–22 May 2011, Marrakech-Morocco.

Çete, M., 2010. The Recent Reforms in the Turkish Cadastre. FIG Congress 2010, 11–16 April, Sydney, Australia.

İnam, Ş., M. Ertaş, Ş. Yalpır, A. Başarır and M. Akca, 2011. Renovation Studies in Turkey Cadastre: Reasons and Project Results. FIG Working Week, 18–22 May 2011, Marrakech-Morocco.

Kahveci, M. and F. Yıldız, 2012. GPS/GNSS Uydularla Konum Belirleme Sistemleri Teori ve Uygulama. (Güncellenmiş 5. Basım), Ankara: Nobel Yayınevi (In Turkish).

Kogure, S., M. Sawabe and M. Kishimoto, 2006. Status of QZSS Navigation System in Japan. *Proceedings of the 19th International Technical Meeting of the Satellite Division of The Institute of Navigation (ION GNSS 2006)*, Fort Worth, TX, September 2006.

Li, M., L. Qu, Q. Zhao, J. Guo, X. Su, and X. Li. 2014. Precise Point Positioning with the BeiDou Navigation Satellite System. *Sensors* 14(1): 927–943.

Majithiya, P., K. Khatri, and J.K. Hota. 2011. Indian Regional Navigation Satellite System – Correction Parameters for Timing Group Delays. *InsideGNSS* 6(1): 40–46.

Meha, M., Q. Kadiri, M. Çaka, R. Murati, A. Ahmeti, B. Shaqiri, F. Kelmendi, Q. Sinani, E. Xani and K. Ahmetaj, 2013. Catalogue for Measuring and Creating Cadastral Units. Kosovo Cadastral Agency.

Mekik, Ç., O. Yıldırım, and S. Bakıcı. 2011. The Turkish Real Time Kinematic GPS Network (TUSAGA-Aktif) Infrastructure. *Scientific Research and Essays* 6(19): 3986–3999.

Montenbruck, O., P. Steigenberger, R. Khachikyan, G. Weber, R.B. Langley, L. Mervart, and U. Hugentobler. 2014. IGS-MGEX: Preparing the Ground for Multi-Constellation GNSS Science. *InsideGNSS* 1(2): 42–49.

Reddy, J., 2010. The Integration of CORS Networks and the Cadastre and its Application in NSW, Australia. FIG Congress 2010, 11–16 April, Sydney, Australia.

Tamrakar, R.M. 2013. Potential Use of GPS Technology for Cadastral Surveys in Nepal. *Nepalese Journal of Geoinformatics* 12: 33–40.

URL-1 The Current Status of the Turkey Cadastral Studies (In Turkish). Available from: http://www.e-tkbm.gov.tr/publisher/projeizleme.htm [accessed 13/03/2015].

URL-2: TKGM 2014 Yılı Performans Programı (In Turkish). Available from: http://www.tkgm.gov.tr/sites/default/files/icerik/ekleri/tkgm-performans_programi_tasari-2014.pdf [accessed 13/03/2015].

URL-3: GPS Constellation Status. Available from: http://www.glonass-center.ru/en/GPS/ [accessed 13/03/2015].

URL-4: GLONASS Constellation Status. Available from: http://www.glonass-center.ru/en/GLONASS [accessed 13/03/2015].

URL-5: Galileo. Available from: http://www.gsa.europa.eu/galileo/why-galileo [accessed 13/03/2015].

URL-6: Galileo. Available from: http://www.esa.int/Our_Activities/Navigation/The_future_-_Galileo/What_is_Galileo [accessed 13/03/2015].

URL-7: Singapore Land Authority – Guidelines and Specifications for GPS Surveys of ISN Markers (Version 2.4), Mar 2011. Available from: http://www.sla.gov.sg/Portals/0/Circulars/att/LS/csdsgsd/Guidelines_and_Specifications_for_GPS_Surveys_of_ISN__Markers_v4.PDF [accessed 13/03/2015].

URL-8: New South Wales Government – Surveyor General's Directions No. 9, GNSS for Cadastral Surveys, May 2014. Available from: http://www.lpi.nsw.gov.au/__data/assets/pdf_file/0006/25944/sgddir9_Ver2.5_May_2014.pdf [accessed 13/03/2015].

URL-9: Guidelines for Cadastral Surveying using Global Navigation Satellite Systems (GNSS), NIMA (National Imagery and Mapping Agency), 2003. Available from: http://www.sssi.org.au/userfiles/docs/QLD Region/documents_1342665635458677117.pdf [accessed 13/03/2015].
URL-10: Large Scale Map and Map Information Production Regulation (In Turkish). Available from: http://www.resmigazete.gov.tr/eskiler/2005/07/20050715-5.htm [accessed 13/03/2015].
Xu, A., Z. Xu, X. Xu, H. Zhu, X. Sui, and H. Sun. 2014. Precise Point Positioning Using the Regional BeiDou Navigation Satellite Constellation. *Journal of Navigation* 67(3): 523–537.

Part II
3D Cadastre with Geo-Technology

Chapter 9
Conceptual Modelling of 3D Cadastre and LADM

Nur Amalina Zulkifli, Alias Abdul Rahman, Muhammad Imzan Hassan, and Tan Liat Choon

Introduction

The cadastral registration system in Malaysia is a parcel based system and it is 2D in nature (Chong 2006; Hassan and Abdul Rahman 2010; Abdul Rahman et al. 2011). This system of cadastral information has served most of the users need for decades. However, in the near future, 2D information may no longer be able to serve the community, especially in more complex situations such as buildings above roads in some large cities and towns. Three dimensional (3D) modelling of cadastral objects such as legal spaces of around buildings, utility networks and other spaces is one of the future aspects for the Malaysian cadastral system. The Malaysian 3D cadastral model could be developed within the framework of the Land Administration Domain Model (LADM) where the generation of the UML model that complies with the concept of LADM for the Malaysian cadastral system can be addressed. The purpose of the LADM is not to replace existing systems, but rather to provide a formal language for describing various sub-systems, so that their similarities and differences can be better understood (ISO 19152 2012).

The LADM covers both the spatial (i.e. LA_SpatialUnit) and administrative (i.e. LA_Party, LA_RRR, and LA_BAUnit) aspects of land administration. The main reason to apply the LADM is to reuse the collective knowledge from many countries in land administration and to have unambiguous definitions of the key concepts (Lemmen 2012). For the Malaysian country profile, the integrated support for both 2D and 3D parcels is very useful. In the LADM, 2D and 3D data are treated in a consistent manner throughout the model. It is important to realise that there is a difference between the 3D physical object itself and the legal space related to the object. The LADM only covers the 'legal space', that is the relevant space for the

N.A. Zulkifli (✉) • A.A. Rahman • M.I. Hassan • T.L. Choon
Department of Geoinformation, Universiti Teknologi Malaysia,
81310 Johor Bahru, Malaysia
e-mail: amalina.jc@gmail.com

T. Yomralioglu, J. McLaughlin (eds.), *Cadastre: Geo-Information Innovations in Land Administration*, DOI 10.1007/978-3-319-51216-7_9

land administration (bounding envelope of the object). To be able to register the 2D or 3D parcels in the cadastral registration, all real estate objects must have a survey document (i.e. LA_SpatialSource), which should make clear to what space the real estate object refers to.

This paper is organized as follows: The next Section describes some legal aspects of Malaysian land policy. The current Malaysian cadastral systems such as *eCadastre* and *eLand* are discussed in third Section. The penultimate Section presents the development of Malaysian LADM country profile (i.e. spatial and administrative parts). Finally, the conclusion is presented in the last Section.

Malaysian Land Policy

Malaysian land registration system requires recording of land rights via the registration of land title. According to the Federal Constitution 1957, land matters are under the jurisdiction of state governments and handled by the respective state registry or district land office, depending on where the document of title is formerly registered and guaranteed by the Federal Constitution 1957 as stated under Article 13 (rights to property).

Land ownerships are governed by the National Land Code 1965 and based on the Torrens System. It is protected by the National Land Code 1965 in Section 340 (Registration to confer indefeasible title or interest, except in certain circumstances). National Land Code 1965 states that land includes: the surface (including air space) of the earth and all substances forming that surface; the earth below the surface and all substances at the surface; all vegetation and other natural products; all things attached to the earth or permanently fastened to anything attached to the earth; and land covered by water.

Strata Rights

The Malaysian strata title registration was first introduced in 1966 by the National Land Code 1965 under Section 355 to Section 374 that dealt with subsidiary titles to each of the parcels within a building having two or more storeys. In order to simplify and overcome the inadequacies of these provisions in the National Land Code 1965, the National Land Council Review Committee deliberated and decided to recommend that a separate legislation on strata titles be enacted, and the existing National Land Code 1965 for subsidiary titles provisions be repealed and replaced by the Strata Titles Act 1985 (Act 318). This legislation came into force in 1985. Although the provisions of strata titles are now in Act 318, this new act is still to be read and construed together with the provisions and rules of the National Land Code 1965. The amendments under Strata Titles (Amendment) Act 2013 include the introduction of the Electronic Land Administration System of Strata Titles, the

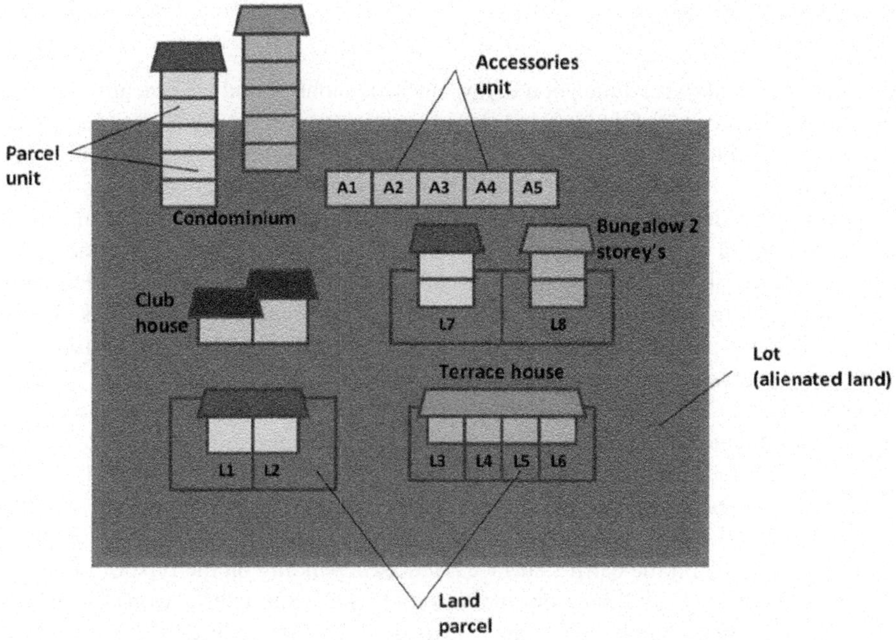

Fig. 9.1 Various cadastral objects related to strata in context of one lot

designation of limited common property, and the creation of one or more subsidiary management corporations to represent the different interests of parcel proprietors.

In Malaysia, a master lot can be subdivided into smaller lots for the purpose of establishing a strata scheme. Subsequently, the strata scheme lot can be subdivided into parcels and land parcels. Each parcel and land parcel can consist of an individual apartment or house. A land parcel means a unit delineated within the lot in which is comprised a building of not more than four storeys held under a strata title, which may have a shared basement, comprises accessory parcels and common property. A building intended for subdivision into parcels means any building or buildings having two or more storeys in a development area and intended to be subdivided into parcels; and any development area has two or more buildings intended to be subdivided into land parcels.

Figure 9.1 illustrates the various types of strata objects in Malaysia. A *parcel* in relation to a subdivided building, means one of the individual units comprised therein (apartment or condominium), which is held under separate strata title. An *accessory unit* means a unit shown in a strata plan, which is used or intended to be used in conjunction with a parcel. A *common property* means so much of the lot as is not comprised in any unit (including any accessory unit). A *limited common property* means common property designated for the exclusive use of the owners of one or more strata lots. A *land parcel* means a unit delineated within the lot (in which is comprised a building of not more than four storeys) which is held under a strata title and which may have shared basement, accessory unit and common property.

Underground Rights

For property deals with dimension below surface, underground land means land that lies below the surface of the earth while stratum means a cubic layer of underground land. Section 44(1)(a) of the National Land Code 1965 states that the extent of the exclusive use and enjoyment of so much of the land below that surface is limited only to such a depth reasonably necessary to the lawful use and enjoyment of the land. According to Section 92B and Section 92E of the National Land Code 1965, the State authority may specify the depth up to which the underground land directly and immediately, below the alienated land may be used, and different depths may be specified in respect of different parts of such underground land. Therefore, the National Land Code (Underground Land) (Minimum Depth) Regulations 2006 in National Land Code 1965 was introduced to specify the minimum depth of such underground lands. For agricultural land use, a minimum depth of six metres from the earth's surface has been suggested as the depth for underground land alienation by the committee for the category of agricultural land use. For the category of building and industrial land use, the fixing of minimum depth depends on the depth of piles for building on the earth's surface. Hence, depending on the type of building, the minimum depth of alienation underground land is ten metres from the earth's surface for residential building and, extending to fifteen metres for industrial constructions.

Native (Customary) Rights

This sub-section explains the customary rights in the Malaysian land administration system. The National Land Code 1965 (Act 56) is only valid within Peninsular Malaysia and Federal territories. Both Sabah and Sarawak states have implemented their own Land Ordinance. Malaysian customary rights are not governed by the Federal Constitution of Malaysia but governed by the state authority as customary rights are regarded as state land matters. Prior to the Malaysian Customary Rights, the states that still implemented these rights are namely Melaka, Negeri Sembilan, Sabah and Sarawak, which are Malacca Customary Land, *Tanah Adat Negeri Sembilan*, Native Lands and Native Customary Land for Sabah and Sarawak respectively. Customary rights are acquired by custom and owned by the natives. Each of these customary rights differs from one another in terms of land administration systems and land management systems due to the their historical differences.

The majority of the indigenous peoples of Malaysia still live in remote areas, although more and more now live in the periphery of the urban areas. Many survive by hunting and gathering, fishing, farming and by trading forest products. There are 28 indigenous groups making up 71.2% of the population of Sarawak state; 13 native people groups in Peninsular Malaysia numbering around 200,000 people (2010 estimate) or constituting 0.8% of the population of the Peninsular Malaysia.

In Sabah, the 39 ethnic groups apparently make up 61.22% of the state's total population.

Various complaints ranging from allegations of encroachment and dispossession of land; land included into forest or park reserves; overlapping claims and slow processing of request for the issuing of native titles or community reserves. A large part of the problem arises from a lack of recognition by the authorities of the concept of customary land of the indigenous peoples, or what constitutes customary land, when much of this land has not been, or is yet to be registered as customary land with the relevant government departments due to ignorance or misunderstanding on the part of the community on the processes involved.

Malaysian Cadastral System

In Malaysia, there are two organizations responsible for managing and maintaining the cadastral system, they are the Department General of Lands and Mines and the Department of Survey and Mapping Malaysia (DSMM). Both departments are within the Ministry of Natural Resources and Environment as shown in Fig. 9.2. The DSMM deals with the cadastral survey to determine the dimension, size and location of the properties. DSMM is also responsible for preparing Certified Plans (CPs), producing and managing the spatial component including the surveying and mapping of the cadastre parcels. The administrative (legal) data is the responsibility of the land offices. The land office deals with ownership registration, i.e. who owns the RRRs.

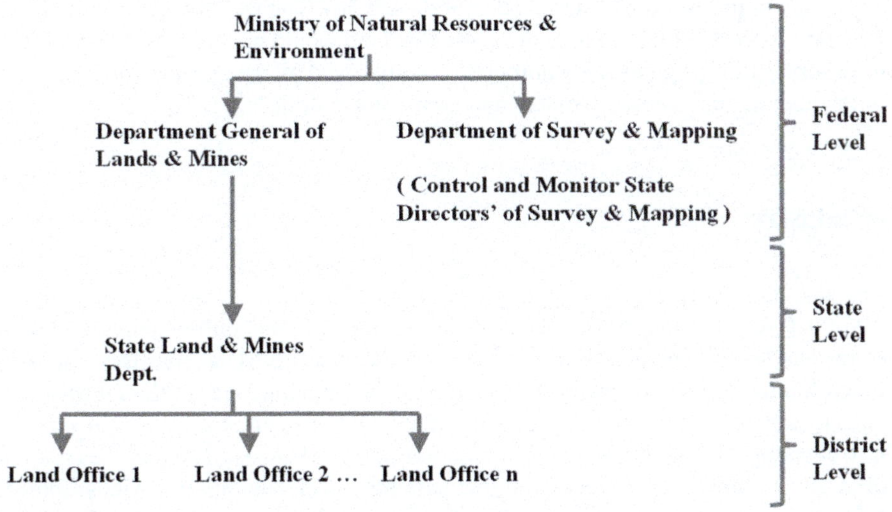

Fig. 9.2 Organisational structure of land administration in Malaysia

Both organizations have their own information management systems – *eCadastre* within DSMM and *eLand* within the land office (Tan and Looi 2013). These are two independent systems and in 2D. The Unique Parcel Identifier (UPI) has been introduced to link the land office and DSMM documents where every cadastral object has a unique identity number to differentiate from other cadastral objects.

The traditional cadastre system that is practised in Peninsular Malaysia is a parcel bound system and provides essential land and property information of the lots and land parcels. The existing Malaysian cadastral survey and mapping registration system and land registration system deal with properties located not only on the surface level, but also above and below the surface level. Therefore, the rights of the proprietor of the surface parcel shall also apply to the air space above and the space underground as well.

The authenticated cadastral map for Malaysia is called the Certified Plan (CP). There are three types of certified plans. Firstly, the Certified Plan (land parcel) is prepared in a standard format that permits the presentation of useful technical data such as bearings, distances, areas, lot numbers, boundary marks, abuttal's of adjoining parcels and the coordinates of two extreme boundary marks as well as other relevant information when dealing with strata and stratum alienations. The Certified Plan (strata building/land parcel) contains the parcel information like floor areas, parcel numbers, boundary dimensions where the buildings reside in a strata scheme and additional information on the height of the building, number of floors/levels and the strata parcels. There is no change to the Certified Plan (strata building/land parcel) after the implementation of *eCadastre* in Malaysia. Likewise, the Certified Plan (Stratum) provides additional information on the depth and mean sea level of the underground volumetric parcels.

The Certified Plan then becomes the source document relating to the creation of the parcel boundaries and specific provisions have been made with regard to its role and status in the National Land Code 1965 (Act 56). The said Act gives much significance to the Certified Plan in the sense that land would not be considered to have been surveyed if the plan is not authorised or approved by the DSMM with evidence of the boundaries, boundary marks and areas shown on it.

eCadastre

The vision of the Malaysian government is to become a developed country by the year 2020 which encompasses the realization of an efficient public delivery system at various levels. The *eCadastre* project is under the 9th Malaysian Development Plan (2006–2010). This has been approved to be implemented by the DSMM and aims to achieve a fully digital Malaysia by 2016. The main objective of *eCadastre* is to expedite the delivery system for land title survey. This would entail the creation of a survey accurate database at the national level suitable for Geographical Information Systems (GIS) users. There are three main components in *eCadastre*, namely Coordinated Cadastral System (CCS), Virtual Survey System (VSS) and

Cadastral Data Integrity System (CDIS). The implementation of CCS is a major part of the *eCadastre* project that includes field and office reengineering to reduce processes and increase the use of digital technology.

Since 1995, DSMM has embarked on a modernization programme that saw the dramatic computerization of both its field and office processes of its cadastral survey division. The digital cadastral database was created by capturing the surveyed accurate information of all land parcels. Under the *eCadastre* project, a comprehensive nationwide readjustment of the mesh-work of the parcels would be carried out based on a new geocentric datum. A dense network model known as the CCS of Real Time Kinematic Global Positioning System (RTKGPS) permanent stations has been established to provide precise geocentric positioning and implemented through the *eCadastre* project.

The current system of cadastral survey is able to capitalize on the advent of satellite based technologies. The new environment allows various cadastral survey processes such as, layout design submission, planning, field data capture, quality control, completed job submission, and approval to be carried out remotely via the mobile telecommunication network. Global Positioning System (GPS) provides real time positioning at centimetre resolution homogeneously to the entire country and coordinates will replace relative measurements as the ultimate proof of boundary mark position. Additional features such as building footprint and space images can be incorporated into the new database in a move towards a multipurpose cadastral.

eLand

Ministry of Natural Resources and Environment (NRE) creates an integrated computerized system, known as the Electronic Land Administration System *(eLand)* to realize the computerization of the overall management and administration of land. *eLand* is designed to improve the delivery of land administration and management services in Malaysia Peninsular using an integrated ICT infrastructure. Currently, the Ministry of NRE has implemented two systems for the administration of land information, which are *'Computerised Land Registration System'* (CLRS) and *'Land Revenue Collection System'* (LRCS). Both systems are already operational in all State Lands and Mines Offices and District Land Offices in Malaysia Peninsular.

The main objective of *eLand* is to develop a comprehensive system in land offices in order to modernize all activities that are related to land and to realize the implementation of electronic government in the public sector. In addition, the mission of *eLand* is to develop and implement a National Land Administration System via ICT towards enhancing the growth of national development. *eLand* is an integrated and a fully computerized system to handle the management and administration of land offices in order to improve the speed and quality of service delivery to the public for all land related transactions. *eLand* also enables the public to make payments online and print the payment receipts, checking details on their own land and so on.

Besides, *eLand* has nine main modules with 85 major business processes in accordance to the existing National Land Code 1965. The business processes supported by *land* maximizes the utilization of the existing ICT infrastructure, taking into account the existing processes and procedures. *eLand* will also be integrated with the existing CLRS and LRCS systems accordingly. The focus of the project is on the major processes that can be implemented without any changes on the existing laws. Any required changes to the existing laws will be done later. However, the modules and *eLand* are designed to be flexible to address possible changes to the system because of the changes in the existing laws.

Generally, the design of the module adheres to the best practices in application development. Emphasizes are given to aspects such as ease of use, traceability, expandability, security and flexibility. These fundamental design aspects of the module and *eLand* in general ensure that the system is able to handle the existing and future requirements on the system.

Malaysian LADM Country Profile

The development of Malaysian country profile is based on the User Requirement Analysis (URA) obtained from DSMM and Land Office officers. The URA of this LADM country profile has been established from workshops and meetings organized by the organization cadastral call group and UTM (together with TU Delft). Many suggestions and views were gathered and discussed during the workshops (i.e. comprehensive, 3D lot, BAUnit, strata objects, 2D topology, intermediate points, spatial source, administrative source, level, abstract class, identifier, code list, constraint in share attribute, indexing, clustering and implicit or explicit encoding of CRS and UoM). All the suggestions were incorporated in the conceptual of the country profile. 'MY_' is the prefix for the Malaysian country profile, covering both the spatial and administrative (legal) data modelling (Zulkifli et al. 2014). All classes in Malaysian model are derived directly or indirectly (via the inheritance hierarchy) from LADM classes. To illustrate the inheritance from the LADM classes, the MY_classes have either in upper right corner the corresponding LA_ class name in italics or have the explicit inheritance arrow shown in the diagram.

Spatial Part

In the Malaysian country profile, spatial units can be in 2D or 3D forms. Traditionally, lots (land parcels) are 2D, but the subsurface of lots do already exist with 3D description with volumetric descriptions. The model has introduced an abstract class MY_GenericLot holding the attributes of a lot and this class has two specializations MY_Lot2D and MY_Lot3D, with their own attributes and structure.

Currently MY_Lot2D is based on 2D topology with references to shared boundaries (MY_BoundaryFaceString).

Note that there are several abstract classes in the Malaysian country profile as indicated in Italics: *MY_SpatialUnit*, *MY_Shared3DInfo* and *MY_GenericLot*. These classes are only supporting the modelling process, representing shared attributes and structures, and these abstract classes will not get any instances (and therefore no corresponding table in the database implementation). For *MY_Shared3DInfo* there is a geometry attribute (of type *GM_Solid*). Normally the 3D geometry in LADM is represented in *LA_BoundaryFace*, but given the fact that no 3D topology is used there is 1-to-1 association with the spatial unit (one of the specializations of *MY_Shared3DInfo*). So, it could be argued that the proposed country profile is ISO conforming, despite the absence of the class *LA_BoundaryFace*.

To make the model comprehensive and future proof, a wide range of spatial units is supported including legal spaces for utilities (3D), customary areas, and reserved land (forest, wildlife areas). It should be noted that reserved land (forest, wildlife) are associated with own RRRs, normally have no overlap, but in some cases overlap can happen depending on state and type. The spatial description of reserved land is by text or sketches, but they may also be surveyed (or a combination with the above).

The various types of spatial units are organized in levels. In this model, *MY_Level* class is used to organize the various types of spatial units. For *MY_Level*, there is a type attribute which describes the level type of the spatial unit. The type of spatial unit includes customary, lots (mixed land and road), building (parts, strata) and utilities. The codelist for this attribute can be referred to *MY_LevelContentType*. Basically, *MY_Level* is a collection of spatial units with a geometric or thematic coherence. The following levels are proposed: level 0 for customary, level 1 for reserved land, level 2 for 2D lot, level 3 for 3D lot, level 4 strata, and level 5 for utility. In the involved classes a constraint has been added (third box in class diagram) to make this more explicit. For an example; *MY_Customary* has a constraint: *MY_Level.name* = 'level 0'. In the class diagram (refer Fig. 9.3), the blue classes refer to part of strata objects for a better readability of the model.

To make the model comprehensive and future proof, a wide range of spatial units is supported including legal spaces for utilities (3D), customary areas, and reserved land (forest, wildlife areas). It should be noted that reserved land (forest, wildlife) are associated with own RRRs, normally have no overlap, but in some cases overlap can happen depending on state and type. The spatial description of reserved land is by text or sketches, but they may also be surveyed (or a combination with the above).

The various types of spatial units are organized in levels. In this model, *MY_Level* class is used to organize the various types of spatial units. For *MY_Level*, there is a type attribute which describes the level type of the spatial unit. The type of spatial unit includes customary, lots (mixed land and road), building (parts, strata) and utilities. The codelist for this attribute can be referred to *MY_LevelContentType*. Basically, *MY_Level* is a collection of spatial units with a geometric or thematic coherence. The following levels are proposed: level 0 for customary, level 1 for reserved land, level 2 for 2D lot, level 3 for 3D lot, level 4 strata, and level 5 for utility. In the involved classes a constraint has been added (third box in class dia-

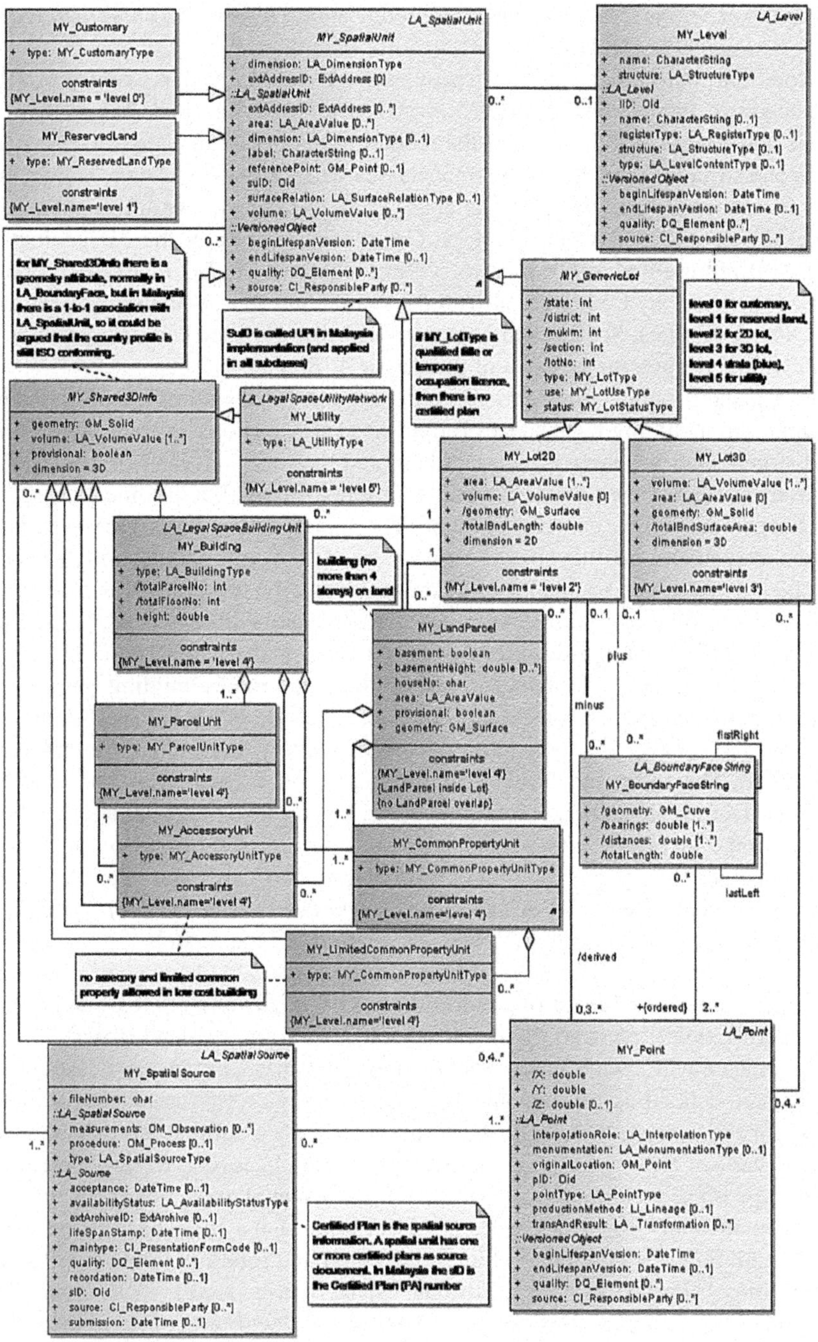

Fig. 9.3 Details of spatial side of the model

gram) to make this more explicit. For an example, *MY_Customary* has a constraint: *MY_Level.name* = 'level 0'.

In case of spatial source documents (usually certified plans) there are links with spatial unit and point tables: *MY_SpatialSource* has association with *MY_SpatialUnit* and *MY_Point*. The LADM Malaysian country profile uses suID for spatial unit and sID for spatial source. Basically, suID in Malaysian country profile is based on Unique Parcel Identifier (UPI). sID for spatial source is the certified plan number. A note has been added in the country profile to indicate this.

In Malaysia there is normally 1 to 1 relationship between BAUnit and spatial unit. However, there are some cases where one BAUnit (with same RRRs attached) has multiple Spatial Units: a combination of farmland with residential house (Group Settlement Act). Also, some status values of *MY_Lot* (e.g. 10, which indicates the charting stage) relate to lots that have yet had RRRs attached; to make this possible, the multiplicity of the association between spatial unit and BAUnit is 0..1 (optional) at BAUnit side. In the future, the Malaysian land administration system can consider more grouping of spatial units with the same RRRs attached via a single BAUnit.

Administrative Part

The legal part of the Malaysian LADM country profile contains Party and Administrative package. The main class of the party package is a *MY_Party* class with its specialisation *MY_GroupParty*. There is an optional association class called *MY_PartyMember*. Basically, a party is a person or organisation that plays a role in a rights transaction. The organisation can be a company, a municipality or a state. A *group party* is any number of parties, forming together a distinct entity. A *party member* is a party registered and identified as a constituent of a group party. This allows the documentation of information to a membership (holding shares in right).

The administrative package concerns the abstract class *MY_RRR* (with its three concrete subclasses *MY_Right*, *MY_Restriction* and *MY_Responsibility*), *MY_Mortgage*, *MY_BAUnit* and *MY_AdministrativeSource*. A *right* is an action or activity that a system participant may perform on or using an associated resource such as ownership, customary, easement and tenancy rights. The rights may be overlapping or may be in disagreement. A *restriction* is a formal or informal entitlement to refrain from doing something. For example, it is not allowed to build a house on a piece of land or not allowed to transfer the title to other parties within some period time. A *responsibility* is a formal or informal obligation to do something such as the responsibility to maintain a monument or a building. The instance of class *MY_Mortgage* is a mortgage. *MY_Mortgage* is a subclass of *MY_Restriction*. *MY_Mortgage* is also associated with *MY_Right* class. The mortgage can be associated to zero or more rights.

A BAUnit is an administrative entity consisting of zero or more spatial units (parcels) against which one or more unique and homogeneous rights, responsibilities or restrictions are associated to the whole entity as included in the Land Administration System. An example of a BAUnit is a basic property unit with two spatial units with same RRRs attached (e.g. Federal Land Development Authority – FELDA). A settler can have two spatial units (i.e. residential and farm land) with same RRRs attached. A BAUnit may play the role of a 'party' because it may hold a right of easement over another, usually neighbouring, and spatial unit.

One of the important foundations of LADM is the fact that all information in the system should originate from source documents and that the association to the source document is explicitly included. In case of administrative source documents (usually titles) there are associations with right, restriction (including mortgage) and responsibility (RRR) and basic administrative unit. *MY_AdministrativeSource* associates with *MY_RRR* and *MY_BAUnit*. The LADM Malaysian country profile uses sID for administrative source. Basically, sID for administrative source is title number.

Except source documents, all classes in LADM (and therefore also all derived classes in Malaysian country profile) are a subclass of VersionedObject and inherit all the VersionedObject attributes (refer Fig. 9.4). The class VersionedObject is introduced in the LADM to manage and maintain historical data. As source documents cannot change, only new source documents can arrive, they are not versioned. The current land administration system in Malaysia does not yet support full history management, so this is a significant change. It is not only an important change for the land administration system itself, but it is also crucial for the future Malaysian information infrastructure, as others might need the functionality to refer to historic versions of land administration objects.

The 3D Cadastre Modelling Within LADM

In 3D cadastre, 3D space is subdivided into volumes partitioning the 3D space. A legal basis, real estate transaction and the cadastral registration should support the establishment and conveyance of 3D right. Hybrid cadastre proposed by Stoter (2004) is an initial step towards implementation of 3D cadastre in Malaysia. The concept of hybrid cadastre is to preserve the current 2D registration and add the 3D component in the registration situation. Figure 9.5 illustrates examples of 3D objects in the cadastral system.

Traditionally, cadastral registration systems are parcel based and it is 2D in nature. However, in the very near future, this 2D cadastral system may not be able to serve more advanced situations and need to be extended to 3D cadastral system as reported by the following researchers (Griffith-Charles and Sutherland 2013; Guo et al. 2013; Karki et al. 2013; Pouliot et al. 2013; Stoter et al. 2013; Vandysheva et al. 2012; Wang et al. 2012). The LADM therefore supports 3D cadastral registration as shown in Fig. 9.5.

In the Malaysian LADM country profile, there are several classes representing 3D spatial unit (i.e. *MY_Building*, *MY_Utility* and *MY_Lot3D*). Both *MY_Building*

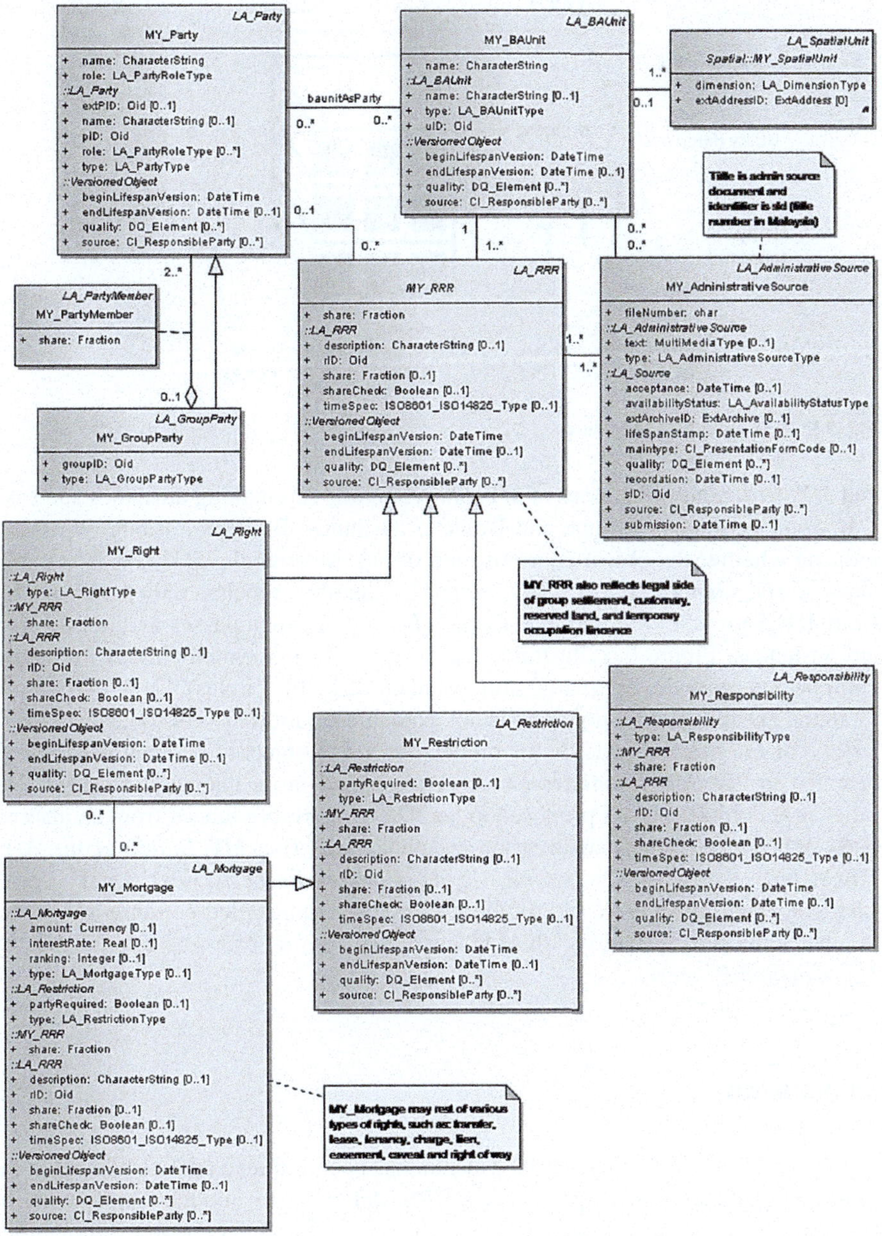

Fig. 9.4 Details of administrative side of the model

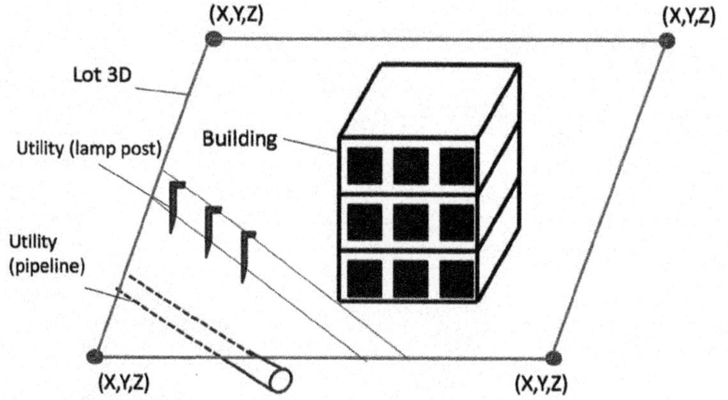

Fig. 9.5 3D cadastre components

and *MY_Utility* are subclasses of *MY_Shared3DInfo*, containing attributes such as a *GM_Solid* geometry, volume and Boolean attributes. Boolean attribute is used to indicate whether the object is provisional or not. Meanwhile, *MY_Lot3D* is a subclass of *MY_GenericLot*. *MY_GenericLot* has another subclass called *MY_Lot2D*. Both *MY_Shared3DInfo* and *MY_GenericLot* are abstract classes and do not have any instances. Figure 9.6 illustrates the overview of associated classes for spatial components (with the 3D spatial unit as indicated by the circles).

In the 3D spatial unit, topology is not available for lot (*MY_Lot3D*), utility (*MY_Utility*) or for strata objects. In the model, one strata object type remains to be represented in 2D, *MY_LandParcel* (with building no more than four storeys). The other strata objects are all proposed to be 3D and therefore inherit from an abstract class *MY_Shared3DInfo*, with strata specializations (i.e. *MY_BuildingUnit*, *MY_ParcelUnit*, *MY_AccessoryUnit*, *MY_CommonPropertyUnit* and *MY_LimitedCommonPropertyUnit*. As there can be several limited common properties in one common property, this is modelled as a part of relationship to *MY_CommonProperty*.

Conclusion

A concept of 3D modelling for 3D cadastre has been initiated and introduced in the Malaysian LADM country profile. The UML details of the model comply with the concept of the LADM has been presented. The presentation of country profile based on the standard is to understand the structure within the individual country land administration system and to show examples of structures that can be useful in building profile for other countries. However, the country profile only proposed 2D topology model and do not include 3D topology model for spatial unit. The potential usage of the 3D topology per building needs further investigation to represent

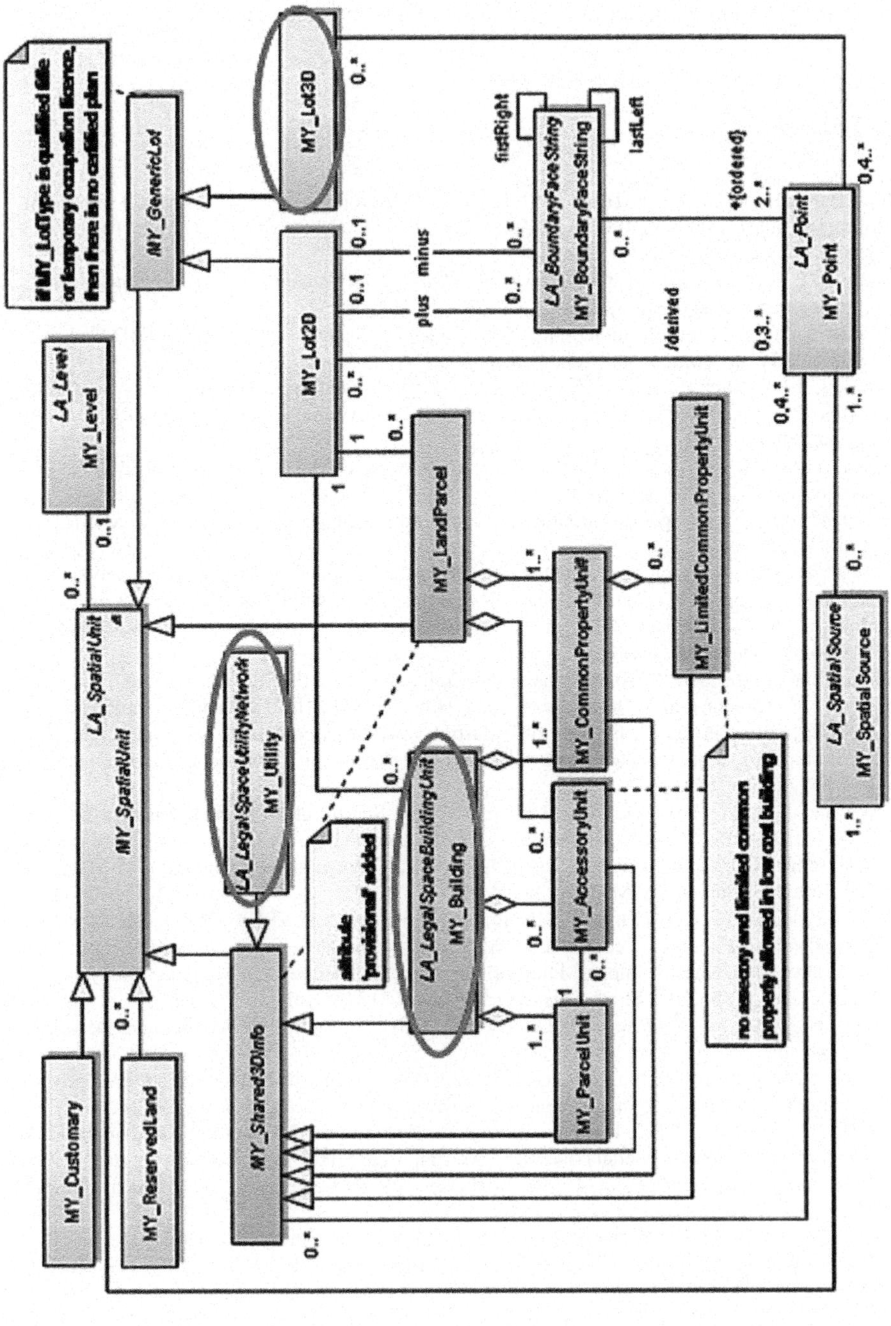

Fig. 9.6 Overview of spatial unit of the model

the various units within the building that share faces. In the near future, the Malaysian LADM country profile should be extended to develop a 3D cadastre prototype that would support and maintain the topology with other geometric elements.

References

Abdul Rahman, A., Chee Hua Teng, and P.J.M. Van Oosterom. 2011. Embedding 3D into multipurpose cadastre. In *FIG Working Week – Bridging the Gap between Cultures*, 18–22 May 2011. Morocco: Marrakech.

Chong, S.C. 2006. *Towards a 3D Cadastre in Malaysia-an Implementation Evaluation*. M.Sc. Thesis. Delft University of Technology, Delft, The Netherlands.

Griffith-Charles, Ch., and M. Sutherland. 2013. Analysing the costs and benefits of 3D Cadastres with reference to Trinidad and Tobago. *Computers, Environment and Urban Systems* 39(2).

Guo, R., L. Li, S. Ying, P. Luo, B. He, and R. Jiang. 2013. Developing a 3D cadastre for the administration of urban land use: A case study of Shenzhen, China. *Computers, Environment and Urban Systems* 39(2).

Hassan, M.I., and A. Abdul Rahman. 2010. An integrated Malaysian cadastral system. FIG Congress 2010. Sydney, Australia.

ISO 19152. 2012. Geographic information – Land Administration Domain Model (LADM), version 1 December 2012.

Karki, S., R.J. Thompson, and K. McDougall. 2013. Development of validation rules to support digital lodgement of 3D Cadastral plans. *Computers, Environment and Urban Systems* 39(2).

Lemmen, C. 2012. *A Domain Model for Land Administration*. PhD Thesis. Delft University of Technology, Delft, the Netherlands.

National Land Code 1965. 2010. National Land Code (Act 56 of 1965) & Regulations.

Pouliot, J., M. Vasseur, and A. Boubehrezh. 2013. How the ISO 191521 and administration domain model performs in the comparison of cadastral systems: A case study of condominium/co-ownership in Quebec (Canada) and Alsace Moselle (France). *Computers, Environment and Urban Systems* 39(2).

Survey and Mapping Director General Secular. 2003. Survey and Mapping Director General Secular (PKPUP), Ref.3, Garis Panduan Amalan Kerja Ukur Kadaster Sehubungan dengan Pelaksanaan Peraturan Ukur Kadaster 2002. As at 22 October 2003. Kuala Lumpur, Malaysia: Department of Survey and Mapping Malaysia.

———. 2006. Survey and Mapping Director General Secular (PKPUP), Ref.3, Peraturan dan Garis Panduan Ukur bagi Pecah Bahagi Bangunan untuk Pengeluaran Hakmilik Strata. As at 29 August 2006. Kuala Lumpur, Malaysia: Department of Survey and Mapping Malaysia.

Strata Titles Act 1985. 2010. Strata Titles Act 1985 (Act 318) & Rules and Order.

Stoter, J.E. 2004. *3D Cadastre*. Ph.D. Thesis. Delft University of Technology, Delft, the Netherlands.

Stoter, J., H. Ploeger, and P. van Oosterom. 2013. 3D Cadastre in the Netherlands: Developments and international applicability. *Computers, Environment and Urban Systems*, 39(2).

Tan, L.C., and K.S. Looi. 2013. Towards a Malaysian Multipurpose 3D Cadastre based on the Land Administration Domain Model (LADM) – An Empirical Study. In *Proceedings of the 5th FIG Land Administration Domain Model Workshop*, 24–25 September 2013, Kuala Lumpur, Malaysia.

Vandysheva, N., S. Sapelnikov, P. van Oosterom, M. de Vries, B. Spiering, R. Wouters, et al. 2012. The 3D cadastre prototype and pilot in the Russian Federation. In *Proceedings of the FIG working week*, May 2012, Rome.

Wang, C., J. Pouliot, and F. Hubert. 2012. Visualization principles in 3D cadastre: A first assessment of visual variables. In: *Proceedings 3rd international workshop on 3D cadastres*, Shenzhen.

Zulkifli, N.A., A. Abdul Rahman, H. Jamil, C.H. Teng, L.C. Tan, K.S. Looi, K.L. Chan, and P. van Oosterom. 2014. Towards Malaysian LADM Country Profile for 2D and 3D Cadastral Registration System. In *Proceedings of the FIG Congress, Kuala Lumpur, Malaysia*, 16–21 June 2014.

Chapter 10
The LADM Based on INTERLIS

Michael Germann, Jürg Kaufmann, Daniel Steudler, Christiaan Lemmen, Peter Van Oosterom, and Kees De Zeeuw

Introduction

Both the conceptual schema language INTERLIS and the land administration domain model (LADM) share the same model driven architecture (MDA) principles. In this paper we explore how INTERLIS and LADM complement each other in actual implementation of land administration system based on the LADM using INTERLIS and its tools.

This paper now first introduces some INTERLIS concepts in next Section. The concepts are then applied to LADM, as explained in third Section, with special attention to data modelling and constraint formulation. The fourth Section compares the LADM/INTERLIS approach with other standards (UML/GML). Finally, future work and main conclusions are provided in last two sections.

INTERLIS Concepts

A Short History of INTERLIS

The first version of the data modelling language INTERLIS was introduced in Switzerland in the late 1980s (Dorfschmid and Kaufmann 1987) and has become a Swiss standard in 1998 (SN 612030). The actual version 2.3 of the standard (COGIS 2006) is an object-oriented conceptual schema language (CSL), which is being used

M. Germann (✉) • J. Kaufmann • D. Steudler
Swiss Land Management Foundation, P.O. Box 5236, CH-3001 Berne, Switzerland
e-mail: jkcons@hotmail.ch

C. Lemmen • P. Van Oosterom • K. De Zeeuw
TU Delft, Delft University of Technology, Delft, The Netherlands

T. Yomralioglu, J. McLaughlin (eds.), *Cadastre: Geo-Information Innovations in Land Administration*, DOI 10.1007/978-3-319-51216-7_10

to precisely define (spatial) data models in textual form with a rigid computer process able syntax. An important characteristic of the language is that it can easily be understood by application and IT experts, thereby bridging the gap between application and IT domains.

While INTERLIS was originally designed and used mainly for land administration, it is not restricted to land administration data modelling. In fact INTERLIS is a general purpose modelling language. Due to its flexibility it has become part of the Swiss Act on Geoinformation (Swiss Confederation 2007) and is currently being used to describe the 160+ data models of the Swiss National Data Infrastructure (NSDI).

INTERLIS Key Features

INTERLIS has a unique set of features which sets it well apart from other modelling standards (i.e. UML or XML-Schema):

- INTERLIS schemas are defined as easy to read text files. The rigid syntax can be directly processed by computer programs;
- The language has built-in geometric data types (point, poly-line, polygon), making it especially suitable for models in the geoinformation domain. Note that at the moment there is not yet a data type of a 3D volumetric solid, such as a polyhedron; and
- Each INTERLIS data model automatically defines a system neutral XML based data exchange format and there are also tools to generate a database schema (SQL DDL).

An interesting aspect of the language is that it is possible to quality check INTERLIS data against INTERLIS data models (including constraints for valid data), thereby enabling fully automated quality control of spatial data.

INTERLIS Tool Chain

The intense use of INTERLIS in Switzerland would not be possible, if the language would not be supported by a wide range of tools. The following list gives a brief overview by naming the most important tools (free and commercial):

- The INTERLIS compiler checks the syntactical correctness of an INTERLIS data model (free);
- The INTERLIS checker can quality check INTERLIS XML data against INTERLIS data models (free);

- The INTERLIS UML editor is used to create INTERLIS models from UML diagrams or to visualize existing INTERLIS data models as UML diagrams (free);
- Data translators can convert data sets from many GIS systems / databases to and from INTERLIS XML (free and commercial); and
- Schema tools can generate database schemata directly from INTERLIS data models (free and commercial).

More information about the INTERLIS language and its tools is available at the official INTERLIS web site at www.interlis.ch.

Integration with LADM

After the introduction of INTERLIS, it should be obvious that LADM and INTERLIS is a perfect match. By applying the INTERLIS data modelling language to the LADM, ISO 19152:2012 (ISO, 2012; Lemmen et al. 2015) standard, we get computer processable model descriptions, which can be used to initialize databases or transfer LADM data via XML. Using INTERLIS for LADM also means that all free available INTERLIS tools such as compiler, checker, UML editor, etc. can be directly applied to LADM derived country profiles.

Model Descriptions

To test the feasibility of a LADM INTERLIS implementation, the Swiss Land Management foundation (SLM) started to describe the LADM ISO 19152 standard with INTERLIS. The core work was completed in February 2014 and the full model can be downloaded freely from the SLM web site www.swisslm.ch. Figure 10.1 shows an example of LADM-UML diagram translated to INTERLIS.

Constraints

The INTERLIS standard includes an OCL-like constraint language. Constraints can be defined on object level (MANDATORY CONSTRAINT) or class level (SET CONSTRAINT, UNIQUE CONSTRAINT). Some of the constraints/invariants of the LADM model can be directly expressed by the INTERLIS constraint language, as the following example shows (the UML pseudo OCL invariant 'if dimension=2D then volume not specified'):

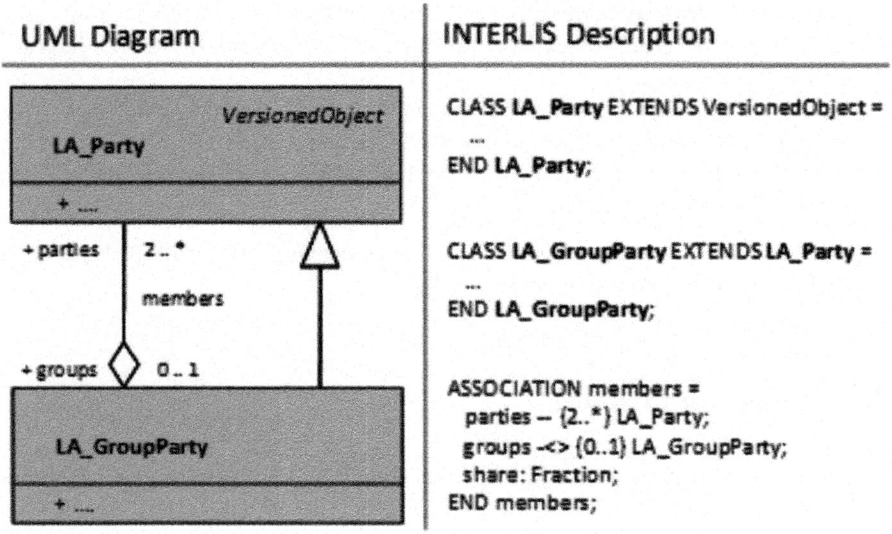

Fig. 10.1 Example of LADM UML diagram translated to INTERLIS (note in the UML Diagram the association class LA_PartyMember is not depicted, but this is included in the INTERLIS description)

```
CLASS LA_SpatialUnit EXTENDS VersionedObject =
 area: LIST {0..*} LA_AreaValue;
 dimension: LA_DimensionType;
 extAddressID: LIST {0..*} LADM_Base.External.ExtAddress;
 label: CharacterString;
 referencePoint: GM_Point;
 suID: MANDATORY Oid;
 surfaceRelation: LA_SurfaceRelationType;
 volume: LIST {0..*} OF LA_VolumeValue;
MANDATORY CONSTAINT
 !! if dimension=2D then volume not specified
 NOT (
dimension == #2D
 )
 AND (
 DEFINED (volume)
 )
 END LA_SpatialUnit;
```

The current version of the INTERLIS checker can process such constraints without additional configuration.

Comparison with Other Standards

After the introduction to the INTERLIS language and its application to ISO 19152 LADM the reader might ask him/herself what makes the INTERLIS/LADM approach so special. After all there are other well established standards that could do the same with another set of tools. In this section we try to answer this legitimate question from our point of view.

Reduced Complexity

INTERLIS was designed from the very beginning to model LIS systems and to exchange data between LIS in a system neutral format. The concentration on the two most important tasks, model LIS system and exchange format, in our application domain leads to significant complexity reductions. The INTERLIS 2.3 reference manual consists of only 160 pages, describing the language and also the data exchange format. The reference manual is self-containing making no references to other standards except XML. This reduced complexity makes it much easier for software developers to create powerful tools (compiler, checker, translators, etc.).

Comparison to UML

UML is a standard mainly intended to document all phases of modern software development (design, development, deployment, maintenance). UML provides many diagram types to support those activities including class, state and behaviour diagrams. As UML has no direct relation to Land Administration or Land Information System in general, applying UML to LIS is sometimes difficult as concepts for even the most basic geometric types (point, line and polygon) are missing in UML. Note that within ISC TC211 there is a family of standards providing both generic base types (e.g. the geometry primitives in ISO 19107), domain models (e.g. LADM in ISO 19152) and many other aspects in between (e.g. temporal type, reference systems, metadata, quality). INTERLIS uses UML class diagrams for visualization of data model structures. There is even a free tool to support the UML / INTERLIS integration (UML-Editor).

Comparison to GML

While GML was originally designed as a data exchange standard it is used today also as a modelling language. While it is possible to use GML for modelling it is somewhat inconvenient as the resulting XML-Schemas are not easily readable by humans without additional tools. Therefore GML should be used as a flexible and model-driven transfer standard only, but not as a modelling language.

The GML standard supports many geometric primitives, making it difficult for software vendors to implement all those types consistently. The Swiss eCH-0118 specification (eCH-0118 2011) therefore defines an INTERLIS/GML mapping to use a subset of GML as an alternative INTERLIS transfer mechanism.

Future Work

Swiss LADM Country Profile

As a next logic step it was decided to implement the Swiss LADM country profile in INTERLIS. This work is sponsored by the Swiss government (swisstopo) and was started in February 2015. The project will also make LADM compatible data from Canton Solothurn available on a public webserver. All provided data will be quality checked (based on more explicit constraints for valid data) by an automatic LADM check service.

INTERLIS 2.4

The LADM/INTERLIS implementation work has directly inspired some additional work on the actual INTERLIS 2.3 language to even better support the LADM standard (i.e. improved constraint formulation, LIST and BAG with basic types). The upcoming INTERLIS 2.4 standard will be published soon.

Complete LADM Described in INTERLIS

After describing the core of LADM in INTERLIS, the next step is a more complete description covering: all possible spatial representation types (spatial profiles: from text to topology), include 2D, 3D and mixed spatial units, describe all constraints mentioned in ISO19152, all specializations of spatial units (such as LA_LegalSpaceBuildingUnit and LA_LegalsSpaceUtilityNetwork), etc.

3D Support

INTERLIS supports 3D point/line and polygons but there are no special 3D-types (i.e. solids) in the standard at the moment. As 3D cadaster becomes more and more common in urban areas the better integration of 3D data types will be an issue in future versions of the language (INTERLIS 3?). In Lemmen et al. (2010) an overview is given of all spatial representation types and the challenges that occur when integrating 2D and 3D spatial units in one environment.

Conclusion

By applying INTERLIS to the LADM ISO 19152 standard we get directly computer processable data models. This approach can significantly speed up the implementation of LADM country profiles. As access to all specifications and important tools (compiler, checker, UML-Editor) is free, the LADM/INTERLIS approach can be easily implemented with minimal financial investments.

References

COGIS. 2006. *INTERLIS 2.3 reference manual*. Coordination, Geo-Information and Services (COGIS), a division of the Swiss Federal Office of Topography, Swisstopo, Switzerland. http://www.interlis.ch/interlis2/docs23/ili2-refman_2006-04-13_e.pdf.

Dorfschmid, S., and J. Kaufmann. 1987. *Detailkonzept: Reform der amtlichen Vermessung*, Swiss Department of Justice. http://www.cadastre.ch/internet/cadastre/de/home/docu/publication/P075.html.

eCH-0118. 2011. *GML Kodierungsregeln für INTERLIS*.

ISO 19152:2012. 2012. *Geographic information – Land Administration Domain Model (LADM)*, version 1 December 2012.

Lemmen, Christiaan, Peter Van Oosterom, Rod Thompson, João P. Hespanha, and Harry Uitermark. 2010. *The modelling of spatial units (Parcels) in the Land Administration Domain Model (LADM)*. In: Proceedings of the XXIV FIG International Congress, April 2010, Sydney.

Lemmen, Christiaan, Peter Van Oosterom, and Rohan Bennett. 2015. The land administration domain model. *Land Use Policy*. http://dx.doi.org/10.1016/j.landusepol.2015.01.014.

Swiss Confederation. 2007. *Federal act on geoinformation*. http://www.admin.ch/opc/en/classified-compilation/20050726/index.html.

Chapter 11
The Underground Space Use Right Registration with the Approach of 3 Dimensional Cadastre Concept

Bambang Edhi Leksono, Andi Ristiawan, Hendriatiningsih Sadikin, and Lucy Meyke

Introduction

Background

Cadastre System (land registry) that prevail in Indonesia today still use land parcels of 2D (two dimensional) as the basic entity of right registration. The advances of developing technology in an effort to overcome land limitation especially in urban area has changed the orientation of development to vertical direction in the form of utilization of space above or below the land. This means the cadastre object has been developed into three-dimensional space that is not located on the land surface so the 3D cadastre approach is necessary in cadastre affair.

Any usage and mastery of land by anyone and for whatever purposes, should be based on a right under national law in accordance with the allocation of land and its usage (Boedi Harsono 1999). National land law is currently more to regulate rights on the land surface (Maharani 2008). This means the law of the land is now only related to the rights existing above the land and has not accommodated the mastery of underground space. Usage of the underground space is not included in the scope of the authority which is based on land rights that is in the national land law today (Hutagalung 2008). Regulation in land such as the Law Number 16 Year 1985 more regulates granting the right settings for the utilization of space for apartment building that are in fact located on the land surface.

B.E. Leksono (✉) • H. Sadikin • L. Meyke
Graduate Program in Land Administration, Institute of Technology Bandung, Jl.Ganesha 10, 40132 Bandung, Indonesia
e-mail: bleksono@gd.itb.ac.id

A. Ristiawan
National Land Agency, Jl.Sisingamangaraja 2, 12110 Jakarta, Indonesia

T. Yomralioglu, J. McLaughlin (eds.), *Cadastre: Geo-Information Innovations in Land Administration*, DOI 10.1007/978-3-319-51216-7_11

Based on the above matter, the usage and mastery form of space in the underground building has not been accommodated by rights stipulated in Law Number 5 Year 1960 and Law Number 16 Year 1985. Therefore, we need a form of new right that can be called Underground Space Use Right (HGRBT) where the implementation needs the 3D cadastre approach.

Problem Formulation

The problem formulation in this research is how the implementation of Underground Space Use Right registration with the approach of 3D cadastre concept?

Hypothesis

This research is conducted based on the hypothesis that Underground Space Use Right registration can be realized with a 3D cadastre concept approach.

Objective

The purpose of this research is to study the implementation of Underground Space Use Right with a 3D cadastre concept approach.

Benefit Research

The result of this research may provide benefit in developing a knowledge in land registration field, particularly regarding the object of underground space. In addition, the result of this study is expected to contribute in improving the land registration policies, particularly those related to the usage and mastery of space below the land.

Expected Results

This research is expected to produce a study regarding Underground Space Use Right registration with 3D cadastre concept approach.

Research Scope

- 3D cadastre in this study is established with the aim to support the implementation of HGRBT registration on the forms of usage and mastery of space at the Blok M mall building.
- Location of case study in this research is Blok M bus terminal located at Jalan Sultan Hasanuddin, Melawai village, Kebayoran Baru district, South Jakarta. Usage and mastery of the underground building has been real in this location in the form of Blok M mall shopping centre.
- The usage of underground space in this research is a form of usage in which the land surface is used as public facilities (Blok M terminal).
- The usage of underground space in this research is not part of land use on it. It means the usage of underground space to be separated with the usage on the land surface.
- The physical data (construction and plan drawings) and juridical data (usage and mastery of building data) were acquired from Blok M terminal manager in 2010. Besides the data above, this research also uses the regulation related to underground space utilization and regulation in the land sector to study the right formation and right registration as a juridical basis for the usage and mastery of underground space in Blok M mall building.
- Software used to create 3D images of Blok M mall building in this study is the AutoCAD software.

Methodology

Preparation

Literature review or literature studies, determining location in accordance with the theme of research, identifying the 3D cadastre object, and prepare research tools.

Data Collection

Data collection consists of physical and juridical data. Physical data are the construction and plan drawing of Blok M mall building. Construction drawing is drawn again using a computer with AutoCad software to obtain construction drawing in digital format. Juridical Data are usage and mastery data of Blok M mall building, the regulations related to underground space utilization and regulations in land sector.

Data Processing

Data processing activities include the selection of 3D cadastre method, 3D building shape design, right form, and right registration form. Construction drawing was digitized to obtain the 2D shape of building which was then undertaken "*extrude*" to obtain 3D building shape in digital format.

From the juridical data in the form of regulation of land rights was established right design with the 3R approach (*right*, *restriction* and *responsibility).* From the juridical data in the form of regulation regarding land rights registration established right registration design with 3D cadastre approach.

Data Analysis

Analysis of 3D building shape of Blok M mall, right form and right registration form are done by comparing the rules in the proprietary right of Housing Units (HMASRS). This analysis considered location, shape, structure and usage, where the mall building was assumed as non-residential flat.

Conclusion

Conclusions from the results of analysis provide suggestions.

Flow Diagram

Flow diagram can be seen in Fig. 11.1.

Results and Discussion

Selection of 3D Cadastre Method

- To determine the 3D cadastre method that will be applied at Blok M mall building, required the consideration of the ease in adjusting to the prevailing system of land law in Indonesia. 3D cadastre method that can be applied to Blok M mall building is a *hybrid cadastre* with an alternative *registration of 3D physical object.*

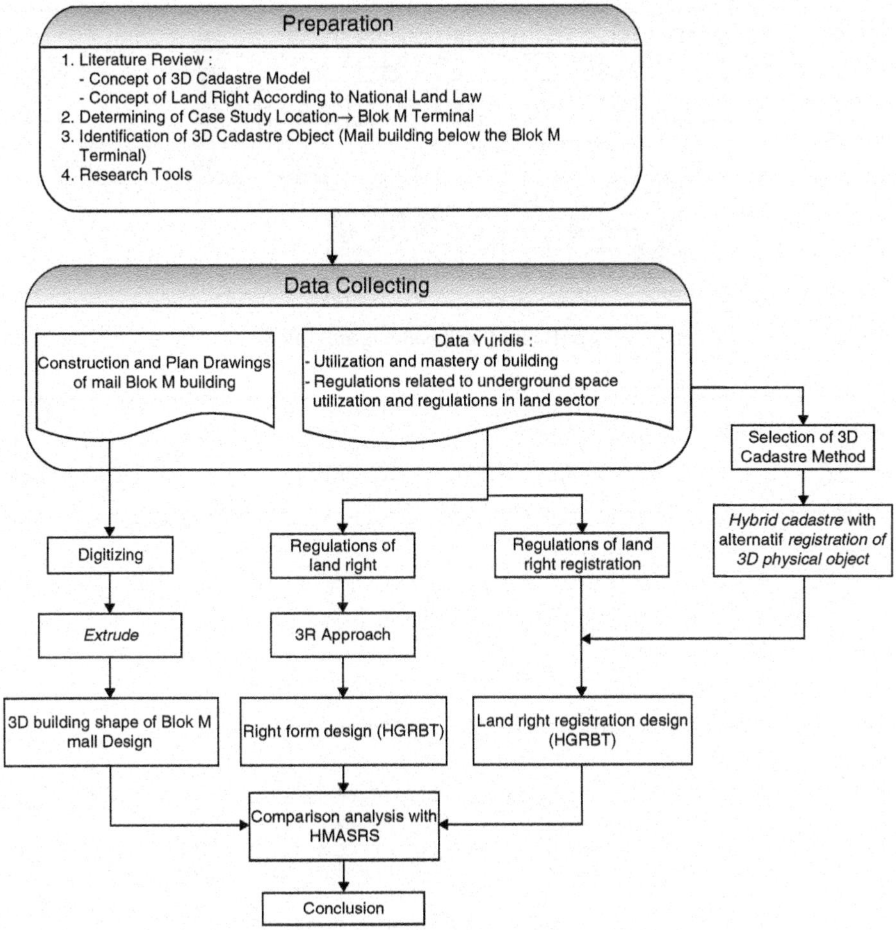

Fig. 11.1 Research phase

- The method of *hybrid cadastre* with alternative *registration of 3D physical object* is selected with the consideration that the current land registration law in Indonesia has not made the 3D space as the basic entity in right registration and still made 2D parcel of land as a basic entity in land registry. Therefore, the formation of 3D cadastre can begin by integrating the 3D property situation on the basis of 2D entity.
- This method is performed by registering the physical object in the form of mall building including spatial information in it. The aim is to describe the 3D situation of Blok M mall building so that information can be used to determine the position of right status and three dimensions situation in the 2D entity.

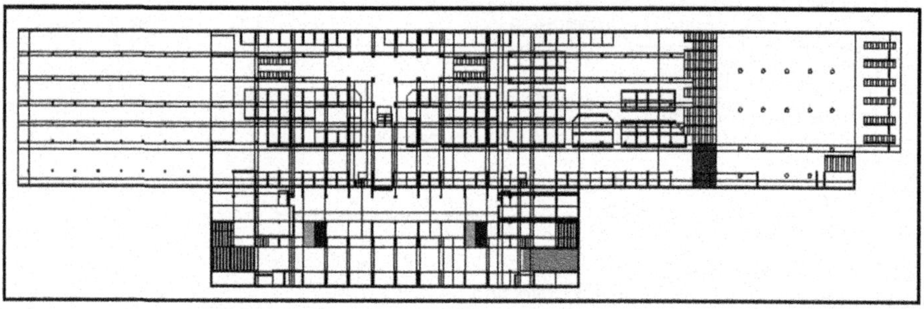

Fig. 11.2 Construction drawing of Blok M mall building at B1 floor

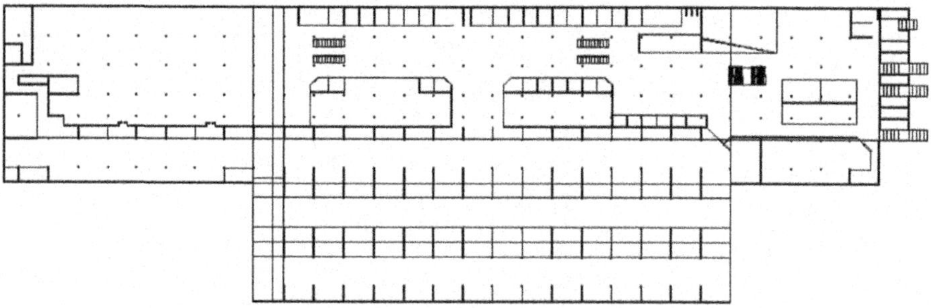

Fig. 11.3 Construction drawing of Blok M mall building at B2 floor

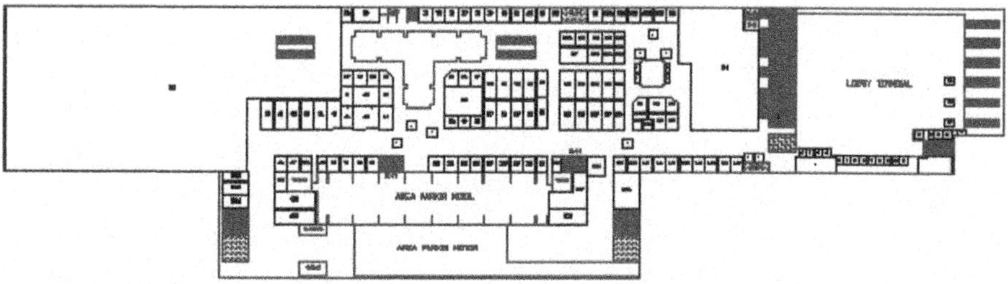

Fig. 11.4 2D building shape of Blok M mall at B1 floor

3D Building Shape

3D shape of Blok M mall building is obtained by processing the physical data in the form of building construction drawings (Figs. 11.2 and 11.3). The construction drawings were digitized by assisting AutoCad software to obtain 2D building shape (Figs. 11.4 and 11.5). While digitizing, geometry images of building used as stores and non stores are directly distinguishable by giving a different colour. Type of usage of building units in the Blok M mall can be seen from the building plans. The building units that had a certain number were identified as a store (see Figs. 11.6 and 11.7).

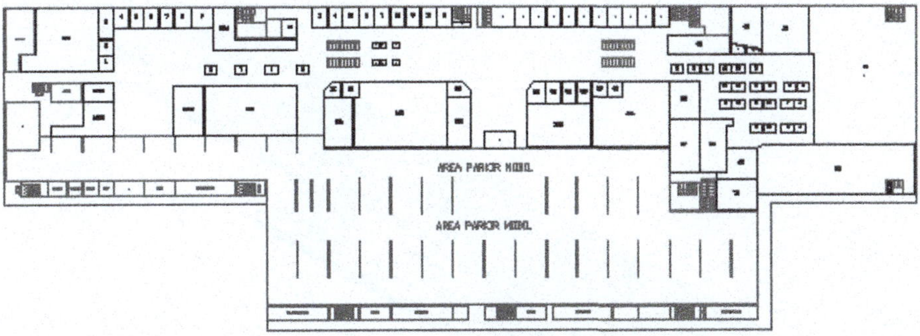

Fig. 11.5 2D building shape of Blok M mall at B2 floor

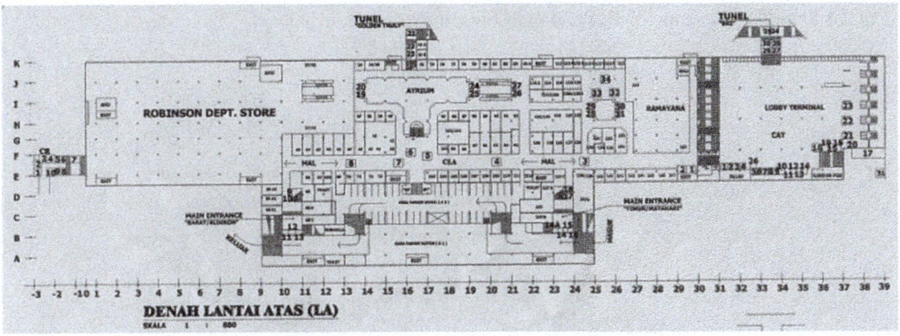

Fig. 11.6 Blok M mall plan drawing at B1 floor

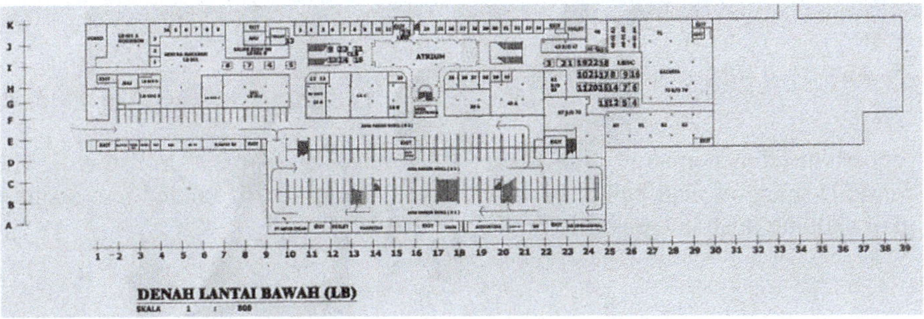

Fig. 11.7 Blok M mall plan drawing at B2 floor

The 3D building shape which is formed as *solid* 3D shape is formed by giving the height (*extrude*) to the 2D building images (Figs. 11.4 and 11.5). The 2D objects were extruded to be *solid* 3D objects in the direction of z axis by helping the "*extrude*" command in AutoCAD software.

Figures 11.8 and 11.9 visualize the *solid* 3D shape of Blok M mall buildings at B1 and B2 floors. In this form, limitation of space usage can be distinguished and clearly visible. Building units used as stores were bordered by blue colours and the

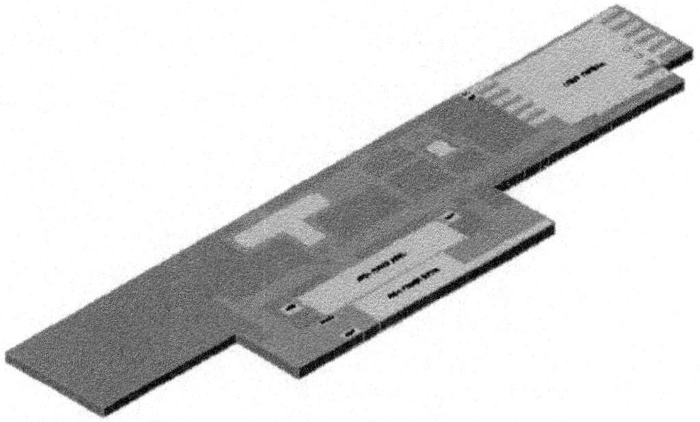

Fig. 11.8 *Solid* 3D shape of Blok M mall building at B1 floor

Fig. 11.9 *Solid* 3D shape of Blok M mall building at B2 floor

property used as non-store were bordered by magenta colours (purplish red). The *solid* 3D shape of mall buildings at B1 and B2 floors can be unified into a single mall building in an integrated form (Fig. 11.10).

Right Form

Indonesia's national land law was stipulated in Law Number 5 Year 1960 on Basic Agrarian Laws (UUPA). Under Article 4 of UUPA, land rights give an authority to use the land, land bodies and water as well as existing space above it, just needed to interests that are directly related to the usage of the land. Outside of those strata, the usage and mastery of the underground space are not included in the authority of the land rights and require specific regulation in a form of new right.

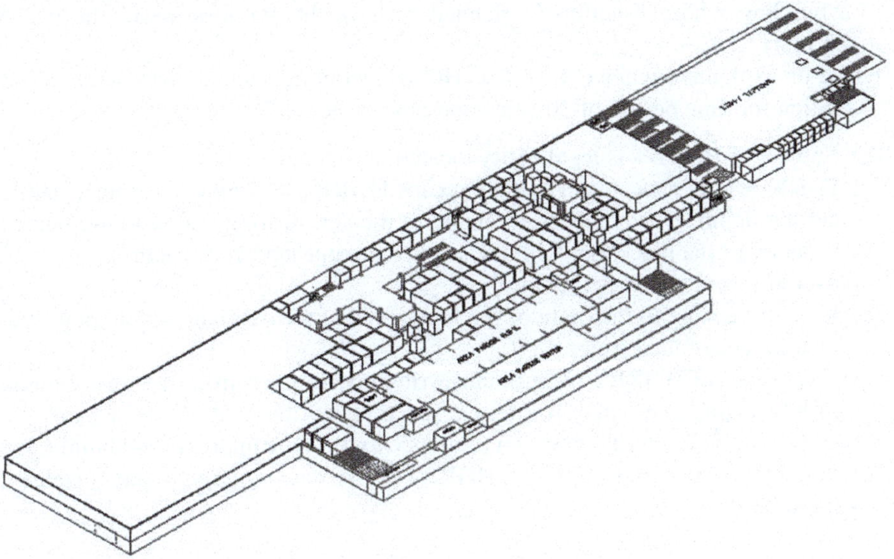

Fig. 11.10 Union of 3D building shape at B1 and B2 floors

The establishment of new right can be done by adopting and modifying the form of rights as stipulated in the regulations concerning land rights. Form of right that is formed can be called with Underground Space Use Right (HGRBT) because of its mastery coverage which includes three-dimensional space located below the land. Designing of HGRBT form was done with the 3R approach (*right*, *restriction*, and the *responsibility*).

Form of authority (*right*) in HGRBT include:

(a) HGRBT gives authority to use three-dimensional space in the body of the land (underground) within a certain time period.
(b) HGRBT gives authority to use part of the land surface above it as a place to build facilities liaison (entrance and exit) to the underground building. Usage of this land surface must be based on a land right.

Forms of right restrictions includes:

(a) Time period of right. The usage of underground space is realized in the form of construction of buildings below the land so that to determine time period of the right, consider the age and strength aspects of building construction.

Under Article 35 of UUPA, land right associated with the construction of the building is Hak Guna Bangunan (HGB) that were given a period of 30 years. Indonesian Institute of Architects, Honorary Chairman, Budi Sukada, had issued a statement that the construction of buildings in Jakarta have been designed with the age of 50 years and the buildings have earthquake resistance until 10 SR (beritajakarta.com, 2009). In addition, Article 5 subsection (3) Government Regulation Number 36 Year 2005, explained that the permanent building means a building whose function is planned to have the services above the age of 20 years.

Based on explanation above, therefore HGRBT granting can be given with consideration for time period of 30 years and renewable for 20 years.

(b) Subject of HGRBT is legal entity located in Indonesia.
(c) Depth of land aspect on HGRBT means HGRBT still may be granted until a certain depth. Depth aspect is related to the soil bearing and the environment capacity of the place that will be used for underground construction.

Form of responsibilities include:

(a) Prevent damage to the body of the land and the environment which may cause disturbance of land usage on it.
(b) The holder of HGRBT must obtain written permission from the owner/holder of land rights that exist on it to use space below the land and to use the land surface to build facilities liaison (entrance/exit) to the underground building.
(c) In the process of granting right, HGRBT holders are required to pay revenue to the state.

Right Resgistration Form

As mentioned previously, the method implemented in 3D cadastre was *hybrid cadastre* with an alternative of *registration of the 3D physical object.* This method was applied at Blok M mall building where the 3D situation of mall building added on basic entity of 2D registration. At the top is *emplasement* of Blok M terminal where at the bottom there is a mall building registered with the land right, namely right of management (HPL) because this parcel is mastered by the Government of DKI Jakarta whereas the mall building is given HGRBT.

Developer who develops Blok M mall is PT. Langgeng Ayom Lestari incorporated as an Indonesian legal entity. To be able to use the land surface associated with the mall building, the developer must obtain licenses and make a written agreement with the Government of DKI Jakarta. On the basis of this agreement, the developer can apply for land right as the basis for the usage of the land surface. As for the land right granted is Building Use Right (Hak Guna Bangunan) (Fig. 11.11)

In the *hybrid cadastre* method with alternative of *registration of 3D physical object,* the physical object of Blok M mall building is registered and then integrated into the land registry map (2D). This is intended to make building have the same coordinate system of land registration map, which is in the coordinate system of TM-3° (x and y axis). For the height (z axis) refers to the local coordinates by making the surface of the terminal that has the same relative height and flat with the land surface as the zero point ($z = 0$).

Design of HGRBT registration form was done by adopting land rights registration form as set out in UUPA and Government Regulation Number 24 year 1997 about Land Registration. The aim is, HGRBT already formed can be registered in

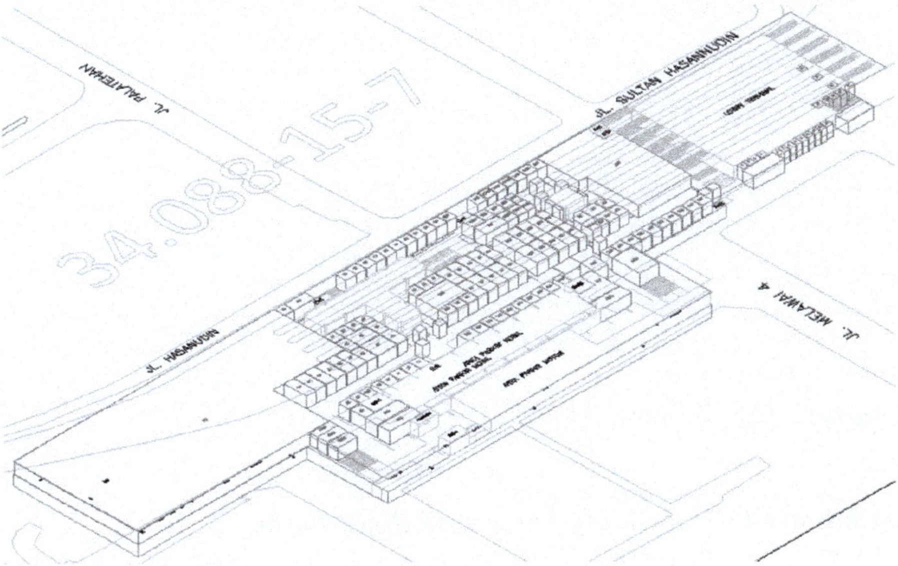

Fig. 11.11 Visualization of Blok M mall building in the land registration map

the cadastre activities. HGRBT registration is conducted through the process of granting right with the decree of granting right decided by the competent authority.

HGRBT registration is made after the prospective right holder qualify technically and administratively. Technical requirements related to requirements of the building while the administrative requirements related to allowance must be met by the prospective HGRBT holder that is local government allowance and written consent (agreement) with the owner/holder of land rights on it.

Product of this granting HGRBT process is document of certificate and Space Book (identical with the Land Book) of HGRBT. Space Book is a document that contains the physical and juridical data of HGRBT object that has been registered. Whereas HGRBT certificate is valid proof for the usage and mastery of space in the underground building. The form of Space Books and certificate of HGRBT can modify the form and format of Land Book and Certificate of Ownership of Land (HAT) as was stipulated in the current land registration laws.

As land rights, physical data of HGRBT object are contained in a document of measurement certificate which became part of the HGRBT Space Book and Certificate. The form and format of measurement certificate can modify the form of measurement certificate on land rights.

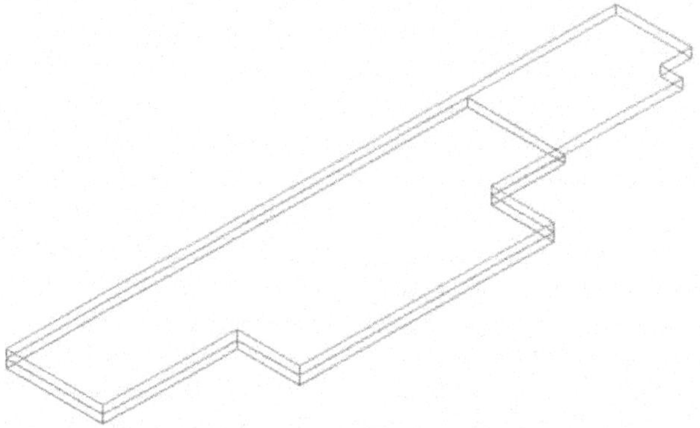

Fig. 11.12 Space boundary for HGRBT at Blok M mall

Analysis of 3D Building Shape and Right Form

Based on Government Regulation Number 36 Year 2005, Blok M mall building as an underground building can be assumed as non-residential flat. Based on this, the analysis is carried out by comparing the 3D building shape, right form, and right registration form towards setting proprietary right of Housing Units (HMASRS).

Basically, for the usage and mastery of the underground building can be given as one HGRBT. This means that HGRBT is attached to the mall building as a whole and owned by a right subject namely PT. Langgeng Ayom Lestari as the developer who built Blok M mall.

In fact, Blok M mall building consists of several building units (underground space units) and most of those can be used individually as a store. The existence of underground space units used by an individual is certainly going to involve many stakeholders as users. Based on this condition, each subject of underground space unit requires a much legally stronger position in using and mastering the underground space unit.

By comparing the rules in the HMASRS, HGRBT inherent in Blok M mall buildings can be separated into several underground space units that can be given right separately and can be called with the proprietary right of Underground Space Unit (HMASRBT). As in HMASRS, before HGRBT is separated into multiple ownership of HMASRBT, developer must make a Deed of Separation approved by the local government. In this mall building, there are units which are used as a store and can be owned with separate right and there are also part of the building which can be used together such as structural columns, walls, toilets, mosque, and parking lots. The used parts for the common good may be called as 'jointly parts'.

Blok M mall building is equipped with means of liaison that connect the underground building with the land surface above it. Liaison facility was built in a certain limitation of the usage of the land surface (Figs. 11.12, 11.13, 11.14).

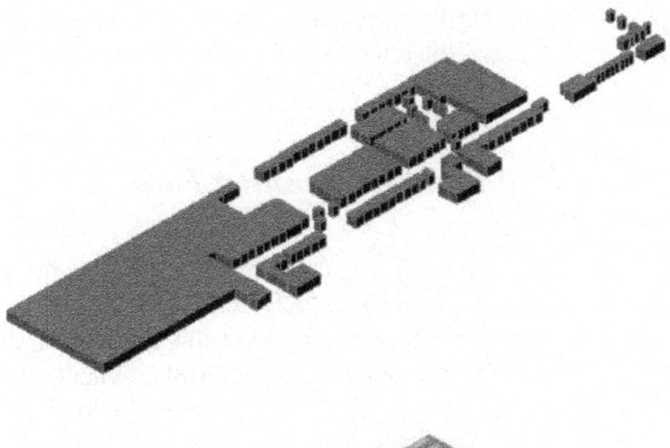

Fig. 11.13 Underground space unit as a store at B1 floor

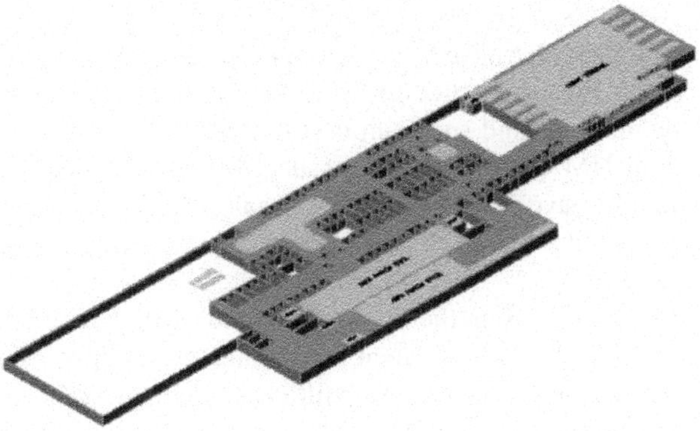

Fig. 11.14 Jointly parts at the B1 floor

Fig. 11.15 Parcels of land surface that are used as facilities of liaison to the Blok M mall

Figure 11.15 shows that the parcels of land surface used as an entrance and exit Blok M mall are the parcels bounded by green colour whereas the area bounded by the dashed red line is the boundary of underground space area used for construction of Blok M mall building.

Basically, underground space unit is part of an underground building that could be used as residential or non-residential. Article 30 Government Regulation Number 36 Year 2005, suggests the construction of underground building that crosses infrastructure or public utilities not addressed to occupancy function/residence place.

Thus the underground space units of Blok M mall can only be used as a non-residential place (business place).

Analysis of Right Registration Form

In the case of Blok M mall, each underground space unit (store) is given HMASRBT. Granting HMASRBT can be done after granting HGRBT of the whole of mall building. By comparing the rules in the HMASRS, before HGRBT at Blok M mall building is separated into multiple ownership of HMASRBT; therefore the HGRBT holder (developer) have to make a Deed of Building Separation which was then approved by the local government.

Deed of Separation which have been made and approved by the Local Government is then registered at the land office to make the HGRBT separation into multiple ownership of the underground space units (HMASRBT). Based on this, certificate and Space Book of HMASRBT are then issued. In this case, the overall certificate is issued on behalf of developer. Form and format of Space Book and certificate of HMASRBT can modify the form and format of Land Book and Certificate of HMASRS.

In the registration process, in the Space Book and certificate of HMASRBT need to attach copy of HGRBT measurement certificate. At the Space Book and certificate of HMASRBT copies of space and building plan drawings from developer are also attached.

The form and format of space and building plan drawings for the issuance of certificate and Space Book of HMASRBT can adopt the form and format of plan drawings at the Head of BPN Regulation Number 4 Year 1989. The form and format of these plan drawings are then modified for HMASRBT registration purpose.

Over a period of time, it is possible that there is a change of ownership of HMASRBT and HGRBT data. Therefore, we need data maintenance activitiy of right registration to record the changes. Changes may occur due to transfer of legal right activities such as buying and selling, grants, inheritance or right encumbrance to ensure debt repayment. In the transfer of legal right process or right encumbrance process, it requires involvement of Maker Official Land Deed (PPAT) for the process of making land deed.

When the right transfer or right encumbrance occurred then it should be noted to the Space Book and certificate. Recording is based on the deed made by PPAT as the basis for changes in registration data. With data maintenance activity of right registration, therefore, any data changes of HMASRBT or HGRBT should be registered, so its status remains *up to date* and to guarantee legal certainty to the parties concerned.

Conclusion

1. Mastering of the form of usage of space in the underground building can be accommodated in form of Underground Space Use Right (HGRBT) where the legalization process of right registration can be implemented by using *the hybridcadastre registration of 3D physical object* method.
2. HGRBT is a right that is given as a basis for the usage and mastery of space for underground building. In the case of Blok M mall, HGRBT holder can separate the space into multiple ownership of underground space unit in the form of proprietary right of Underground Space Unit (HMASRBT).
3. HGRBT is still associated with the usage of land surface and land rights above it. In this case, its required written consent (agreement) with the owner or land rights holder on it should be included and mentioned in the decree granting right.
4. Space Book and certificate of HGRBT can provide assurance of certainty of the position and clear boundary of space usage to the HGRBT holder.

Suggestions

1. Granting HGRBT and HMASRBT including right registration activity for legalization of mastery of space in underground building need to be regulated by law in order to guarantee legal certainty to the owners or interested parties.
2. For the registration purpose, it's required further technical study related to delimitation of underground space parcel and its mapping on the registration map. In this case also it needs further study on the standardization of 3D object representations for right registration.
3. In order to provide complete information to the parties concerned about the object (physical data) and subject (legal information) of HGRBT or HMASRBT, the land information system development is needed regarding the existence of 3D cadastre objects.
4. Just as in the case of land rights, HGRBT or HMASRBT can be lost if the underground building was damaged or destroyed. Therefore, further study is needed about the form of building (property) insurance in order to protect the interests of right holders from losses due to natural disasters or disasters caused by human negligence that caused destruction of building.

References

Anonymous. 2009. *Multi-storey building in Jakarta Earthquake up to 10 SR.* http://www.beritajakarta.com/. Access to 25 Apr 2010 at 8:40.

Harsono, Budi. 1999. *History of the formation basic agrarian law, content and implementation.* Jakarta: Djambatan.

Hutagalung, Arie S. 2008. *Land use rights in upper room, underground and bodies*. Jakarta: Seminar, Juridical Research Institute of Registration Bodies and Technical Policy Right to Space and Water.

Maharani. 2008. *Right to land space and underground space use rights*. Jakarta: Seminar, Juridical Research Institute of Registration Bodies and Technical Policy Right to Space and Water.

Chapter 12
Towards 3D Land Registry in Hungary

Gyula Iván and András Osskó

Introduction

Hungary has a long tradition in cadastral surveying and land registration. The first cadastral survey of the country started at the mid of the nineteenth Century, in the ages of the former Austro-Hungary. Surveying and registration of lands finished at the end of the Century. It means all the lands in Hungary have been surveyed and registered since the end of the nineteenth Century.

At this time, the main goal of the Cadastre was the taxation of land. Cadastral surveying and land registry were separated. Registration of Condominium Units (CUs) has started in the 1930s. Developed registration process was quasi 3D registration of CUs. Identification of CUs linked the cadastral parcel number and the building where the unit was. Each floor of the building had a layout plan, on which the CUs were delineated. This process remained the same until now. In registration of cadastral parcels based on a typical 2D registration method, no 3D aspects had been taken into consideration.

Operation of Land Registry and Cadastral Mapping was persistent during the socialist ages, because private ownership of land and real properties was extinct.

In 1972, because of the large number of discrepancies between the Land Registry and Cadastral Map, Hungary introduced a new Unified Land Registry, in which Land Registration and Cadastral Mapping is the responsibility of the same organization, i.e., Land Office Network. Land Offices have to update Land Registry and Cadastral Maps together, in an integrated way. Land Offices' tasks are also Agricultural Land Valuation, Land Protection, Land Consolidation and Land Lease Registration altogether.

G. Iván (✉) • A. Osskó
Institute of Geodesy Cartography and Remote Sensing,
5 Bosnyák tér, 1149 Budapest, Hungary
e-mail: ivan.gyula@fomi.hu

T. Yomralioglu, J. McLaughlin (eds.), *Cadastre: Geo-Information Innovations in Land Administration*, DOI 10.1007/978-3-319-51216-7_12

Institute of Geodesy Cartography and Remote Sensing (FÖMI) developed a standard (MSZ 7772:1–1996, Digital Base Map, Conceptual Model, DAT Standard) on Cadastral Map Database. DAT Standard is an object-oriented data model for cadastral map database. Based on DAT Standard, new (mandatory) instructions have been developed for cadastral surveying and mapping activities in Hungary, which was introduced in 1997. DAT Standard was approved by the Hungarian Standards Institution in 1996 (Iván et al. 2004).

New standard and instructions totally reformed the Hungarian Unified Land Registry. Object-oriented database view of Cadastre inspired many software and IT system developments. IT developments of Hungarian Land Administration are the responsibility of FÖMI. Until now FÖMI developed many IT systems for Land Office Network:

- TAKAROS (IT system for the management of Land Records and Legal Data, Application management, Statistics, Reporting, finished in 2000)
- FÖNYIR (IT system for Land Lease Register, finished in 2000)
- TAKARNET (Network of Hungarian Land Administration, Unified Land Registry Services for the users, finished in 2003)
- DATR (Integrated Unified Land Registry Management IT system, Management of Land Records and Cadastral Map by integrated way, finished in 2010)
- MENYÉT (IT system for Farmer's Registry, finished in 2014).

Development of an IT system has never finished. Continuous update, customization of the systems for the changing technical and legal environment is a very strict requirement. Since all the codes were developed by FÖMI en-bloc, this development task is a permanent duty of our institute.

All these IT systems are based on the DAT Standard. Reliable operation of the systems shows the success of our developments. Core data model of DATR system acts as a country profile in Land Administration Domain Model (ISO 19152:2012) Standard.

After the political changes at the beginning of the 90s, the newly established local governments had not enough financial resources; therefore, they sold out the flats and other properties for the citizens. This action caused a huge increase of the number of condominiums and registration of them.

Development in urban infrastructures (e.g. roads, railways etc.) and privatization of infrastructure companies (e.g. utilities, water supply, sewerage, electricity etc.) increased the value of properties and the number of privately owned infrastructure elements as well. Fast development led to intersecting, overlapping objects, which could not be registered in a traditional 2D Land Registry. Utility networks, owned by different companies, also rose registration problems. Hungary has 22 wine-growing regions with a lot of cellars in the hills. Under the traditional 2D Cadastre regulations registration of cellars as individual properties within a cadastral parcel cannot be executed, because it is a part of the cadastral parcel. In the case of cellars, registration of them is a very important issue for wine-makers. Samples can be counted, but these were the reasons, why Hungary changed Act on Surveying and

Mapping Activities and Act on Land Registry introducing 3D Land Registry in 2012.

New Act on Surveying and Mapping Activities

The goal of Act on Surveying and Mapping Activities is to determine the tasks of the State in surveying and mapping and to establish a condition system, which provides map databases in a cost-effective way for the whole economy and society.

The main issues, which are handled in this Act, are the following:

- State works and State Data
  - State Databases
  - Data Services
  - Control Networks
- State Databases
  - Database of State Boundary
  - Database of Control Points
  - State Cadastral Map Databases
  - State Topographic Databases
  - State Remote Sensing Databases
  - Databases for State Defence
  - National Gazetteer
  - Archive Databases
- Surveying and Mapping Activities
  - Mounting and measuring surveying marks
  - Ownership of surveying marks
  - Protection of surveying marks
- Ownership of Surveyed Data
- Institutional Issues in Surveying

New concept on surveying and mapping activities changed from the old, map-based regulation to database fundamentals. State Cadastral Map Database is the geometric part of the Unified Land Registry Database, which is defined in the Act on Land Registry. Unified Land Registry Database contains two main parts, Cadastral Map Database and the Database of Land Records, which must be integrated.

Topographic mapping activities are shared between the public (FÖMI) and military mapping agencies. Large scale (1:10,000) topographic mapping is the responsibility of public (FÖMI), while smaller scale topographic mapping belongs to the military mapping agency.

State Remote Sensing Databases are Orthophotos, Satellite Images, LIDAR (including Terrestrial LIDAR technics), photogrammetric products, where production financed by the State are also regulated. There is a very strict statement in the new Act, which really helps the renewal, updating and production of State Map Databases: "A copy of any map database product, of which production is fully or partly financed by public funds, must be provided for FÖMI, without any financial and natural compensation." This means that every map database, produced in Hungary, of which production is financed by public money, can be used for State level map database renewing establishment. It is very important, if the financial resources in State Budget are generally low or not existed (like in Hungary). This statement really helps the Hungarian Mapping Agencies in their work.

New concept also includes 3D Cadastral issues, which determine that 3D parcels, related to Land Registry, should be stored in State Cadastral Map Databases. Because of the importance of 3D Cadastre solution (e.g. Condominiums in Hungarian Land Registry are registered since 1930s as quasi 3D parcels), Hungarian Land Administration Sector now is elaborating the legal restrictions of 3D Cadastre.

Act on Land Registry defines a Unified Land Registry, which means the legal and geometric (cadastral map) part of the Unified Land Registry compose one system. All geometric characteristics of Land Registry components (parcel boundary, area etc.) are derived from the State Cadastral Map Database. The Database of Land Records and State Cadastral Map Database compose one database, i.e., unified Land Registry Database.

This Act defines cadastral parcel, which is a continuous part of the Earth's surface, on which the ownership relationships are homogenous. There are other types of real properties beside cadastral parcels, which are the components of Land Registry (e.g. buildings, condominiums, flats, shops etc.).

Former Act defined some 3D Cadastre issues. These are the registering of condominiums, flats, shops, other areas within the condominium, cellars with access to public domain etc. Registering of these 3D situations are based on the 2D cadastral map. For example flats and shops are not the part of Cadastral Map Database; these are described by the floor layout plans, which act as cadastral map of condominium units. Change management of them is a very hard work. Cellars are described only by the line of access of them in the Cadastral Map Database.

Because of the above situation, and introducing a real 3D Cadastre in legislation, the renewed Act on Land Registry defines a new type of property, which opens the doors to 3D Cadastre. This new type of property is defined as follows:

> Under-ground and above-ground passes, objects, structures, which have homogenous ownership relationships should be taken into account as property, which must be registered in Land Registry.

With the help of this definition utilities, over-crossings and other objects (e.g. cellars) should be registered as property and 3D legal relationships should be modeled in Hungarian Unified Land Registry.

Vis-a-vis with some 3D Cadastre solution, the Hungarian concept registers 3D object in space. Connecting legal space of 3D object should be derived from the geometry of the objects itself and other regulations (e.g. spatial planning regulations).

Legal space required for 3D Cadastre object is defined in different laws and regulations related to land use and land development in Hungary. This means if 3D objects and their legal space should be registered in Land Registry, the required legal space must be modelled based on regulations.

The Act authorizes FÖMI to elaborate the required legal and technical regulations for the implementation of 3D Cadastre in Hungarian Unified Land Registry. Therefore 3D elements of the Act will not come into force immediately, but only after these regulations and technical conditions are ready.

Legal Issues

Civil Code of Hungary defines the extension of ownership on real-property as follows: "Ownership right on a real-property extends to the air-space above and to the subsurface below it until utilization is possible". It is a typical "from-heaven-to-hell" situation as in many countries all over the world.

Fortunately the word "utilization" can limit the infinite extension of property rights. The main legal problem in the introduction of 3D Land Registry is the legal handling of 3D situations, for example in Fig. 12.1.

In Fig. 12.1, a typical situation is shown. There is a cadastral parcel (number: 023), and under the parcel there is a tube, whose owner is different from the cadastral

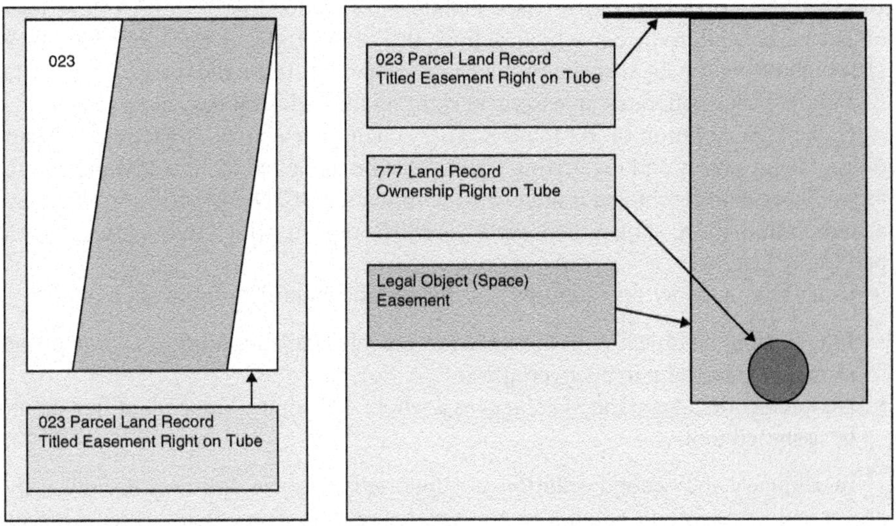

Fig. 12.1 2D and 3D Land registry situation

Fig. 12.2 3D object overlaps cadastral parcels

parcel's owner. Based on the regulations on tubes, a tube must have safety zone, which is defined by the certain distance from the axis of the tube.

In 2D situation the tube is surveyed, and the safety zone is defined. If the safety zone intersects a privately owned cadastral parcel (023), the safety zone must be inserted into the cadastral map database and an easement right must be registered on this cadastral parcel in Land Registry.

In 3D situation, the tube itself is a property. It has a land record in Land Registry. But the tube generates a legal space (safety zone) around it as well. A new, legal object must be introduced, which defines the safety right of the tube. This legal space intersects the legal space of the cadastral parcel (023), the intersection of the two legal spaces generates an easement right on the cadastral parcel.

In the legal solution of 3D Land Registry such intersections, touching, overlapping of legal spaces and the arising rights must be modelled technically and legally as well. Legal and technical issue is the identification of 3D properties. In Fig. 12.2, a cadastral map is shown and there is a 3D object, which overlaps cadastral parcels.

Generally there are two solutions for the unique identification of such object:

- Intersecting the object with the cadastral parcels, and split it into as many pieces as many cadastral parcels overlapped.
- No intersection, keep the 3D object as a whole, but some unique identifier should be assigned to it.

In Hungary the second solution is supported, because splitting dissolves the object and its registration and lose main meaning of it. The object, which overlaps parcels (under or below the surface of the earth) must connect to a physical object within a cadastral parcel (e.g. pillars of a bridge, transformer house of an electricity cable, distributor station of gas pipeline etc.). For the unique identification of 3D object, these "starting" cadastral parcel identifiers should be used with some other identifiers, which help in uniquely identifying the 3D object. This identification can connect 3D object identifiers to the traditional 2D identification. The whole concept of such "octopus" identification has not been materialized yet.

Geometric Model

There are different recommendation, solutions for modeling legal spaces in 3D Cadastre. One of the most known was published in LADM, which was using boundary face concept for 3D representation of legal spaces. It has a great advantage, because mixed representation of 2D and 3D legal spaces is also available using boundary face string and boundary face solutions (ISO 2012).

The author recommended the usage of homogenous coordinates for managing unbounded legal spaces, or the tessellation of legal spaces by using tetrahedrons (Iván 2012). This solution could be really flexible, but the management of homogenous coordinates could cause computational problems. Tessellation of space requires that all parts of the space should be tessellated, which requires a lot of computation and storage, and mixed 2D and 3D situations cannot be handled in this way. In the future, taking into account the development of ICT technology, this solution could be introduced.

The Hungarian Cadastral Map Standard and the IT system of the Unified Land Registry is using the concept described in DAT Standard. DAT Standard is using 3D points as default for the representation of geometry of the Cadastral Database. Until now only the 2D capacity of the model was used, generally the "Z" coordinate of a point was set to zero (except in the case of vertical ground control points, which require the "Z" coordinate). The sketch of DAT Standard geometry can be found in Fig. 12.3.

The base is the class Point, which describes the positions (3D as default). If an object is a point type object, it is directly linked to Point class. Line is a class, which stores Points describing a line. Connection type is line as default, but curved connection is also available. If an object is a linear object, it is linked to class Line. Boundary_Line class is similar to Line, except it must be a part of a boundary. Boundary is a set of connected Boundary_Lines and it must be closed. Face is a composition of boundaries. Face must have only one outer and zero or more inner boundaries. Polygon type objects are directly linked to Face, describing their geometry. This concept is very similar to any GIS systems concept as it is a 2D situation.

DAT Standard based solution for modelling 3D legal spaces is based on the condition that the legal spaces are not unbounded or can be closed in the finite. Boundaries of legal spaces are represented by planes (no curved surfaces are allowed, Fig. 12.4).

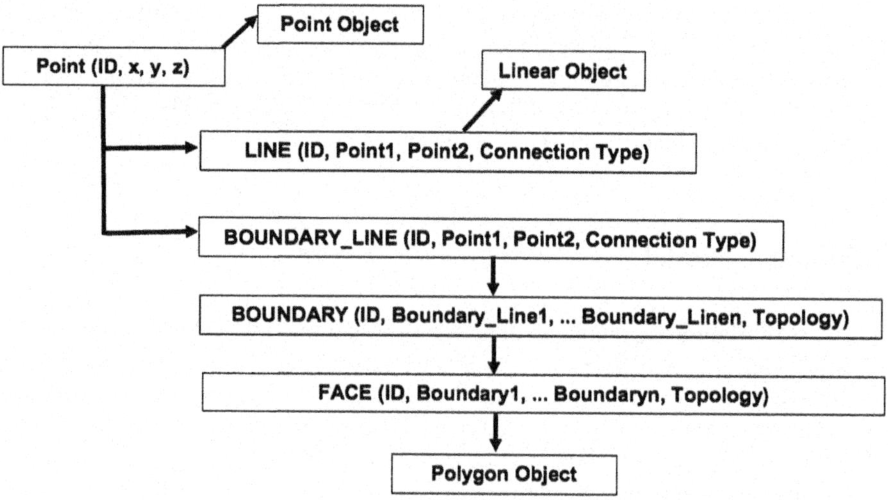

Fig. 12.3 Geometric concept of DAT Standard

Fig. 12.4 Legal spaces delineated by planes

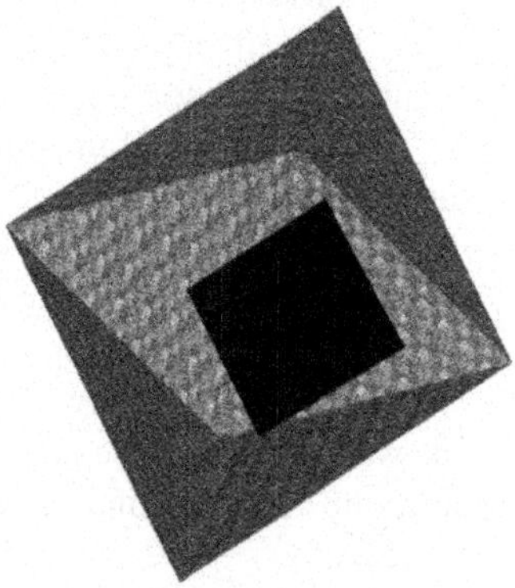

In Fig. 12.4, two legal spaces are shown delineated by planes. The outer (dark+light grey) legal space contains an inner legal space (black), which is not a part of the dark and grey one (hole in the space).

If the original geometric construction of DAT Standard is used, Face class should describe the planes, which delineate the legal space. In Face class holes within a face are also allowed. For the usage of Face in 3D a co-planarity constrain is required as defined in the following formula:

$$\mathrm{Det}\begin{vmatrix} X_1 & Y_1 & Z_1 & 1 \\ X_2 & Y_2 & Z_2 & 1 \\ X_3 & Y_3 & Z_3 & 1 \\ X_i & Y_i & Z_4 & 1 \end{vmatrix} = 0$$

where:

- X_1, Y_1, Z_1 – are the coordinates of point 1 on the face, which cannot be collinear with points 2 and 3,
- X_2, Y_2, Z_2 – are the coordinates of point 2 on the face, which cannot be collinear with points 1 and 3,
- X_3, Y_3, Z_3 – are the coordinates of point 3 on the face, which cannot be collinear with points 1 and 2,
- X_i, Y_i, Z_i – are the coordinates of point i on the face, $i = 4...n$, n is the number of vertices of the face.

Constrain: Faces should have only one outer boundary, no multiple faces are allowed.

DAT Standard geometry has been expanded with two geometric primitives: shell and legal space. Shell is the polyhedron, which bounds a legal space. Legal space is a set of shells (or one shell), which describes the 3D land registry entity. The following definitions and constrains were defined (Iván 2014):

- Outer shell of legal space is the shell, which bounds the legal space. There is no same legal space (regarding to Land Registry) outside this shell.
- An edge of a shell is the intersection of only two, and only two Faces.[1]
- Legal space has only one outer shell.
- Outer shell cannot intersect or touch oneself.
- Inner shell of legal space is a shell, which is entirely within the outer shell of the legal space, and bounds a legal space, which is different from the legal space bounded by outer shell (regarding to Land Registry, see Fig. 12.4, red space).
- Inner shell cannot intersect or touch outer shell.
- Inner shell cannot intersect or touch oneself.

An example of the usage of these rules is shown in Fig. 12.5.

By the usage of above expansion of DAT Standard geometry model 3D legal situations can be modelled in the existing Hungarian Land Administration IT environment. It is a flexible solution, because 3D Land Registry is required anywhere. At that place the original 2D model can be used and mixed use of 2D and 3D geometry also available, because 3D solution is based on the 2D primitives.

[1] This constrain is nearly similar to Euler's Polyhedron Formula $V - E + F = 2$, where V is the number of vertices, E of the edges and F of the Faces, but not the same, because holes are allowed not the Faces.

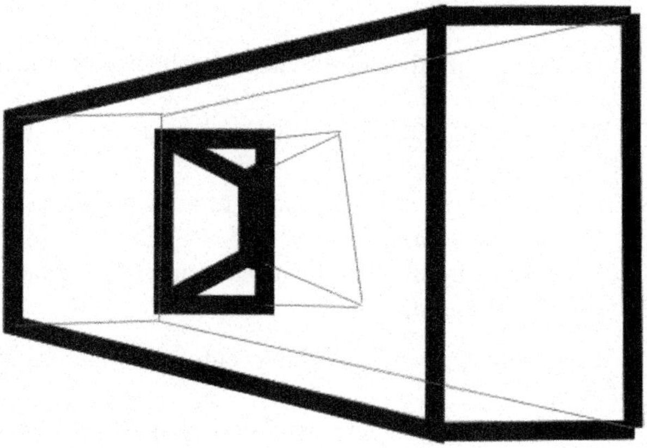

Fig. 12.5 Legal space description by shells

Conclusion

Hungarian Unified Land Registry has enough resources for the introduction of 3D Land Registry in Hungary. Introduction must be very careful, an agenda should be planned, because not only human and IT resources are needed, but a huge financial and surveying capacity are required as well. FÖMI is working on the implementation of the system both on legal and technical level as well.

References

ISO. 2012. International Standard Organization Geographic Information – Land Administration Domain Model, ISO 19152:2012.

Iván, Gyula. 2012. *3D cadastre developments in Hungary*. Proceedings of FIG working week, Knowing to manage the territory, protect the environment, evaluate the cultural heritage – Rome, Italy 6–10 May 2012.

———. 2014 (in Hungarian). *A háromdimenziós ingatlan-nyilvántartás geometriai modelljéről* (About the Geometric Model of 3D Land Registry). Geodézia és Kartográfia, 2014. 3–4., Hungarian Association of Surveying Mapping and Remote Sensing, Budapest, Hungary, 2014.

Iván, Gyula, Mihány, Szabolcs, Szabó, Gábor and Weninger, Zoltán. 2004. *Standards and new IT Developments in Hungary*. Proceedings of Joint 'FIG Commission 7' and 'COST Action G9' Workshop on Standardization in the Cadastral Domain, Bamberg, Germany, 9 and 10 December 2004.

XLVI. *Act on surveying and mapping activities*. Hungary.

Chapter 13
Germany on the Way to 4D-Cadastre

Markus Seifert, Ulrich Gruber, and Jens Riecken

The Demand for 3D-Building Information

Energy Turnaround

In Germany the government targets at climate and environmental protection currently lead to extensive changes in the energy sector, the so-called energy turnaround. This includes the end of the use of nuclear energy by 2020, the reduction of greenhouse gases and other objectives (BlmSchG, 2012). As a result planning processes especially have to take into account the use of photovoltaic technology, geothermal energy, wind energy and the energetic isolation of buildings.

From the process view, data must be available to provide actual information of the environment and all energetically relevant topics. Very often this leads to a data collection or at least to a data processing task. Having the required information, the analysis and the evaluation will give a sustainable picture of the energy balance, including possible savings the use of renewals energies and energetic isolations of buildings (Fig. 13.1).

M. Seifert (✉)
DVW WG Geoinformation, Bavarian Agency for Digitization, High Speed Internet and Surveying, Bavarian, Germany
e-mail: Markus.Seifert@ldbv.bayern.de

U. Gruber
DVW WG Geoinformation, Recklinghausen, Germany

J. Riecken
DVW WG Geoinformation, Bavarian Agency for Digitization, High Speed Internet and Surveying, Bavarian, Germany

DVW, Cologne District Government, Cologne, Germany

T. Yomralioglu, J. McLaughlin (eds.), *Cadastre: Geo-Information Innovations in Land Administration*, DOI 10.1007/978-3-319-51216-7_13

Fig. 13.1 Photovoltaic map of the city of Dusseldorf

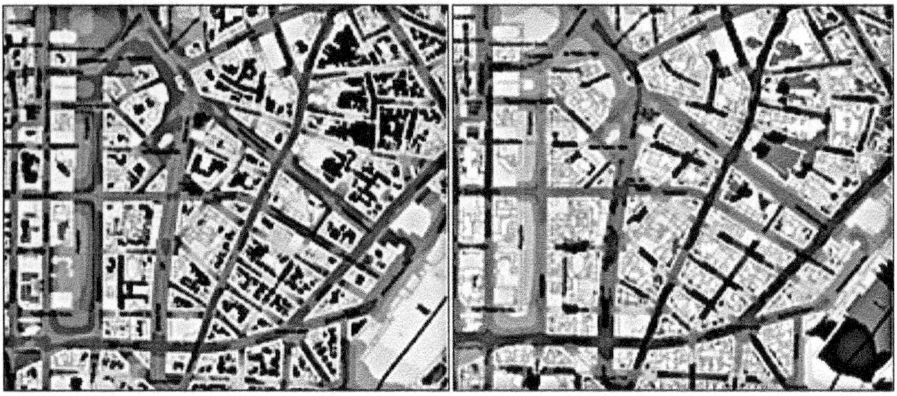

Fig. 13.2 Noise map of the city of Dusseldorf (*left*: during day time, *right*: at night)

Noise Protection

The 3D-geometry and semantics, particularly of buildings, are very important for simulating and mapping of noise expansion (Fig. 13.2). By a European directive every 5 years the member states of the European Union are obliged to determine and to document noise pollution in cities. In addition the progress of noise-reduction is checked.

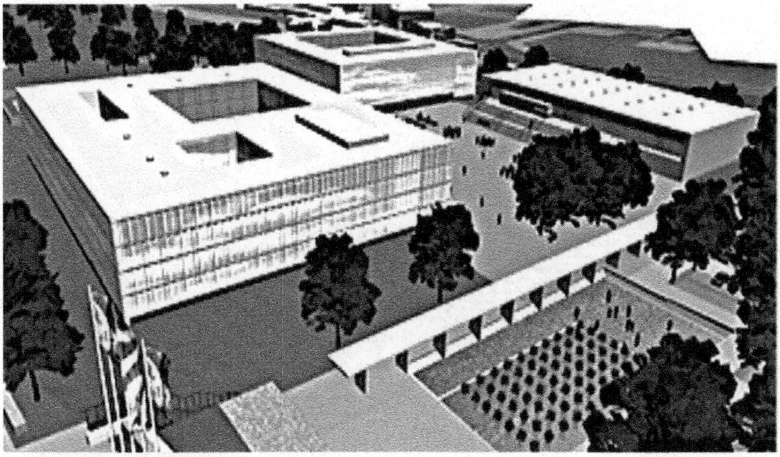

Fig. 13.3 Planned school in the county of Recklinghausen

Urban Planning

The use of cadastral information for urban planning was always essential in the 2D-world, especially to consider the property distribution. Nowadays 3D-information is a basic demand of the urban planning sector (Fig. 13.3). Demographic effects and other restrictions could be visualized in planning alternatives (Riecken and Seifert 2012).

Disaster Management

Increasingly 3D-information is used in the simulation of disasters, for example for evacuation and flood scenarios (Fig. 13.4).

Requirements for 3D-Building Information

Level of Detail (LoD) in the Cadaster

While 3D-building information in the LoD 1 are sufficient for applications like noise mapping (Fig. 13.2) many other applications like the aforementioned photovoltaic map (Fig. 13.1) at least need a higher LoD 2 resolution (Oestereich 2014). As a consequence, so-called "city-models" were built up in many cities in Germany. Their basic goal was to support or even allow a visualization of special application scenarios (examples: Figs. 13.3 and 13.4). On the other side these models had not special quality or updating mechanisms. Often they used the cadaster as a data

Fig. 13.4 Air rescue – county of Recklinghausen

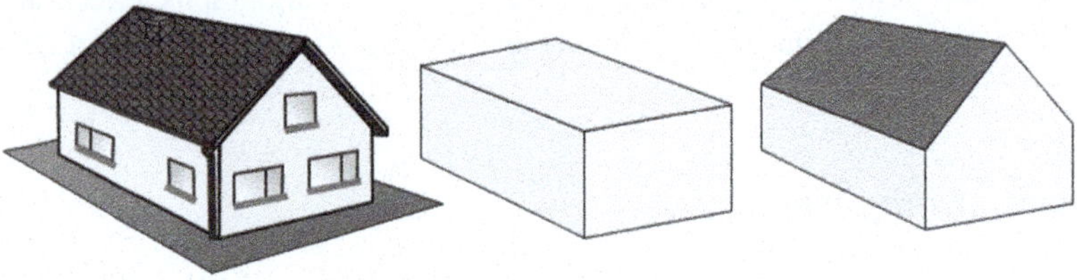

Fig. 13.5 Real World (*left*); LoD 1 (*centre*) and LoD 2 (*right*)

source (exact location/2D building information), but they never became part of the cadaster. To overcome this lack a proposal came up to expand the official cadastral AAA® - data model towards the third dimension (Fig. 13.5).

Approach

Several investigations have proved that only a few additional information is needed to build up a 3D-spatial data set out of the existing 2D-spatial cadastral data and to keep the information up to date. What is needed are the number of floors, ridge direction, and the building height. Most of this information already exists in the planning process; additional data is collected during the cadastral survey. With this

Fig. 13.6a City model, Recklinghausen

Fig. 13.6b 3D-spatial data, Recklinghausen

approach and the integration of the aforementioned information, a future 3D-cadaster could be implemented.

These days the 3D approach is a "topographic" extension of the cadaster in Germany. The demand for taxation was the reason for the establishment of the cadaster in the beginning of the nineteenth century. One hundred years later the property cadaster was established. In the last decades the cadaster was increasingly used for other necessary mapping and planning issues – it became a so-called multi-purpose cadaster, at that time restricted to 2D. With the AAA® model modern technologies, XML-descriptions' suitable software came up (Hawerk, 2002) and today the link to CityGML takes place (Gröger et al. 2012). With this the 3D-ability is included.

The pictures (Figs. 13.6a and 13.6b) show the additional contents of a city model compared to 3D-spatial data. While city models are often based on visualization, the AAA®-3D-spatial data are focusing on analysis. After the implementation of

AAA®-3D-spatial data, city models might be developed automatically as cadastral applications.

Standards

CityGML-Profiles and AAA®-3D-spatial Data

The following step-by-step approach is applied to realize nationwide 3D-geodata set in Germany.

Interim Solution CityGML-Profiles

Already today, there is a demand for 3D spatial information. The currently used AAA®-data model (version 6) is not able to store and to provide the expected 3D-information. The expanded AAA® version 7.0 will not be available before 2018 all over Germany.

Therefore the existing OGC standard CityGML (Gröger et al. 2012) for the representation and exchange of 3D-information is used. In March 2012, CityGML 2.0 was published as an international standard by the Open Geospatial Consortium (OGC). To realize the above-mentioned interim solution profiles were created from GML and CityGML taking into account the needs of 3D spatial information of the cadastral and surveying administration. As a result, the classes, attributes and values have been reduced to the maximum extent permitted by the product definition (Gerschwitz et al., 2011).

Figures 13.7a and 13.7b show that the AdV profile uses only parts of the CityGML-schema, especially mandatory requirements and quality indicators. The profiles are logical restrictions to CityGML-schema. The updating process of the described interim solution will be done by reprocessing of the existing/original data. An object based actualization does not exist yet.

AAA®-concept

The AAA®-concept is national standard for official spatial information in Germany. It was built up completely by specialization of international standards (Fig. 13.8) (AdV 2008a, 2008b). The AAA®-schema is a GML-application schema which represent the national standard for geospatial data of the surveying and cadastral administration in Germany. The model and external schema are completely embedded in existing standards of ISO and OGC.

According to size (number of citizens), Recklinghausen is the biggest county district in Germany and therefore comparable to a city like Cologne. In 2011 about

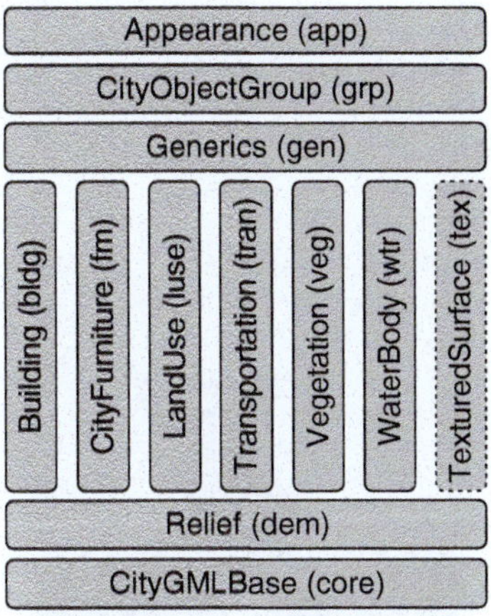

Fig. 13.7a CityGML Version 1.0

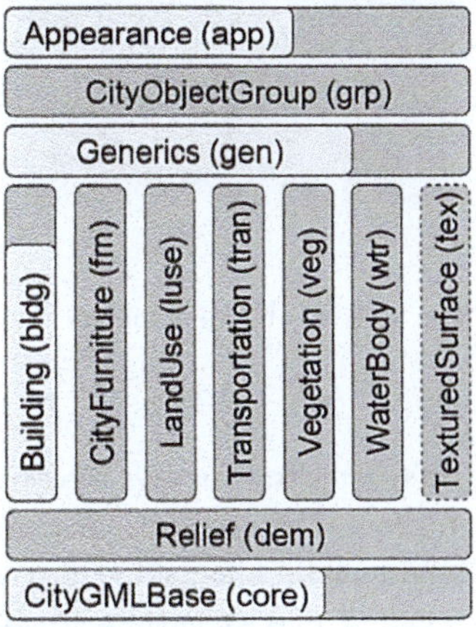

Fig. 13.7b CityGML-Profile of AdV

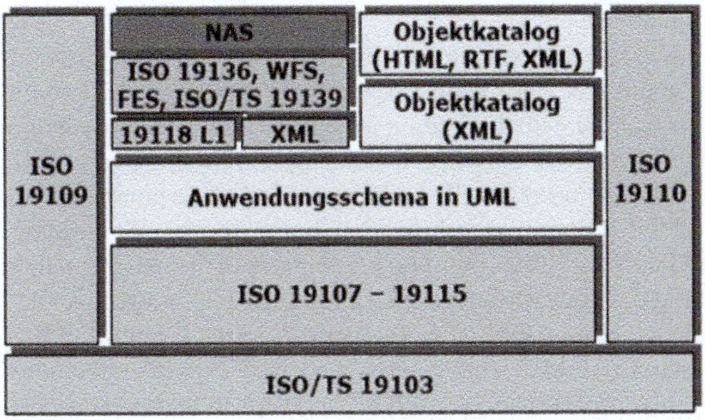

NAS = exchange interface, Objektkatalog = feature catalogue,
Anwendungsschema = application schema

Fig. 13.8 AAA®-embedding existing international standards (AdV 2008a, b). *NAS* exchange interface, *Objektkatalog* feature catalogue, *Anwendungsschema* application schema

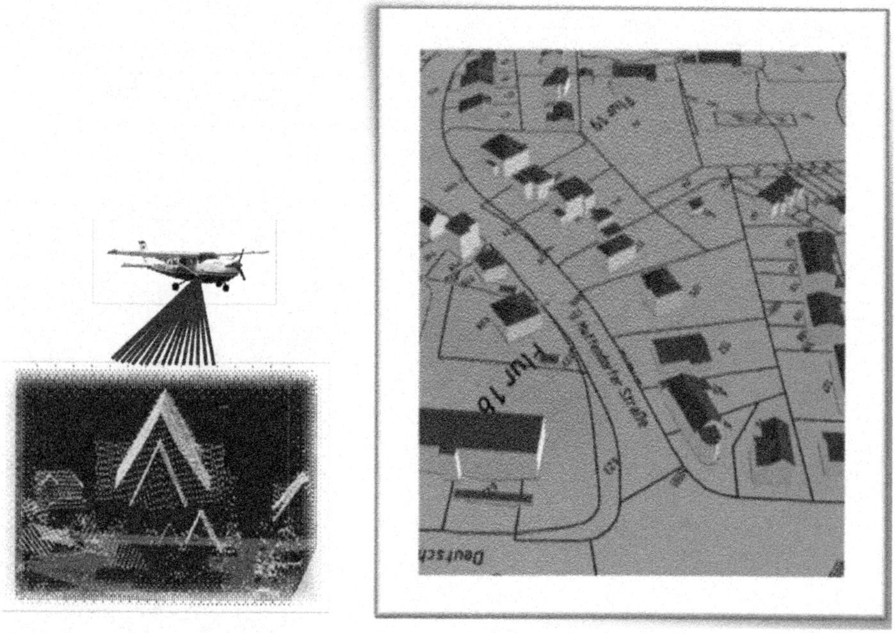

Fig. 13.9 3D-building model

1600 cadastral surveys took place with respect to buildings. For Recklinghausen, as in general for the German cadaster with over 50 million buildings, it is therefore of fundamental interest to store actual 3D-building information conform to the AAA® standard and consistent to 2D- and 3D-cadastral object information (in general: 2D-property building layer identical to 3D building footprint) – the so-called "vertical integration concept".

This "vertical integration concept" takes into account the source of the data and the production process. The "legal" 2D-property building layer as a major cadastral information is merged with the 3-dimension from laser scan as a topographic source. The result is a "legal" 3D-building model (Fig. 13.9).

It defines the AdV product "3D building model". As a consequence, the demand, especially of the economy, for official (administrative) 3D-building information could be fulfilled. In addition this data participates in the existing national and international spatial data infrastructure (SDI), for example through simple export to the defined INSPIRE topics (INSPIRE Thematic Working Group Building 2012).

In contrast to CityGML, which is designed as an external interchange format and for the easy use of 3D-data, the AAA®-concept defines a standard: application schema, feature catalogue and exchange interface (Fig. 13.10).

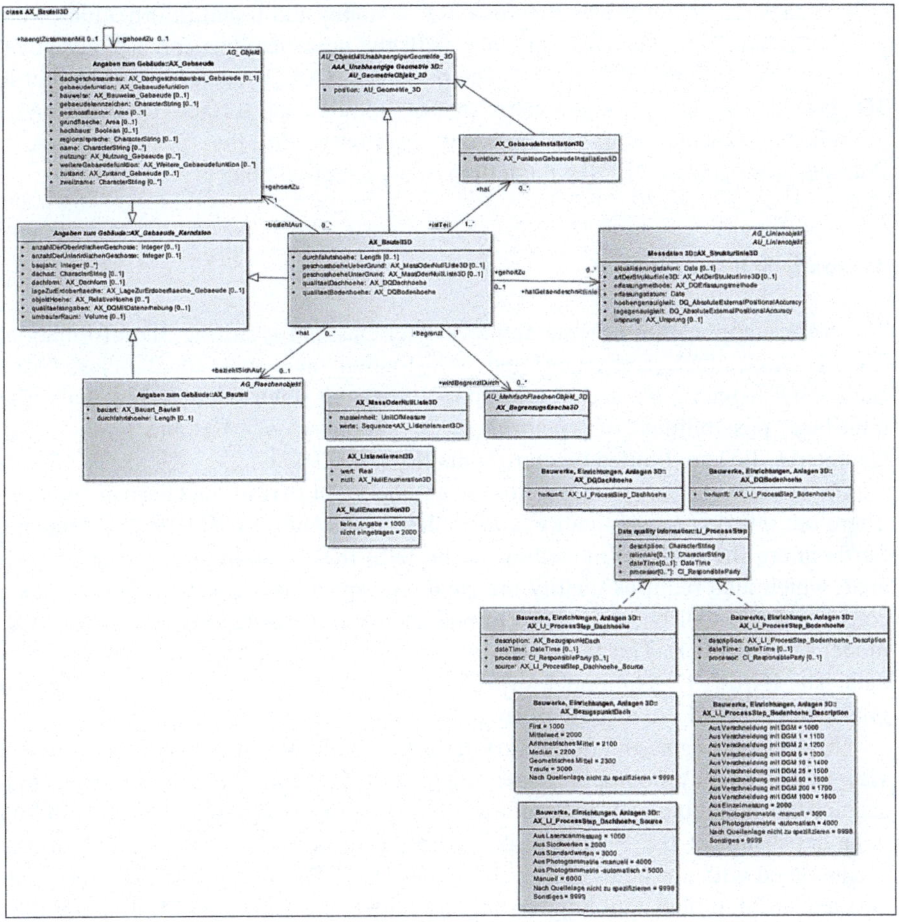

Fig. 13.10 Specification to the building 3D-AAA®-schema

Modelling Aspects

Basic Schema

The AAA®-schema is logically divided into several packages, essentially into the thematic independent basic schema and the thematic schema, which is based on the basic schema. 3D-classes, which are necessary, are integrated into AAA®-schema in three new packages:

- AAA_SpatialSchema 3D,
- AAA_Unabhaengige Geometrie 3D, and
- AAA_Praesentationsobjekte 3D.

The package "AAA_SpatialSchema 3D" contains additional information of the existing AAA®-schema in accordance with the specifications for 3-dimensional objects of the ISO-norm "191XX". The package "AAA_Unabhaengige Geometrie 3D" provides all necessary geometric shapes (dot, line and surface) for the AAA®-3D-schema objects with independent geometry. In the package "AAA_Praesentationsobjekte 3D" the modelling of presentation objects is described.

Thematic Schema

The AAA®-application schema defines object classes for storing 3D-information: The 2D-classes "AX_Gebaeude" and "AX_Bauteil" as well as the 3D-class "AX_Bauteil3D" which have a common upper class "AX_Gebaeude_Kerndaten". The multiface possibilities of occurrence of geometry of 3D-objects in "AU_Geometrie_3D" are limited by constraints (SIG 3D 2012).

The storage of quality information is an important part in the German cadaster. Therefore information of quality is modelled conforming to ISO-19115 Metadata. Furthermore the relevant modelling in the INSPIRE[11]-building-topic was considered, which also requires quality information, especially the source of data. As a consequence it will be possible to provide semantics match between the AAA®-model and INSPIRE. This allows the realization of the exchange and conversion of data. The INSPIRE data model, especially the profile extended3D, is one special profile of CityGML, in a similar way to the AAA®-3D-e

The 3D-building model of the AdV describes buildings in terms of the cadastral view as well as for topographic surveying (LoD 1-3, chapter 2). It does not take into account the modelling of interior rooms (LoD 4), or city topography. The 3D-building is an expansion of the "Hausumringe" (hose foot prints) in the third dimension, accumulated with attributes of associated cadastral 2D-objects. Currently the product standard describes building resolutions conform to LoD 1 and LoD 2 (Gruber, 2012).

The Fourth Dimension

Traditionally, in the German cadastre every change of a parcel (e.g. subdivision) is documented by surveying sketches and textual documentations. The development of the cadastral map is continuously monitored and every change over time can be restored in case of cadastral disputed, but usually using non-digital paper documents. Therefore, modern possibilities for inquiries were also a technical requirement for the AAA®- standard. Besides this more internal cadastral

[11]Infrastructure for Spatial Information in Europe: A European Directive, see http://inspire.jrc.ec.europa.eu/

use-case, there are lots of further requirements for the time-related cadastral information, such as:

- Monitoring the development of cities and villages over time
- Statistic of changes of land use and land cover
- Planning purposes
- Historical archiving
- Monitoring cultural heritage.

The AAA®-data model requires for each object a unique identifier together with a designated time stamp for creation and deletion of an object. However, once an object has to be deleted during an updating process the object will not be physically removed from the data base. Only the life cycle of the thematic relevance has ended, but not the existence of the object as an instance. A "deleted" object is then considered as a historical information which can be easily distinguished from the actual information. Sometimes there are changes in an object which do not require the deletion of the object (e.g. only a name of the person changes). In that case also the different versions of an object can be stored. Within the AAA®-data model this approach is therefore called "versioning concept". Since every object carries life cycle information the storage of historical objects and versions of objects is not limited to any specific object type.

Within the AAA®-data model this approach is used for providing historical information as well as for the incremental updating of secondary used information systems.

Conclusions

Availability of LoD 1 und LoD 2 in Germany

Due to the constitutional responsibility of surveying and mapping, the responsibility for cadastral data is on the state level. As mentioned above the Working Committee of the Surveying Authorities of the States of the Federal Republic of Germany (AdV) defines nationwide cadastral standards AdV (2009). In addition a nationwide access point was established in North Rhine-Westphalia to distribute about 21 million house coordinates (coordinates of buildings with an official address), about 50 million 2D-"Hausumringe" (house foot prints), LoD 1 and LoD 2 - data for Germany (for more information: www.adv-online.de).

Economy, science and administration have an increasing demand for official three-dimensional spatial information (3D-geodata) as a base for multiple applications. The surveying and mapping administration in Germany has accepted this demand as a challenge to develop and realize sustainable conceptions for 3D-geodata, focusing on quick and economic solutions. In this context, national and international standards, infrastructures and activities had to be considered. The German AAA®

cadaster standard takes into account the international standardization of ISO and OGC to include 3D-geodata as an economic solution for guidance and continuation. The approach of the vertical integration of 3D-geospatial into the cadastral standard guarantees an interface to the German and European spatial data infrastructure. Especially consistent regulation of modelling, actualization concepts and the quality management are activities which have to be finished in the next years.

References

AdV. 2008a. Working Committee of the Surveying Authorities of the States of the Federal Republic of Germany: GeoInfoDok (ALKIS®) main document Version 6.0.

———. 2008b. Preliminary report "Digital Surface Models, architectural models" of the working groups "Geotopography" and "cadastre" for 120th Committee Sessions of the Working Committee of the Surveying Authorities of the States of the Federal Republic of Germany on 10 to 11 September 2008.

———. 2009. Final minutes of the 121st Committee Sessions of the Working Committee of the Surveying Authorities of the States of the Federal Republic of Germany on 9 to 10th September 2009.

BImSchG. 2012. Law to protect against harmful effects of air pollution, noise, vibration and similar phenomena, as amended by the 03.15.1974 26.9.2002, as last amended by Article 2 of G v. 06/27/2012. (Gesetz zum Schutz vor schädlichen Umwelteinwirkungen durch Luftverunreinigungen, Geräusche, Erschütterungen und ähnliche Vorgänge, vom 15.3.1974 i.d.F. vom 26.9.2002, zuletzt geändert durch Art. 2 G v. 27.6.2012).

Gerschwitz, A., U. Gruber, and S. Schlüter. 2011. Die dritte Dimension im ALKIS®. In *Das deutsche Vermessungs- und Geoinformationswesen 2012*, Hsrg. K. Kummer, and J. Frankenberger, 279–310. Berlin/Offenbach: Wichmann Verlag.

Gröger, G., T.H. Kolbe, C. Nagel, and K.-H. Häfele (Hrsg.). 2012. OGC City Geography Markup Language (CityGML), Encoding Standard, Version 2.0, OGC Doc No. 12-019, Open Geospatial Consortium, 2012.

Gruber, U. 2012. Die dritte Dimension im amtlichen Vermessungswesen Deutschlands. In *Geoinformatics "Mobility and Environment"*, Hsrg. M.-O. Löwner, F. Hillen und R. Wohlfahrt. Proceedings of "Geoinformation Conference 2012" of 28-30.03.2012. Braunschweig.

Hawerk, Winfried. 2002. *Cadastre 2020 – New trends in Germany's Cadastre?!* Washington, DC: FIG XXII International Congress.

INSPIRE Thematic Working Group Building. 2012. *D2.8.III.2 Data specification on building – Draft guidelines*. Version 3.0rc3.

Oestereich, Marco. 2014. *Das 3D-Gebäudemodelle im Level of Detail 2 des Landes NRW*, Növ 1/2014.

Riecken, J., and M. Seifert. 2012. Challenges for the multi purpose cadastre. *FIG Working Week*.

SIG 3D. 2012. *Quality concept*. http://www.sig3d.org/indexphp?catid=2&themaid=8777960. Last visit at 11.12.2012.

Chapter 14
Automated Extraction of Buildings from Aerial Lidar Point Cloud and Digital Imaging Datasets for 3D Cadastre – Preliminary Results

Pankaj Kumar, Alias Abdul Rahman, and Gurcan Buyuksalih

Introduction

3D extraction of buildings is required for many applications such as cadastral, city modelling, infrastructure mapping and urban growth analysis. Moreover, an accurate information about location and dimension of building features provides crucial input for the fire-safety analysis and managing other hazards in an urban environment. Traditionally, building boundaries are delineated based on manual or semi-automated reconstruction from close-range and satellite images. These processes are time-consuming and limited to 2D reconstruction of building objects. The lack of automated methods can be attributed to problems in finding an appropriate information from the data and the complexity in the scene (Elberink 2008).

Advances in geospatial data acquisition techniques have transformed the concept of 2D building modelling to 3D. Light Detection And Ranging (LiDAR) enables 3D modelling of real world environment by measuring the time of return of an emitted light pulse (Kumar et al. 2013). Laser scanning systems use this technology to acquire an accurately georeferenced set of highly dense LiDAR point cloud data (Kumar 2012). These systems provide high level of automation during data acquisition and have an ability to capture data beneath tree's canopy. The applicability of laser scanning systems continue to prove their worth in geospatial mapping due to the rapid, continuous and cost effective 3D data acquisition capability (Barber et al. 2008). LiDAR data records a number of attributes including elevation, intensity, pulse width, multiple echo and range information, all of which can be used for extracting various features (Kumar et al. 2015). The methods developed for

P. Kumar (✉) • A.A. Rahman
Department of Geoinformation, Universiti Teknologi Malaysia, 81310 Johor Bahru, Malaysia
e-mail: pankajkumar@utm.my

G. Buyuksalih
BIMTAS, IBB, Istanbul, Turkey

T. Yomralioglu, J. McLaughlin (eds.), *Cadastre: Geo-Information Innovations in Land Administration*, DOI 10.1007/978-3-319-51216-7_14

segmenting LiDAR data are mostly based on the identification of planar surfaces and the classification of point cloud data based on its attributes (Vosselman 2009).

Automated extraction of building objects has been a topic of intensive research since last few years. Several approaches have been developed over the past decade for extracting urban building features from LiDAR data. Mumtaz and Mooney (2009) developed a semi-automated approach for extracting building objects from the integration of airborne LiDAR and digital imaging datasets. In their approach, Normalized Digital Surface Model (NDSM) was generated from LiDAR and a Normalized Differential Vegetation Index (NDVI) was developed from digital image. Both the NDSM and NDVI values were thresholded and then morphological operations were applied to binary image for extracting building features. However, in their approach, some of the large vehicles and industrial installations were incorrectly identified as buildings while smaller buildings were missed. Oda et al. (2004) proposed a method to extract building features from aerial LiDAR data in which Digital Surface Model (DSM) was segmented and then Hough transformation was applied for extracting building boundaries.

Finally, 3D building model was created by attaching vertical walls from aerial image to each of the extracted building polygon. The proposed method did not address the problem of extracting inclined roof. Pu and Vosselman (2006) presented an approach to automatically extract building features from terrestrial laser scanning data. LiDAR point cloud was segmented using the planar surface growing algorithm and then several human-knowledge driven feature constraints such as size, position, direction and topology were applied to extract building features. Mancini et al. (2009) used multi-source aerial LiDAR and multispectral dataset to automatically extract urban building and road objects. They involved multi-class supervised pixel classification using adaptive boosting algorithm to classify buildings, grass, land and tree objects. Finally, filtration and Hough transformation techniques were applied to extract linear road and roundabout features. Rutzinger et al. (2009) extracted vertical walls from mobile and airborne laser scanning data. A region growing segmentation technique based on 3D Hough transform was applied to extract planar surfaces from point cloud data and then the extracted segments were analyzed based on their inclination, size and dimension.

Most of the approaches developed for extracting buildings require semi-automated or manual intervention. The developed methods are also associated with the misclassification of large vehicles, trees and other features as building objects. There is a need to develop an operational and automated approach for extracting building features. LiDAR data provides elevation, intensity, pulse width and multiple echo attributes which can be a useful source of information for extracting building objects. The integration of multispectral digital images with LiDAR data will provide more efficient and accurate extraction of buildings. The use of LiDAR data provides to distinguish between high and low rise objects while multispectral data helps to distinguish canopies from the building objects. In this paper, we present some preliminary results based on automated extraction of building objects from the integration of aerial LiDAR and multispectral digital imaging datasets. In the next

Section, we detail our methodology to extract buildings while in the third Section, we test our approach on aerial LiDAR and multispectral digital imaging dataset. Finally, we discuss the test results and make conclusions in the last Section.

Methodology

Our methodology is based on the integration of aerial LiDAR and digital imaging dataset to extract buildings. A workflow of the automated building extraction approach is shown in Fig. 14.1. We make a use of digital imaging dataset to remove canopies from the data. The available multispectral digital image consisted of blue,

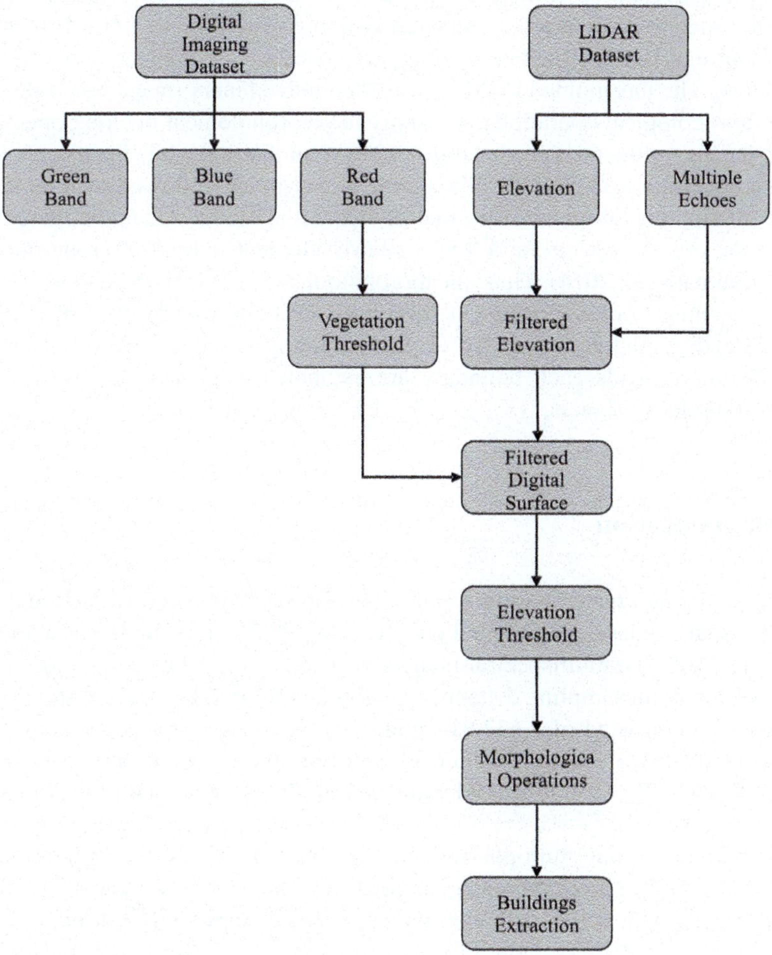

Fig. 14.1 Automated buildings extraction approach

green and red bands which represent brightness information of the targets. We utilize a low reflectance property of the vegetation in the red band to suppress them in the data. We apply empirically estimated T_1 threshold value to the red band image in order to remove the vegetation area.

LiDAR data provides multiple echo information which refers to multiple return of echo pulses from the targets. We filter out multiple reflected points and retain those points that record a single reflection. These single return reflection points belong to buildings, roads and other solid objects. The filtered points are used to generate Digital Surface Model (DSM) using the maximum elevation value of points within the cell while natural neighbourhood interpolation method is used to estimate values for cells that do not have points within their extent. We use the thresholded red band image to remove vegetation area from the DSM. We apply empirically estimated T_2 threshold value to the DSM in order to remove ground level objects such as roads, parking areas etc. and retain high rise building features in the data. In order to complete the extracted buildings and remove noise that is introduced through the use of thresholding, we make a use of binary morphological operations. The thresholded DSM is converted into a binary image and is processed using morphological operations. We apply the morphological opening operation in which the binary image is eroded followed by their dilation while the morphological closing operation is applied by dilating the binary image followed by their erosion. In the dilation operation, a binary matrix element is used to dilate the image pixels and in the erosion operation, a binary matrix element is used to erode the image pixels (Kumar et al. 2014). Thus, the morphological operations applied to the binary image are able to extract inherent shapes of the building objects and to remove noise. Finally, contour boundary of each building object is identified and then LiDAR points inside each boundary are estimated to provide 3D generation of extracted building objects.

Experimentation

We tested our automated building extraction approach on aerial LiDAR and multispectral image datasets of Istanbul city covering 89.76 km^2 which was acquired in October, 2012. The multispectral image consisted of three bands i.e. red, blue and green with ground sampling distance of 0.1 m and 8-bit radiometric resolution. The LiDAR data consisted of 2,647,912 points with 0.18 m spacing. The point cloud was associated with elevation, intensity and multiple echo attributes. The empirically estimated $T_1 = 130$ threshold value was applied to red band image in order to remove vegetation.

After filtering out multiple reflected points, the LiDAR data consisted of 2,376,200 points. The DSM was generated from the maximum elevation value of filtered points with 0.1 m cell size. We applied the empirically estimated $T_2 = 45$

threshold value to the DSM. The morphological opening and closing operations were applied using 3 × 3 matrix element. The tested multispectral image is shown in Fig. 14.2a while the automated extracted 2D and 3D building objects are shown in Figs. 14.2b and 14.3 respectively.

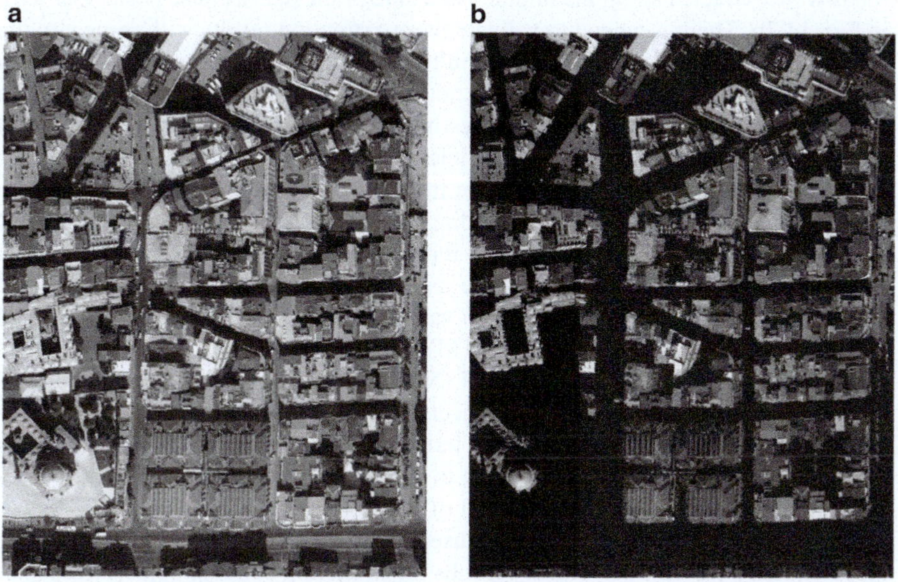

Fig. 14.2 (**a**) Input multispectral image and automated extracted; (**b**) 2D building objects

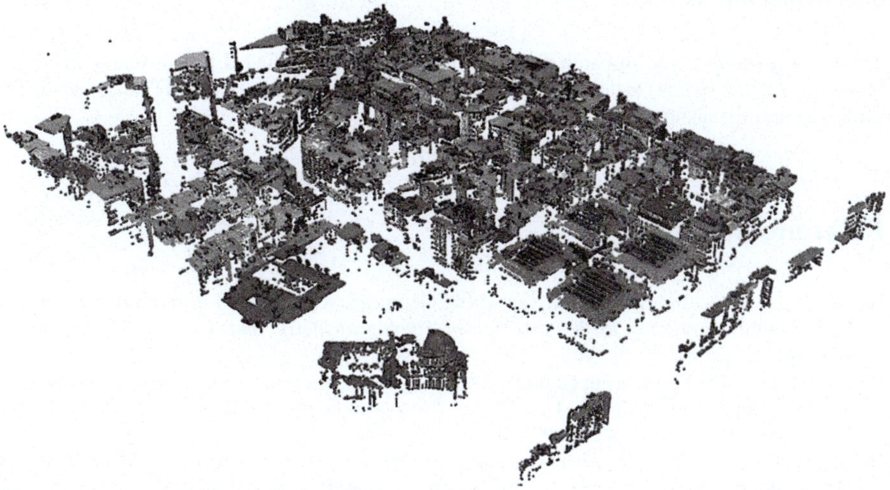

Fig. 14.3 Automated extracted 3D building objects

Discussion and Conclusion

Our automated approach was able to successfully extract the building objects from aerial multispectral digital imaging and LiDAR point cloud datasets. Some of the building objects along the lower-left side of the data were missed while some of the roads were extracted along the middle-right side of the data as false positive as seen in Fig. 14.2b. There is a need to validate these extraction results with respect to the ground truth. We used red band in the multispectral image to remove canopies from the data; however, this information was not adequate. The use of both the near-infra red and red bands would provide us to estimate Normalized Differential Vegetation Index (NDVI) which would be more efficient in removing vegetation areas from the data. The use of multiple echo attribute in the LiDAR data was further helpful in retaining the points that belong to buildings, roads and other solid objects. LiDAR data provides intensity attribute that represents the maximum amplitude of a reflected pulse. Intensity values can be used to differentiate buildings from other terrain objects. The minimum elevation value of points within the cell can be used to generate Digital Terrain Model (DTM) which can be further used to estimate Normalized Differential Surface Model (NDSM). NDSM values can be more efficiently used to remove ground level objects and retain building objects as they represent absolute height values of the terrain objects. The opening and closing morphological operations were applied to complete the shapes of extracted buildings and remove noise. There is a need of their inclusive use in which the dimensions of the extracted objects can be used to remove non-building objects. This research study presents preliminary results for extracting building objects from integrated aerial LiDAR point cloud and digital imaging datasets. In future, we intend to develop more comprehensive approach for automated and operational extraction of building objects.

Acknowledgements The research presented in this paper was funded by Universiti Teknologi Malaysia (Vot No. 01E64) and the experimental datasets were provided by BIMTAS, Istanbul. The authors gratefully acknowledge this support.

References

Barber, D., J. Mills, and S. Smithvoysey. 2008. Geometric validation of a ground-based mobile laser scanning system. *ISPRS Journal of Photogrammetry and Remote Sensing* 63(1): 128–141.

Elberink, S.O. 2008. Problems in automated building reconstruction based on dense airborne laser scanning data. *International Archives of Photogrammetry, Remote Sensing and Spatial Information Sciences* 37(XXX): 93–98.

Kumar, P. 2012. *Road features extraction using terrestrial mobile laser scanning system*. National University of Ireland Maynooth. http://eprints.maynoothuniversity.ie/3995, 300 p., 25 Jan 2015.

Kumar, P., C.P. McElhinney, P. Lewis, and T. McCarthy. 2013. An automated algorithm for extracting road edges from terrestrial mobile LiDAR data. *ISPRS Journal of Photogrammetry and Remote Sensing* 85: 44–55.

———. 2014. Automated road markings extraction from mobile laser scanning data. *International Journal of Applied Earth Observation and Geoinformation* 32: 125–137.

Kumar, P., P. Lewis, C.P. McElhinney, and A.A. Rahman. 2015. An algorithm for automated estimation of road roughness from mobile laser scanning data. *The Photogrammetric Record* 30(149): 1–16.

Mancini, A., E. Frontoni, and P. Zingaretti. 2009. Automatic extraction of urban objects from multi-source aerial data. *International Archives of Photogrammetry, Remote Sensing and Spatial Information Sciences* 38(3/W4): 13–18.

Mumtaz, S., and K. Mooney. 2009. A semi-automatic approach to object extraction from a combination of image and laser data. *International Archives of Photogrammetry, Remote Sensing and Spatial Information Sciences* 38(3/W4): 53–58.

Oda, K., T. Takano, T. Doihara, and R. Shibasaki. 2004. Automatic building extraction and 3D city modeling from LiDAR data based on Hough transformation. *International Archives of Photogrammetry, Remote Sensing and Spatial Information Sciences* 35(B3): 277–280.

Pu, S., and G. Vosselman. 2006. Automatic extraction of building features from terrestrial laser scanning. *International Archives of Photogrammetry, Remote Sensing and Spatial Information Sciences* 36(5): 25–27.

Rutzinger, M., S.J.O. Elberink, S. Pu, and G. Vosselman. 2009. Automatic extraction of vertical walls from mobile and airborne laser scanning data. *International Archives of Photogrammetry, Remote Sensing and Spatial Information Sciences* 38(W8): 7–11.

Vosselman, G. 2009. Advanced point cloud processing. *Proceedings Photogrammetric Week*, Stuttgart, 7–11 September.

Part III
Cadastral Cases

Chapter 15
Modelling PGIS for Multipurpose Cadastre in Ghana

E.A. Gyamera, E.E. Duncan, and J.S.Y. Kuma

Introduction

Cadastre is the pivot of the Land Administration (LA) system, and it describes the right, restrictions and responsibilities (RRRs) associated with land (Njuki 2001). The current cadastre system package is made up of the following: a cadastral plan (geometric dimension of the parcel), the interest, right (ownership), and values. Other vital information such as cultural attribute, physical attribute, utility data, and legal attributes are not considered. Though no one can claim absolute ownership of land, every individual traces a lineage to land. This means that every one owns right to land with respect to and in accordance with the local laws, practice and encumbrances. Multipurpose Cadastre which combines geospatial data to cultural, physical and legal attributes with utility data through technical integration is very essential for developing countries like Ghana. The model has a potential to support spatially enabled government, private sectors, society and to expand Information Technology Communication (ICT) support in the process of visualization, organization and management of useful land information (Bin Taib 2012). Figure 15.1 illustrates the potentials of MPC.

According to Untong (2013), multipurpose cadastre is a land information system that incorporates legal (property right or cadastre), physical (topology, artificial features etc.), and cultural (land use, demographics etc.) information about land into an accurate framework. PGIS facilitates the presentation of local people's spatial knowledge using two or three dimensional maps (Corbett et al. 2006). PGIS practice is geared towards community empowerment through tailored, demand-driven

E.A. Gyamera (✉)
Department of Soil Science, School of Agriculture, University of Cape Coast, Cape Coast, Ghana
e-mail: gyamengineering@yahoo.com

E.E. Duncan • J.S.Y. Kuma
Faculty of Mineral Resource Technology, University of Mines & Technology, Tarkwa, Ghana

T. Yomralioglu, J. McLaughlin (eds.), *Cadastre: Geo-Information Innovations in Land Administration*, DOI 10.1007/978-3-319-51216-7_15

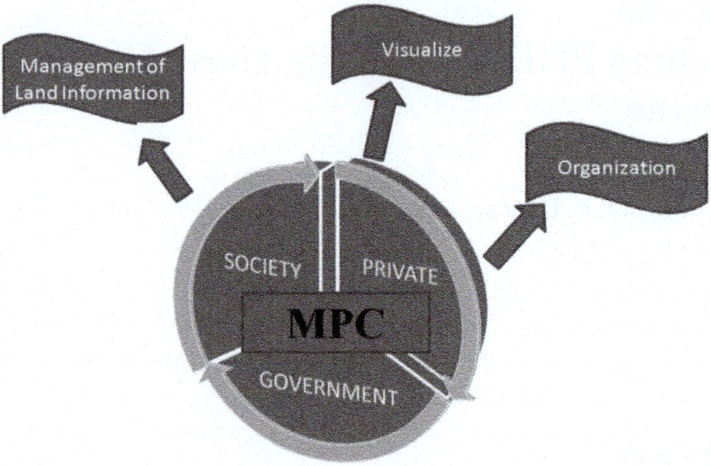

Fig. 15.1 Potentials of multipurpose cadastre

and user-friendly applications of these geospatial technologies. Good PGIS practice is flexible and adaptable to different socio-cultural and biophysical environments. It often relies on the combination of expert skills with local knowledge. Unlike traditional GIS applications, PGIS places control on access and use of culturally sensitive spatial data in the hands of those communities who generated it. Involvement of the local community in developing a multipurpose cadastre in Ghana will be good for sustainable development. Candler et al. (2006) concluded that since the first Traditional Use Studies (TUS) in the Treaty 8 area in 1974, PGIS has proved itself as a valuable tool in communicating first nations relationship to interests on land. Through changing political, social, cultural and ecological environments, PGIS has been able to adapt and maintain an important role.

Countries like Italy, Sweden and the Netherlands have already started MPC and deriving enormous benefits. In Ghana, MPC is implemented in the form of Land Title Registration (LTR). LTR was established purposely to replace the deed registration which was flawed due to the occurrences lands conflicted that created a lot of tension in the country. Multiple sales of land were a major failure of the deed registration. Meanwhile, the LTR also has a lot of short falls due to the limitation of geospatial data in its composition. The current LTR which is being used for taxation and financial transaction, have the following deficiencies among others; inadequate archival process, foot prints are not being included in the cadastral plan for most areas and also, the process is not fully automated (Arko-Adjei et al. 2010).

The need for Ghana's MPC is long overdue, hence, the purpose of this work. MPC when adopted and implemented in Ghana by the use of PGIS will improve upon the current LTR by a number of ways including: the provision of fully automated geospatial database with good archival records. All attributes related to land parcel information will be captured in to the database. The GIS involvement will aid in detailing every foot print in the preparation of a site plan. This will then minimize land conflict due to stakeholders' active participation in the initial process of MPC which is the participatory mapping and survey.

PGIS as a tool for resolving land and resource related conflicts have been widely used. Reyes-García et al. (2012) used a randomized evaluation to assess the effects of participatory mapping in conflicts resolution. It was concluded that participatory mapping can help in conflict resolution or contribute to conflict generation or exacerbation depending on the political and socio-economic context. Tudor et al. (2014) investigated four different cases of land-use conflicts in Switzerland and Romania by the use of PGIS. For conflict-resolution to be successful, the findings indicated that it is important to foster on long-term ecological benefits and to take into consideration, people's needs. Brown and Raymond (2014) used PGIS to describe and evaluate alternative methods for identifying land use conflict potential. Jankowski (2009) explored the potentials of using PGIS as a tool to help the public become meaningfully involved in decision making processes affecting their communities and promoting the sustainable use of natural resources. GIS-generated maps need highly skilled facilitators to generate and interpret, and this posed a fiscal constrain for small community to developing GIS capability (Rambaldi et al. 2005). Bojorquez-Tapia et al. (2001) also used a GIS-based multivariate application for land suitability assessment taking into account issues and concerns of stakeholders, and employed a multivariate statistical procedure for classifying land units into land suitability. PGIS had some difficulties, which included varied understanding of the assessment's objectives, inability of the some stakeholders to grasp the analysis because of lack of formal education. Cinderby et al. (2011) discussed the development and piloting of PGIS techniques to facilitate decision making that integrates the knowledge of local stakeholders from individual communities with information from environmental managers at the watershed scale, the environmental decision making scale. It was noted that there is a significant potential that PGIS aids informed-decision making.

White et al. (2010) reviewed the emerging lessons from utilizing PGIS in Shoreline Management Planning. The research also examined the potentials for improving stakeholder dialogue and involvement in flood risk management with access to mapping. It was concluded that incorporating PGIS and other visual approaches offset the weaknesses and maximizes the usable responses from the stakeholders. It is also important to engage citizens in PGIS process which enhances their collaboration with the policy making community and their participation in policy making. Bemigisha et al. (2009) investigated the utility of evidential belief functions (EBFs) and Dempster's rule of combination to represent classification uncertainty and integrated the PGIS-based grazing intensity maps. Tracing the social history of PPGIS, Sieber (2006) argues that PPGIS has been socially constructed by a broad set of actors in research across disciplines and in practice across sectors. Finally, Duncan et al. (2012) modelled 3D for multipurpose cadastre. An initiation of 3D modelling for 3D cadastre was established with an introduction of a concept for volume parcels.

Upon all the research works that have been done on PGIS and MPC, none has been able to model PGIS for MPC in Ghana, and that is what this paper seeks to achieve. The concept is expected to benefit government and society. Among a lot of the benefits are increased sharing of datasets, public transactions of data and reduction of administrative costs.

Components of MPC

The component of an accurate MPC is predominantly the formulation of the Ghanaian Digital Cadastral Database (GDCD) that has been populated, and have undergone strict adjustment and quality checks at every formation level with large scale geospatial data sets. The MPC can be described as a spatially enabled system that consist of the integration information system which contains survey accurate cadastre, topology, manmade features and cultural (e.g., land use, demographics) in a common and accurate reference framework (Bin Taib 2012). Figure 15.2 presents the components of MPC that enhances delivery system to the public and connected governmental realization.

The main constituent of MPC is the addition or the combination of NDCDB and Large Scale Geographic Features (LSGF). A Unified Modelling Language (UML) class diagram for MPC constituent is shown in Fig. 15.3.

Traditionally, MPC is made up of five to six phases as shown in Fig. 15.4. The first and essential phase, which is NDCDB, provides a survey accurate fundamental layer in MPC. This is followed by large scale data acquisitions by the use of MTLS. Large scale spatial features such as building, road, utility, vegetation, etc. will be detailed during the survey. MTLS is a main source of spatial data for MPC. Geospatial Data Centre (GDC) dataset that is made up of GIS layers and large scale topographic map will then be captured. This is followed by the application of

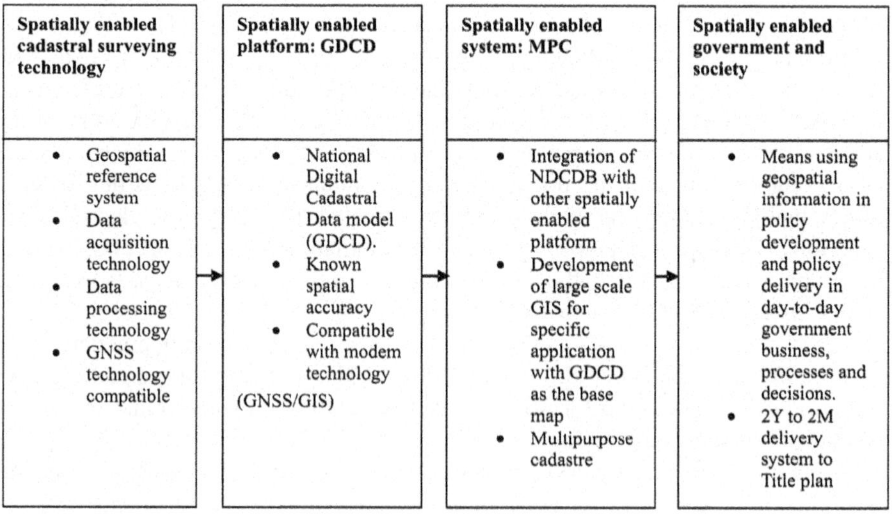

Fig. 15.2 MPC component

Fig. 15.3 Class diagram for MPC constituent

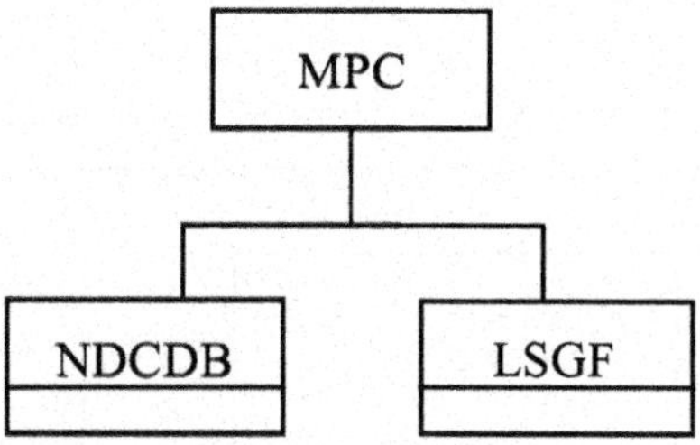

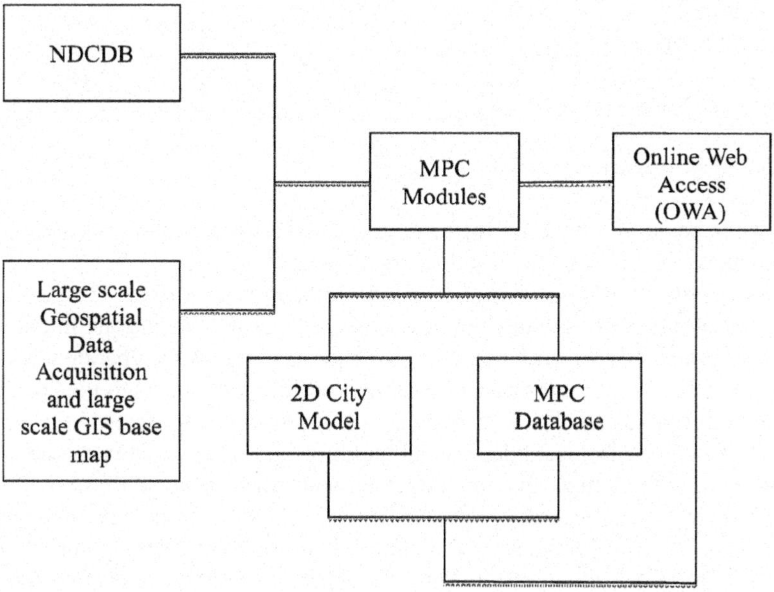

Fig. 15.4 UML class diagram for MPC phases

modules for integrating multiple data sources, validating MPC data base and the update of new spatial features. MPC database is then created with the provision of Online access Web (OWA) so that it can be accessed on the internet. The UML class diagram for MPC phases is shown in Fig. 15.4.

MPC Core Dataset

The core dataset for MPC are as follows:

1. *National Digital Cadastre Database (NDCDB):* To maintain a homogenous spatial accuracy of cadastral boundary coordinate to a better unit is the main objective of survey accurate NDCDB which is the most important element in the development of large scale geospatial database and will facilitate the development

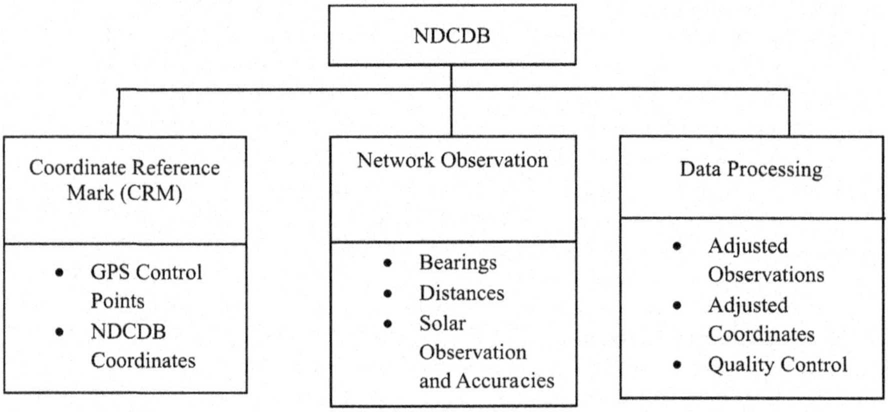

Fig. 15.5 UML diagram for MPC ingredient

of large scale Spatial Data Infrastructure (SDI). Figure 15.5 presents the main ingredient of NDCDB in a UML class diagram.

2. *Large Scale 2-D MTLS based Geographic Features Dataset:* This is the main source of MPC. It includes point clouds data that has undergone initial adjustment by post-processed kinematic GNSS procedures from separate GNSS base by a local transformation to a point that is well defined through the project area to produce the final station placed throughout a particular area.
3. *3D City Model:* Relevant details and buildings in 3-D representation have added opportunity to include semantic information about facade and thus not limited to geometric data only. Areas of buildings covered by dense and accurate measurement are used to model spatial features like transportation network, grounds, water bodies, buildings, city furniture, electric power lines and land cover. Thus, 3-D city models are now becoming the essential basis for city planning, development and control (Bin Taib 2012).
4. *Utility Dataset:* Utility features are features like water mains, sewerage system, telephone lines, power lines, gas lines etc. and they are needed to be mapped well in order to contribute to the benefit of MPC. It will critically aid city authorities to plan, maintain, and control infrastructure development. Repairs and replacement of utility lines become easy and efficient with GIS/Utility mapping system. Damages due to exaction can be easily avoided. Location and characterization of features can be systematically stored with easy access to information such as utility lines dataset and base maps. Updating, extraction and analysis become efficient and flexible with easy map distribution via digital or hard copy and internet access. A good link to data register and consumer information is added advantage.
5. *Street Address:* This provides a very pertinent data support for MPC to function well. Geocoding process can systematically be conducted by the use of cadastral lot numbers, UPI, and OID. The data can then be linked to zip, postal to generate street addresses based on MPC GIS base map integration.

6. *State Geospatial Data Centre (SGDC):* By the use of SGDC dataset as a core geospatial data impute, peri-urban MPC can be developed.

PGIS-MPC Conceptual Model Development

Multipurpose cadastre consists of multiple independent, interrelated layers commonly used to describe the graphic component of a GIS database (Organization Of American States, 2010). The fundamental constituent of the multipurpose cadastre is a cadastral overlay delimiting the current status of property right. The individual building block for the overlay is called cadastral parcel, which is also defined as a unit of land within which unique property interests are recognized. The overlay will consist of a series of maps showing the size, shape and location of all cadastral parcels within a given jurisdiction. The modern cadastral system is a step towards the Multipurpose Cadastre concept. In his view, Kaufman (1998) defined a Multipurpose Cadastre as a systematical arrangement of public register of data concerning all legal land objects in a certain country or district, based on a survey of their boundaries. The Multipurpose Cadastre is an extension of the modern cadastre to include other land information registers (Jamil et al. 2013). These registers might include databases with planning and valuation information.

From the Committee on Geodesy, 1980 report, the basic requirements for MPC are listed below:

(i) The development of technical standards and specifications and the means to enforce these;
(ii) The development of linkage mechanisms in order to relate other land information to the basic components;
(iii) An emphasis on gradual, phased reorganization and quality control of existing governmental functions, rather than creation of new functions and agencies;
(iv) A focus on the county level as the place where much of the work in developing and maintaining a multipurpose cadastre will occur, with appropriate support by state and federal governments; and
(v) The development of qualified personnel through encouragement and support of university research and education.

A conceptual model of PGIS for MPC has been derived and presented in Fig. 15.6. It begins with the initial field inspection and subsequent participatory mapping by stakeholders of land management and administration in Ghana. The stakeholder's composition is made up of the Local Community, House of Chiefs, Government institutions and department (e.g. Lands commission, Town and Country planning, Stool lands administration, the police, Judicially, CHRAJ etc.), interested NGOs, and individual investors who matter. The mapping is done by any GIS tools. This is followed by technical integration of a geoscientist who will make sure that statutory requirement and standardization is achieved. Geospatial data and utility features will be captured together with legal (land right), physical (topology) and cultural

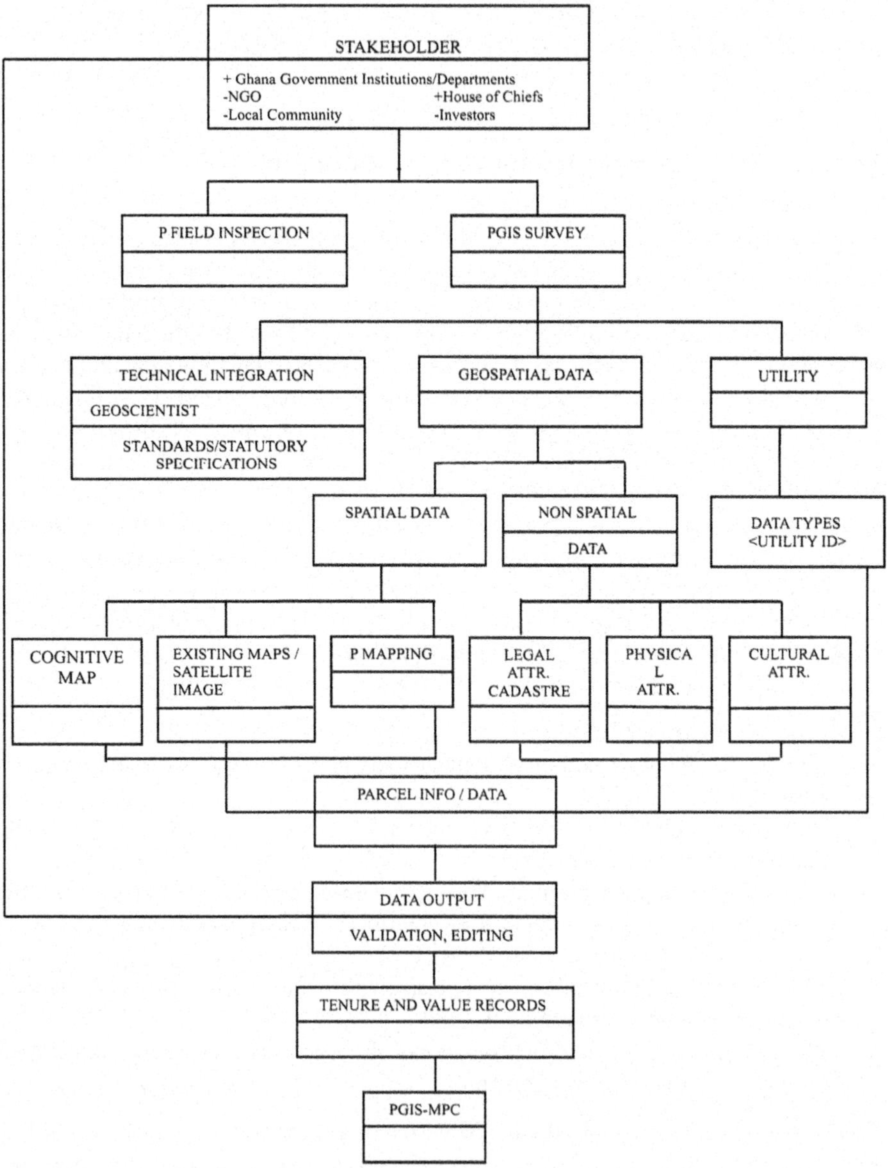

Fig. 15.6 PGIS model for MPC

(demographic) attributes. A parcel information database will then be produced after strict adjustment, computation and coordination of the geospatial and the utility data. The output will then be subject to scrutiny, validation, editing, formatting by stakeholder representatives to produce a valid, and reliable tenure and value record.

Implementation Concepts

The entire process will follow the following steps:

(i) Stakeholders' consultation meeting to define the task and the need to work together.
(ii) Meetings with the community for site analysis and also to elicit knowledge on historic to present geospatial information.
(iii) Participatory mapping where the community will be thought how to GPS and GIS instrument which are simple, user-friendly and very accurate.
(iv) Technical integration whereby the geoscientist will produce the cadastral map or base map with boundary information's impute from the local authority having considered all lay down statutory requirements. The geoscientist with a shared computer display to selected and trained community members.
(v) Legal (property right, or cadastre), physical (topographic and relevant feature), and cultural (land use and demographics) details or attribute will then be incorporated.
(vi) Data processing (Validation, editing, translating and formatting of output).
(vii) System output

Conclusions

Participatory Geographic Information System model for Ghana's MPC has been proposed. An integration of spatial data (land parcels) with cadastre information has been gathered with community involvement through participatory mapping and interviews. Adopting PGIS MPC in Ghana by 2016 will help in land tenure and management system in the country. Due to continually changing humankind to land relationship, demands and resource management, these will largely depend on the ability to adopt and integrate PGIS model for MPC in an enhanced land administration system. The PGIS model for MPC is driven by the needs of users that critically require accurate surveyed data. PGIS-MPC will prop up effective land development and administration, increase sustainable economic development activities, agricultural productivity, and environmental management. It can also empower different levels of stake holders at communal authority, state authority and national authority for participatory decision making processes which will enhance delivery system to the public. The pilot project towards the development of MPC for only one region in Ghana will provide informative insight on the future direction in implementing nationwide MPC and new cadastre management system.

References

Arko-Adjei, A., D.J. Jitske, Z. Jaap, and T. Arbind. 2010. *Customary Tenure Institutions and Good Governance*. FIG Congress 2010, Facing the Challenges Building the Capacity. Sydney.

Bemigisha, J., A.K. Skidmore, C. Polce, J. Carranza, M. McCall, and H.H.T. Prins. 2009. Representation of uncertainty and integration of PGIS-based grazing intensity maps using evidential belief functions. *Transactions in GIS* 13(3): 273–293.

Bin Taib, A.K. 2012. *Developing a multi-purpose Cadastre in Malaysia*. Malaysia Geospatial Forum 2012. Melaka.

Bojorquez-Tapia, L., S. Diaz-Mondragon, and E. Ezcurra. 2001. GIS-based approach for participatory decision making and land suitability assessment. *International Journal of Geographical Information Science* 15(2): 129–151.

Brown, G., and C.M. Raymond. 2014. Methods for identifying land use conflict potential using participatory mapping. *Landscape and Urban Planning* 122: 196–208.

Candler, C., R. Olson, S. DeRoy, and K. Broderick. 2006. PGIS as a sustained (and sustainable?) Practice. First Nation Experiences in Treaty 8 BC, Canada. *Participatory Learning and Action* 54: 51–57.

Cinderby, S., A. de Bruin, B. Mbilinyi, V. Kongo, and Barron. 2011. Participatory geographic information systems for agricultural water management scenario development: A Tanzanian case study. *Physics and Chemistry of the Earth* 36: 1093–1102.

Corbett, J., R. Giacomo, K.A.P. Kyem, D. Weiner, R. Olson, J. Muchemi, M. Mccall, and R. Chambers. 2006. Mapping for change – The emergence of a new practice. *Participatory Learning and Action* 54: 6–10. Date accessed: January 2015.

Duncan, E.E., A. R. Abdul, P. O. Van, T. H. Chee, K. S. Hafiz, N. Azri, and M. I. Hassan. 2012. 3D modelling for multipurpose cadastre. Shenzhen: International Workshop on 3D Cadastres, Developments and Practices. http://pubs.iied.org/pdfs/G02154.pdf.

Jamil, H., M.J. Mohd Yusoff and N.Z. Abdul Halim. 2013. *Discovering possibilities of implementing multipurpose cadastre in Malaysia*. FIG working week, environment for sustaiability. Abuja.

Jankowski, P. 2009. Towards participatory geographic information systems for community-based environmental decision making. *Journal of Environmental Management* 90: 1966–1971.

Kaufmann, J., and D. Steudler. 1998. *Cadastre 2014: A vision for a future cadastral system*. Berne: Rüdlingen and Bern.

National Research Council. 1980. *Need for a multipurpose cadastre: Panel on a multipurpose cadastre, Committee on geodesey, Assembly of mathematical and physical sciences, National Research Council*. Washington, DC: National Academy Press.

Njuki, A.K.. 2001. *Cadastral system and their impact on land administration*. Paper presented at the spatial information for sustainable development international conference, Nairobi, Kenya. Date accessed January, 2015.

Rambaldi, G., K.A.P. Kyem, P. Mbile, M. McCall, and D. Weiner. 2005. *Participatory spatial information management and communication in developing countries*. Paper presented at the mapping for change international conference (PGIS'05), Nairobi, Kenya.

Reyes-García, V., M. Orta-Martínez, M. Gueze, C.A. Luz, J. Paneque-Gálvez, M.J. Macía, and J. Pino. 2012. Does participatory mapping increase conflicts? A randomized evaluation in the Bolivian Amazon. *Applied Geography* 34: 650–658.

Sieber, R. 2006. Public participation geographic information systems: A literature review and framework. *Annals of the Association of American Geographers* 96(3): 491–507.

The Organization Of American States. 2010. *Modernization of cadastre and land rights infrastructure in Haiti*. Project proposal, A rapid and inclusive approach.

Tudor, C.A., C.I. Ioja, I. Pătru-Stupariu, M.R. Nit, and M.A. Hersperger. 2014. How successful is the resolution of land-use conflicts? A comparison of cases from Switzerland and Romania. *Applied Geography* 47: 125–136.

Untong, H.J.S.B. 2013. *Developing multiPurpose cadastre in Sabah*. Asia Geospatial Forum. Putra World Trade Centre, Kuala Lumpur, Malaysia.

White, I., R. Kingston, and A. Barker. 2010. Participatory geographic information systems and public engagement within flood risk management. *Journal of Flood Risk Management* 3: 337–346. www.fig.net/pub/fig2014/papers/ts08d/TS08D_dabiri_7004pdf.

Chapter 16
Aims and Actual Outcomes of Tuscany Castore Project: A Final Balance

Massimiliano Grava, Maurizio Trevisani, Umberto Sassoli, Andrea Peri, and Fabio Lucchesi

The Castore Project

Sponsored by the Region of Tuscany (*Regione Toscana*, RT henceforward), the CASTORE project started in cooperation with Tuscany State Archives on the basis of an agreement signed with the Ministry of Cultural Heritage and Activities and Tourism (MIBACT) in July 2004, and two main objectives were originally established. On the one hand, local government authorities would be provided with a historical cartographic basis enriching cognitive frameworks of spatial organisation, landscape management, and environment planning; on the other hand, the actual value of the first cartographic representations drawn via scientific methods would be definitely enhanced by way of systematic online publication – thus also improving fruition and preservation of original documents at the official archives and facilitating access by private citizens and students as well as technicians and researchers.

Considering the overall profile of the CASTORE project, several complex operational phases envisaged filing, digital reproduction, and georeferencing of primary historical cadastral maps: the General Cadastre of Tuscany (also known as *Catasto Ferdinandeo-Leopoldino*), covering a wide part of current regional continental and insular territory; the complementary Bourbon and post-unification cadastres of the province of Lucca; the Este cadastre of the territories of Massa and Carrara; and the French cadastre, integrating some of the gaps of the General Cadastre of Tuscany (Pazzagli 1979).

M. Grava (✉)
CFS Department, University of Pisa, via P. Paoli 15, 56126 Pisa, Italy
e-mail: assimiliano.grava@cfs.unipi.it

M. Trevisani • U. Sassoli • A. Peri
Regione Toscana, via di Novoli 26, 50127 Florence, Italy

F. Lucchesi
DIDA Department, University of Florence, via della Mattonaia 14, 50121 Florence, Italy

T. Yomralioglu, J. McLaughlin (eds.), *Cadastre: Geo-Information Innovations in Land Administration*, DOI 10.1007/978-3-319-51216-7_16

During the first operational phase, namely the study of both consistency and distribution of cartographic heritage preserved in the State Archives of Tuscany, each map was univocally given a code, selected, and described in a single board indicating the main documental and content features. The advanced technical and methodological procedures of this first phase were carried out by the Department of Historical and Geographical Studies at the University of Florence (Sassoli 2013).

In the second phase, original maps preserved at the State Archives underwent digital capture and elaboration. For this purpose, large-format scanners were used to acquire each document in a single scan at high resolution; these operations were carried out at the premises of the State Archives, in accordance with appropriate security protocols.

Thirdly, the over twelve thousand digital maps were georeferenced in order to display as a *continuum* the coverage of the area considered in the Cadastre. To this end, the most advisable procedure should minimise geometric inconsistencies in correspondence to the edges of the maps: on the methodological point of view, homologous checkpoints necessary for georeferencing were firstly located on ancient maps and matched with corresponding elements in present geographical sources, then chosen in concurrence with 'persistent' topographic elements – such as intersections, junctions of hydrographic elements, and presence of more prominent and relevant buildings – and eventually identified on maps.

Furthermore, the process of georeferencing employed for each map was replicated on contiguous maps and finally adopted for greater 'blocks' of maps with the aid of additional homologous connecting points, thus respecting geometric matching to the edges of maps and assembling a consistent cartographic 'mosaic'. The operations of georeferencing, carried out by experts in the use of GIS (Geographic Information System) technology, led to the creation of a vector chart of the boundaries in cadastral maps: through successive geometric operations of aggregation, it was eventually possible to reconstruct both a mesh of cadastral sections and the boundaries of nineteenth-century Communities.

Eventually, in the spring of 2007, the results of the three phases were integrated into a single information system designed to support publication on the Internet. This system, developed and managed by the technical units of RT, integrates the three main areas: research, display of original maps, and navigation in WebGIS environment (URL-1 2015). The system has also been implemented with a WMS (Web Map Service) allowing users to freely access both the map archive and cartographic work directly via a GIS software (URL-2 2015).

Historical Sources

The historical sources examined by the CASTORE project provide data of basically fiscal nature: in addition to cartograms – whose systematic scanning and georeferencing have been leading to online publication via WMS – quite a few public registers compiled by State employees and preserved in provincial archives still record

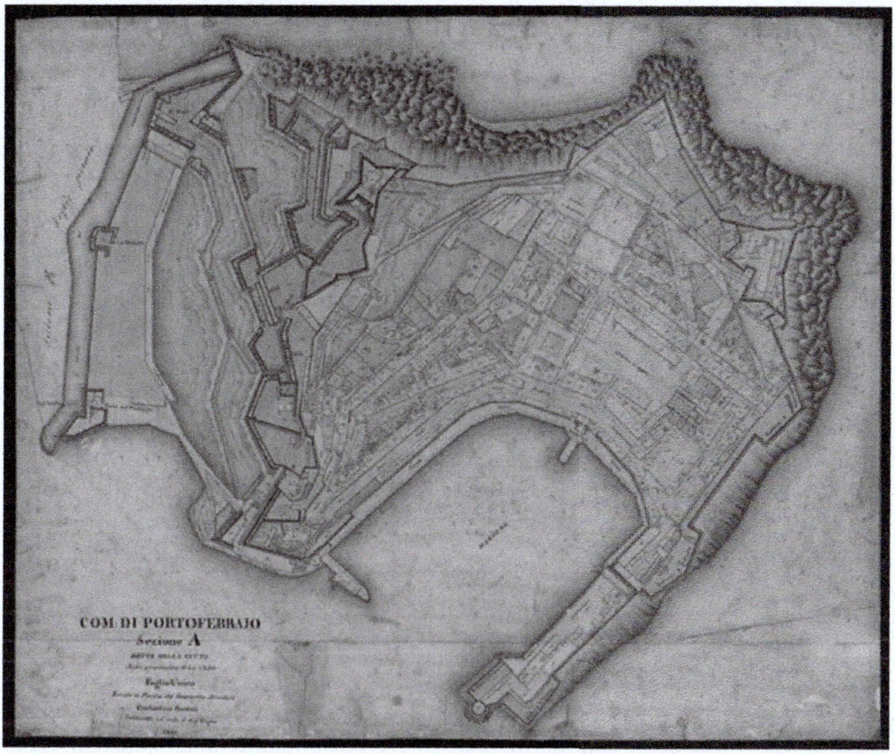

Fig. 16.1 Community of Portoferraio – General Cadastre of Tuscany Islands

all due information about taxes and fiscal withdrawals, in the form of either praedial levies or extraordinary imposts (Campana 2003). In particular, the following pages will draw the attention to the most ancient of the three documental series pertaining to the whole area of current RT, namely the *Catasto Ferdinandeo-Leopoldino* (Gregory and Healey 2007).

The General Cadastre of Tuscany (*Catasto Generaledella Toscana*, or *Catasto Ferdinandeo-Leopoldino*) is a nineteenth-century land register geometrically organised into land parcels covering the whole area of the Grand Duchy (*Granducato*): for the first time in Tuscany, it was based on geodetic principles, thus representing a modern and most effective fiscal tool produced through implementation of innovative land survey principles – just like its French model (Fig. 16.1) (Mirri 1980).

The work for the *Leopoldino* cadastre started in 1819 and ended with the completion of the land register in 1835. At that time, the Grand Duchy included 242 Communities, each consisting of a variable number of sections. Maps were drawn to a scale ranging from 1:1250 (built-up area) to 1:5000; the original unit of measure for areas was the 'square Florentine ell', while taxable income was "expressed in coin of account (*lira*) and in actual currency (*fiorino*)". In order to keep to proportions, it had been necessary to both subdivide the whole area and to allot each land unit to several single map sheets. Furthermore, an Index Map was drawn showing on a single sheet the relief of the entire area covered by each Community to a scale varying between 1:20,000 and 1:30,000.

On the other hand, the issue concerning evaluation of land properties was slightly more complex. In fact, a controversial and long-standing debate had opposed the members of the Deputation for the Creation of the New Cadastre as to whether "to tax the land either on the basis of its potential productivity or on the basis of how it was used at the time of the register compilation; in other words, on the basis of either potential or effective income" (Biagioli 1975). The final decision of the *Deputation* was a parcel-by-parcel evaluation of properties taking into account the effective production of the year 1817, when the land register had been 'commissioned'.

During the inspections, surveyors relied on clear drawings created on the model of the first triangulations by Father Giovanni Inghirami from the Ximeniano Observatory (author of the 1830 *Carta Geometricadella Toscana*), and total parcels, land measurements, and agricultural uses were eventually recorded in the *Giornali di Campagna*. At a later stage, data and drawings were used by the appointees of the land register in the Journals, Copy-books, *Value Relationships*, *Tithable Samples*, and the *Response of the Gonfalonieres to the Agrarian Questions* to almost arbitrarily calculate tax rates of each land parcel (Biagioli 1981).

The most important documents attached to the cadastre are the final *Community Land Register Samples* and the *Guide Tables* showing owners and their respective properties, and the maps. In particular, the *Community samples* resemble actual registers recording deals and transactions of each Community in alphabetical order. In fact, properties could be registered under the name of one or more 'persons' (either individuals or corporations), and for this very reason each document is close to a business census: "For every business transaction in the *Samples* there is the indication of surname, name, family name, and sometimes either noble ranks and titles (knightly, military or ecclesiastical) or professional qualifications (doctor or lawyer)". The samples were compiled with double-entry accounting: on the left-hand page, land parcels either involved in the transactions or acquired after establishment of the cadastre are indicated (*Dare*, that is, debits), while on the right-hand page there is the indication of properties ceded and purchased by others after the establishment of the cadastre (*Avere*, that is, credits). On the other hand, the *Guide Tables* progressively reported the number of land parcels, register sections, different uses, and the names of the owner.

Data

The emergence of the Internet as an effective platform for sharing also geographic data and information is certainly one among the most noteworthy events of the last years, and both the possibility to accede to the platform and the appeal to a wider and wider number of users prompted to the planning of new cartographic services increasingly based on shared access to geographical information and its interoperability.

Such a scenario immediately drew the attention to the urgent need to overcome the old elaboration paradigm, demanding preliminary specific treatment of data for each single use and obliging professionals to copy and elaborate them for further adaptation to different new needs and uses within off-line GIS.

Confronted with both maintenance and updating requirements, this paradigm eventually revealed to be a weak and rather problematic spot within the system. For this reason, the concept of interoperability in treating data and geographic information came to the fore. In particular, this idea principally implied the possibility to access the datum in itself (and the piece of information it provides) after combination with other data and information: this would be possible by means of devices abiding by approved and versatile standards in order to consider any possible future scenarios of use. In this context, RT launched a programme of publication, access and sharing of geographical data produced in the execution of its institutional duties through interoperable platforms conform to the standards approved by the Open Geospatial Consortium (OGC) and valid also in ISO context.

More specifically, RT started in 2008 the publication on WMS systems of its topographic products (namely technical cartographies to the scales of 1:2000 and 1:10,000 and cadastral maps; pictures from aerial shots and orthophotomaps; historical maps; several thematic charts such as geological maps, charts of use and covering of the soil, pedological charts, and maps of landscape and territorial restrictions) with the aim of providing not only the scientific and technical community, but also an even wider number of common users with as great interoperability of geographical information as possible.

Furthermore, data were distributed anonymously and free of charge via dedicated WMS systems: this connection mode relieves users of mandatory technical skills for structural elaboration of the single datum stressing the informative aspect, and such a rich informative basis led to planning and realisation of new and otherwise unimaginable applications.

Availability of both modern (twenty-first century) and ancient (nineteenth century) maps is a most incisive example. On the technical point of view, data access via WMS technology is based on simple HTTP calls, that is, on browser interrogation: this very system was used by RT to activate a WebGIS client (GEOSCOPIO) for easy navigation through georeferenced ancient cadastral maps and comparison to modern ones.

Another noteworthy interoperable service offered by the CASTORE project is the consultation of scans of original maps abiding by the specific International Image Interoperability Framework (IIIF). As it is the case with WMS, the system allows access to a specific cluster of information by means of simple browser calls. RT activated another web client supporting information and consultation of single scans.

To conclude, the interoperable approach is characterised by same protocols as many other services, by web clients directly created by RT, and by free access to completely open databases.

Laws and Land Government

Land government, that is, the cluster of activities aiming at organising, planning and defining different territorial uses and transformations, demands complete and reliable reference frames on the topographical and environmental as well as landscape, thematic, and historical points of view. Article 56 of the Regional Law no. 65 lists the fundamentals of the *Base Informativa Geografica Regionale* as follows:

(a) a topographical, geological and pedological informative basis accompanied by charts of use and covering of soil, orthophotomaps, aerial and satellite shots, and historical maps;
(b) a thematic informative basis mainly focused on the condition of territorial heritage; and
(c) a current legal informative basis pertaining to the effects of territorial and urban planning as well as of direct land government.

In the last few years, RT has paid particular attention to documental sources, with the aim of rescuing and collecting as much information as possible and eventually providing new references for reconstructing historical landscape evolution.

Recovery of historical information recorded in maps, registers, documents and archive paper material must confront with the demanding issue of reconstructing a historical and reference frame. To this end, multidisciplinary competences should be combined with several variables: acquisition, cataloguing and usability of sources; planning of integrated databases providing digital information (scans, tabs, texts); and use of both divulgation and promotion strategies for informative contents.

As far as public administration is concerned, departments appointed to production, collection, and organisation of databases supporting evaluative, decisional, and planning processes required by territorial authorities do not have the due scientific and research competence to manage collection and digitisation of historical documental sources. For this purpose, the Regional Council (*Giunta Regionale*) of Tuscany created a specific protocol of understanding and cooperation connecting Tuscany universities with RT University Centre for Territorial Sciences (*Centro Interuniversitario di Scienza del Territorio*, CIST).

In particular, the protocol envisaged a few research projects:

1. Documentation and digital capturing of selected historical cartographic collections dating from the fifteenth up to the nineteenth century, thus integrating the cartographic database of nineteenth-century cadastres within the CASTORE project.
2. Realisation of a digital archive supporting historical comparison of Tuscany place names, that is, systematic collection of toponyms recorded by both cartographic sources used by the CASTORE project and present maps (regional technical cartography and cadastral cartography).
3. Reconstruction of the hydrographic net (with particular focus on water courses indicated in the lists of public waters following the dispositions of the Regio

Decreto no 1775 of 11 December 1933) and of the location of manufacturing plants by examination of the maps of the Catasto Leopoldino, accompanied by investigation on the use of soil and land parcels in some Communities of Tuscany as documented by historical cartography and nineteenth-century cadastral registers.

In addition, further projects on environmental subjects in collaboration with CIST have started: acquisition of data pertaining to the distribution of natural areas deserving preservation within the Sites of Community Interest (*Siti di Interesse Comunitario*); capturing, measuring, and digital processing of data regarding technical property of lands by means of specific surveys *in situ* and in laboratories, with the aim of creating a database of lands of all typologies of geological surfaces interested by hydrogeological instability (Trevisani 2014).

Some of these projects have already been completed, while some others are still being carried out. In particular, reciprocal collaboration and confrontation of academic researchers in territory sciences with regional employees gave a strong impulse in defining strategies and solutions to informative requirements of authorities appointed to both administration and planning of shared resources; to this end, also the potentialities of up-to-date computer devices and instruments for the elaboration of geographical data were used.

Territorial Planning Between Time and Identity

The rich supply of information about the historical evolution of places is a most precious tool for professionals in the preparation of documents regulating urban planning, territorial organisation, and landscape structuring. Furthermore, cultural approaches and cognitive paradigms related to urban and territorial planning underwent a deep change in the last quarter of the twentieth century, and territory has progressively gained an active role in determining transformation choices (Lucchesi 2002). In particular, a clear overturning of trends has occurred. On the one hand, the traditional cognitive model, sometimes labelled as 'functionalist' paradigm, interprets human request and needs as determiners of choice perspectives, so that the territory itself must support fulfilment of such needs – at least within the limits of its own 'carrying capacity'. On the other hand, a new model claims direct influence on the quest for change of the 'existing territory' and its supply of both assorted places and consolidated vocations.

This cognitive reversal is often attributed to the imposition of the change sustainability principle; nevertheless, a close connection between sustainability and time should be considered too. Actually, persistent historical relations between settlement organisation and environment represent a repertoire of tested epistemological rules, and this repertoire might work as a heuristic tool for estimating the impact of any changing intervention on future long-lasting contexts. In other words, change sustainability can be evaluated by examining consistency of active transformations with durable relation rules connecting human action and environment.

More specifically, territorial identity is the expression of the dynamic relation between human action and the environment, that is, the resulting 'residue' of both alternations of peculiar social and productive structures and progressive metabolising of successive arrangements, eventually delivering new interpretations of the spatial table to the crucial test of time. As a matter of fact, each culture intervenes on the territorial palimpsest and evaluates its suitability with respect to current leading value systems. In contrast with the primacy of efficiency and functional performance, the sustainability principle focuses on the evaluation of the durability of territorial structures, namely on the acknowledgement of the influence of history in determining territorial transformation projects. Consequently, territorial identity originates from long-lasting historical sedimentation.

As far as operational practices are concerned, the capability to recognise and interpret territorial identity is strictly connected to the acquisition of technical skills supporting assignment of 'time dimension' to spatial description. In this particular case, access to the plentiful nineteenth-century cartography of Tuscany allowed the integration of information pertaining to the 'durability' of topographical signs with institutional cartography aimed at providing tools for territorial planning. Moreover, the possibility to elaborate thematic cartographic synthesis facilitates the evaluation of both structural entity and morphogenetic quality of described elements, this representation being a possible answer to the urging demand by planners for updated technical tools.

However, the above considerations do not imply elaboration of images with immediate normative value. Neither does the process aim at preparing a detailed list of past manufactured products and signs transmitted in the course of time, nor at keeping these elements unaltered in the future by way of different restrictions to transformation. On the contrary, it rather points at representing the final structure derived from sedimentation of all territorial features resisting to transformation. Therefore, the underlying causes of persistent historical paradigms of relation connecting anthropic elements and environmental conditions could only be understood by evaluating the sustainability and future effects of territorial transformations.

Conclusions

The exponential increase of accesses to CATORE online platform and to the assorted informative series of Tuscany map library (SITA *Cartoteca*) proves the excellence of services offered by the regional digital help desk. As far as web publication of geographical services is concerned, it turns out to be a most effective reference tool, both in Italy and maybe in the whole European scene. The many accesses not only to WebGIS, but also to portals supporting online visualisation of thematic cartography clearly show the great importance of geographical knowledge and data sets produced by Tuscany Public Administration for a wide user base.

In this context, the flow of accesses to CASTORE historical cartography should be particularly taken into account: amounting to over 20% of total accesses to the information system. It is mainly used by experts in urban planning and researchers in territory studies in quest of historical data. Actually, the maps of the CASTORE project describe a mainly rural (over 80% of total area) historical region. The still sparsely anthropised territories of the Grand Duchy of Tuscany counted no more than 1,393,334 inhabitants in 1883 (in contrast with 3,704,152 of 2014), and wealth derived from possession of land properties rather than from manufacturing proto-industry, this last being confined only to areas where energetic (and especially water-) sources could be found. The two-fold merit of the CASTORE project is a direct consequence of the accessibility of such a rich supply of cartographic and fiscal material: on the one hand, a representation of three historical States was possible thanks to documentary series describing the whole territory parcel by parcel; on the other hand, a new platform connecting historical and current data and eventually providing new tools for a better knowledge of territory in all its components was developed.

In conclusion, the combination of the CASTORE project with Territorial Informative System is the result of an extraordinary convergence, that is, a new functional elaboration by RT employees and managers of an excellent nineteenth-century source through institutional initiatives and regulatory frameworks: in other words, an excellent example of mutual communication and interaction between University and Public Administration for the development of GeoData, as expected by INSPIRE directive.

References

Biagioli, G. 1975. *L'agricoltura e la popolazione in Toscana all'iniziodell'Ottocento*, 17. Pacini, Pisa: Un'indaginesulcatastoparticellare.

———. 1981. Il Catasto in Toscanafra '700 e '800. Direttivecentrali e documentilocali: la città di Pisa. In *Città e proprietàimmobiliare in Italia negliultimi due secoli*, ed. C. Carozzi and L. Gambi, 374. Milano: Franco Angeli.

Campana, S. 2003. Catastoleopoldino e GIS technology: Metodologie, limiti e potenzialità. In: AA.VV., Tramenellospazio, *Quadernidigeografiastoricaequantitativa*, LaboratorioInformatico di Geografia, 1, Siena.

Gregory, I., and R.G. Healey. 2007. Historical GIS: Structuring, mapping and analyzing geographies of the past. *Progress in Human Geography* 31(5): 638–653.

Lucchesi, F. 2002. Nuovetecnologie per la descrizione e la rappresentazione del territorio. In: *Città e insediamenti*, ed. Cusmano, G. (a cura di) Milano: Franco Angeli.

Mirri, M. 1980. La fisiocrazia in Toscana: Un tema da riprendere. In: *Studi di storiamedievale e moderna per Ernesto Sestan*, II, 703–760. Firenze: Etàmoderna, Olschki.

Pazzagli, C. 1979. *Per la storiadell'agricolturatoscananeisecoli XIX e XX*. Dal catastoparticellarelorenese al catastoagrario del 1929, Fondazione L. Einaudi, Torino; Conti E. 1966, I catastiagraridellarepubblicafiorentina e ilcatastoparticellaretoscano (secoli XIV. XIX). Roma: Istituto Storico Italiano per il Medio Evo.

Sassoli, U. 2013. I Catastistoricidella Toscana e ilprogetto CASTORE, Rassegnadegli Archivi di Stato, VII, 113–119; De Silva, M., G. Tarchi, C. Ciampi, M. Ercolini, E. Loi, F. Lucchesi,

F. Nardini, and I. Scatarzi. 2012, *Valorizzazionedellefonticartografichestorichesull'uso del suolo per ilgoverno del territorio in Toscana.* Vicenza: XVI Conferen zaNazionale Asita.
Trevisani, M. 2014. Gliopendata e la Regione Toscana. *Geomedia* 6: 36–40.
URL-1. 2015. http://www.regione.toscana.it/-/castore-catasti-storici-regionali. Accessed 1 Jan 2015.
URL-2. 2015. http://www.opengeospatial.org/. Accessed 1 Jan 2015.

Chapter 17
Cadastral Renewal and Automation Project in Cyprus

Nihat Şahin, Baransel İzmirli, Sinan Çolakoğlu, and Rabia Bovkır

Introduction

Cadasters are often seen as being one of the pillars of modern land administration systems alongside efficient land registration, property valuation, real estate taxation, and land use management systems. An effective land administration system, it is conventionally argued, is seen as being an essential prerequisite for an efficient property market. Such a system should include ways of ensuring that property rights are protected and that trading in land, the transfer of property rights, and the raising of capital by pledging property as security can take place efficiently. In order to achieve these ends modern land administration systems have land registration to record property rights, their ownership and transfer, and cadasters to map property rights and record their geo-co-ordinates.

Land registration is a process of official recording of rights in land through deeds or title (on properties). It means that there is an official record (the land register) of rights on land or of deeds concerning changes in the legal situation of defined units of land. It gives an answer to the question "who" and "how" (Fig. 17.1).

Cadaster is a methodically arranged public inventory of data concerning properties within a certain country or district, based on a survey of their boundaries. Such properties are systematically identified by means of some separate designation. The outlines or boundaries of the property and the parcel identifier are normally shown on large scale maps which, together with registers, may show for each separate

N. Şahin (✉) • S. Çolakoğlu
Land Registration Office, Turkish Republic of Northern Cyprus, 06930 Ankara, Turkey
e-mail: nihatsahin61@gmail.com

B. İzmirli
Company of Sebat Mapping, 61830 Trabzon, Turkey

R. Bovkır
Istanbul Technical University, 34000 Istanbul, Turkey

T. Yomralioglu, J. McLaughlin (eds.), *Cadastre: Geo-Information Innovations in Land Administration*, DOI 10.1007/978-3-319-51216-7_17

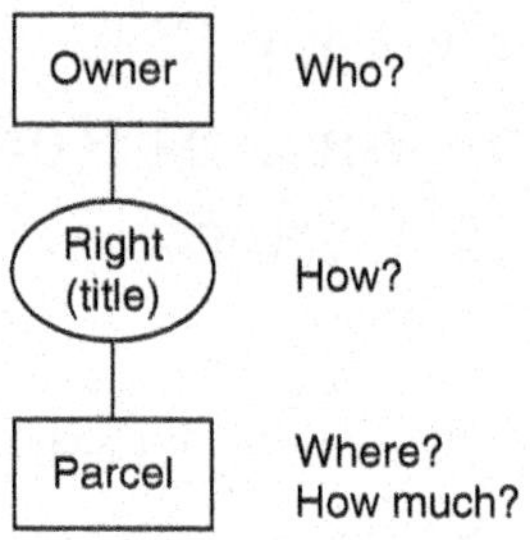

Fig. 17.1 Core entities connected (Zevenbergen 2002)

property the nature, size, value and legal rights associated with the parcel. It gives an answer to the questions "where" and "how much".

In many cases the land registration and cadastral functions are organized independently, and regularly they are not co-operating in the most effective way. Improvements, both technological and others, in quite some cases only mend one or a few links in the chain formed by land registration and cadaster. Often not even the weakest link. To fully understand land registration and cadasters, however, it should be treated as an integrated system, and be studied, analyzed and improved in its wholeness (Zevenbergen 2002).

The ownership structure in Turkish Republic of Northern Cyprus (TRNC) dates back to the Ottoman Empire which dominated the island between the years 1570–1878. Then the island came under British rule in 1878. The land classification and the first ownership determination for tax purposes were worked out in that period by using methods such as simple triangulation, chain measurements or plane table. However, only blueprint copies of the said maps come until today and most of which being unusable because they do not have measurement, bench mark and coordinate values.

Therefore, Land Registration is currently being executed under inappropriate conditions in TRNC. Although a register of property owner is kept, it is not possible to have access to owner information in a sound manner. With a view to remedying said deficiencies, protocols were signed from 1998 onwards between the relevant ministries of TRNC and Turkey, spelling out the principles of working geared towards under a contract awarded by the General Directorate of Land Registry & Cadaster of Turkey. At the beginning of 2013, works have commenced, covering up to 25% of the project, envisaging renewal of 125 thousand parcels in the form of three packages.

Duration of the project is one year, and estimated cost of the project is approximately USD 5 million. The first stage of the works covering 61 villages has been completed and related announcement has been posted. Related works will be completed in five months, and the cadastral maps newly prepared in metric system, digital and certified outputs thereof and the books of real estate registers will be done. The CORS-TR system surface network covering TRNC has been created, ensuring conversion from the British Imperial System into universal metric system. The paper covers the said works and the results obtained.

Background

Cyprus is the third largest and third most populous island located in the south-eastern region of the Mediterranean Sea, with an area of 9251 km^2 and a population of approximately 793,000. Topographically Cyprus consists of two mountain masses (on the north – The Kyrenia Range and south - the Mesaoria Plain) and central lowland. Administratively Cyprus is divided into six districts named after the island's principal towns: Nicosia, Limassol, Larnaca, Paphos, Famagusta and Kyrenia. Nearest neighbours are: on the 70 km north Turkey, on the 100 km east Syria, on the 170 km southeast Lebanon and Israel, on the 370 km south Egypt and on the 950 km northwest Greece (Fig. 17.2).

The recorded history of Cyprus extends to the 8th century BC. At a strategic location in the Middle East, it was subsequently occupied by several empires: Roman and Byzantine administration (30 BC–1191 AD), the Crusaders, the Frankish House of Lusignan (1192–1489), the Republic of Venice (1489–1571), the Ottomans (1571–1878) and finally the British (1878–1960).

Island was conquered by Ottoman Empire in 1571 and stayed over three centuries of Ottoman rule between 1571 and 1878. However, in the aftermath of the Russo-Turkish War (1877–1878), Cyprus was leased to the British Empire which de facto took over its administration in 1878 (though, in terms of sovereignty, Cyprus remained a de jure Ottoman territory until 5 November 1914) in exchange for guarantees that Britain would use the island as a base to protect the Ottoman Empire against possible Russian aggression. In 1923, under the Treaty of Lausanne, the new

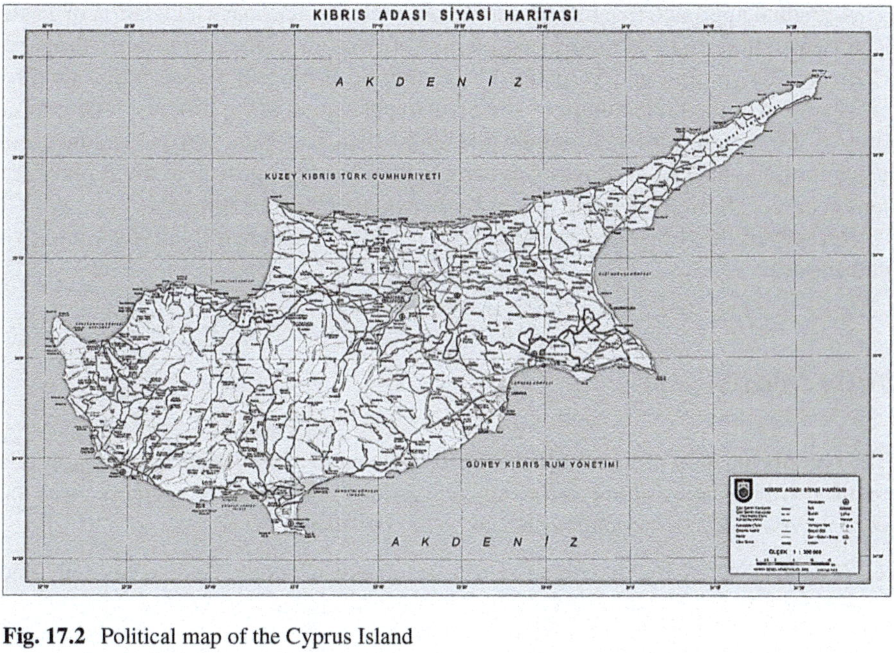

Fig. 17.2 Political map of the Cyprus Island

Turkish republic relinquished any claim to Cyprus, and in 1925 it was declared a British crown colony. British period ended on August 1960 with the Zürich and London Agreement between the United Kingdom, Greece and Turkey and Cyprus gained its independence.

Land use and tenure in Cyprus was on a totally communal basis before the year 1050 BC. The first idea of individual ownership was evolving after the period of this year. An inscription of the fifth century reveals certain important characteristics of land ownership in ancient Cyprus. It indicates that there were then not only royal lands but also private ownership, mentioned by owner's names. It also indicates the existence of land taxation and of inheritance in those days.

During the period 1571–1878, Cyprus was under Ottoman Rule, therefore, the most important legal provisions were contained in the Medjelle and the Ottoman Land Code (1858) (arazi kanunnamesi humayunu), which set land ownership on a more rational basis. There were five categories of land: Mulk, Arazi Mirie, Arazi Mevkoufe, Arazi Metrouke and Arazi Mevat. The Ottoman System may be described as a system of registration of deeds combined with the registration of title. The parcel boundaries are stated without any reference to a map or plan.

During the nineteenth century, the pattern of land tenure in Cyprus developed so as to serve the purposes of land use, within the framework imposed by the customs and traditions, and the economic, social, religious and political needs and circumstances of that time. In 1878, an agreement between Great Britain and Turkey transferred administration of Cyprus to Great Britain. In 1907, the Immovable Property Registration and Valuation Law, was enacted (kanunlaştırmak). This Law introduced the registration of title and set the basis for the modern registration of all immovable property, based on the plan in use.

A gradual approach to a reform started with the revolutionary legislation of 1946. The Legislation of 1946 comprised mainly of two laws: The Immovable Property (Tenure, Registration and Valuation) Law and the Wills and Succession Law. The legislation was the first attempt by the State to put an end to the process of deterioration of the structure of land tenure that until then had been left to decline from generation to generation. It was a dynamic and positive approach to the problem of land tenure, offering ways and means to owners to achieve a remedy.

Records of the tittle deeds belonging to pre-1974 are generally divided into two categories.

Disordered System

The records made in this system have been made on owner of immovable property's request without being connected to any regulations. As a result of such practices the deeds given to the owner called 'Kocan' were two-tier (Fig. 17.3):

1. The ones given by the organization according to study results without field survey.

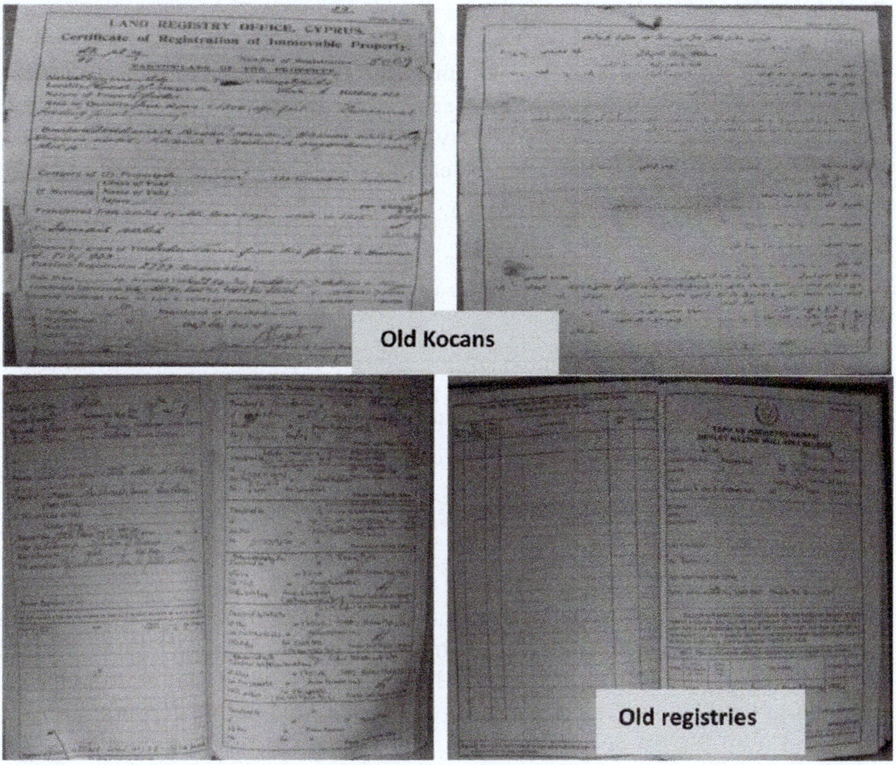

Fig. 17.3 Old Kocan and records examples

2. The ones given after the fieldwork. These feeds were recorded in 'Village Register Books' in the villages by the field officers.

General Registry System

Studies that was in the scope of this system were conducted by the British legislation and based on generally maps in particular plane table maps. The books that these studies recorded were called 'Field Book' or 'Land Register'.

Until today Cyprus has undergone four major stages in terms of land registry and cadaster works. These are:

Registered Period

Written cadastral works were carried out in the period of Ottoman Rule (1570–1910) in the Cyprus island. During this period, determination and registration of the real estates were performed. Legal validity of these registration records equals to the title deeds in Turkey. However, after the cadaster, land register entries based on the cadastral maps replace them.

Deed Records with Sketch

Beginning of the British dominion on the island (years of 1911–1912) in some areas, it is known that identification of the parcels was performed using scaled sketches. The only differences separating these works from written ones is sketches. However these sketches can not be replaced with cadastral maps.

Chain Survey

It is known that cadastre of the large part of the island was performed by using chain survey method which is preferred mostly by the British. This method was conducted in the period of the British Rule years 1912–1929 and all field measurements were recorded in field books. Cadastral plans on scales from 1:500 to 1:2500 were prepared and used for the general registration. While doing these studies instead of triangulation polygons were used. In the course of time because most of these polygons are lost application has become impossible in most places.

Plane Table Survey

To speed up field studies between the years of 1929–1960 British rule, applied plain table measurement was based on triangulation method. Plain table method is coarser than the chain method and using this method 1/5000 scale maps are produced (Fig. 17.4).

Current Cadastral System

Cadastral systems are organized in different ways throughout the world, especially with regard to the Land Registration component. Basically, two types of systems can be identified: the Deeds System and the Title System. The differences between the two concepts relate to the cultural development and legal setting of the country. The key difference is found in whether only the transaction is recorded (the Deeds System) or the title itself is recorded and secured (the Title System) (Fig. 17.5).

Fig. 17.4 Measurements with plane table (Louca and Fani 2008)

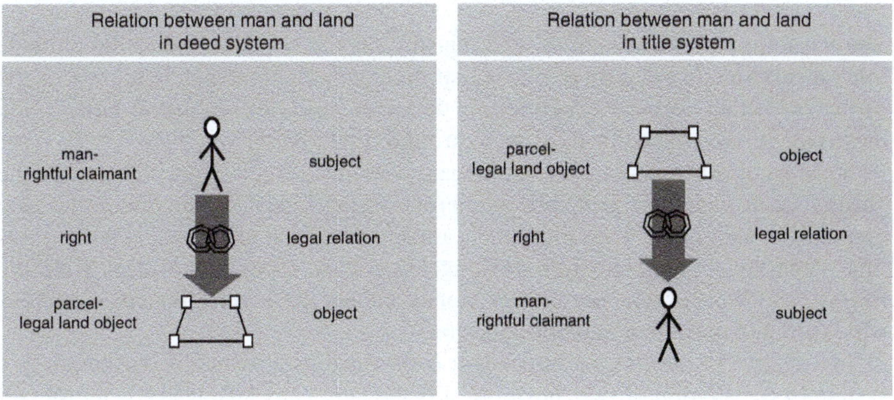

Fig. 17.5 Basic approaches of the two basis of Human-Land relationship: (**a**) Land Registration and (**b**) Land Recording

The Deeds System is basically a register of owners focusing on "who owns what" while the Title System is a register of properties presenting "what is owned by whom". The cultural and judicial aspects relate to whether a country is based on Roman law (Deeds Systems) or Germanic or common-Anglo law (Enemark 2006, 2010).

Deeds Registration

Deeds registration system is merely a registration of all important instruments related to that land. In order to establish one's title to the land, a person (or usually their purchaser's attorney) will have to ascertain, for example:

- all the title documents are properly executed;
- "a chain of title" is established, i.e. the proper ownerships from the granting of the land from the government to the present owner; and
- there are no encumbrances on the land that probably will harm the title of the land.

What is registered is not the title but only the 'evidence' of title, namely instruments purporting to transfer or deal with various interests. Therefore a would-be purchaser has to decide, by examining these instruments, and by inspection of the property, whether or not the vendor is the owner and has the right to sell. On the other hand an advantage of deeds registration is that the procedure for accepting the deed by the registrar can be very quick. Only a short check might be made to see if the deed meets the formal requirements.

Title Registration

The most appropriate system of land registration can be described as the ultimate title registration. In a system of title registration one can immediately see who is the owner of certain property. The register therefore needs to be 'parcel based', and these parcels are well defined (usually through 'title plans').

Systems of title registration exist in many varieties. Especially with regard to the "insurance or guarantee principle" numerous variants exist. Furthermore there are great differences in the ways the parcels are described and identified. In some cases title plans are just copied from existing large scale topographic maps (like the English and Welsh Ordnance Survey maps). Others use precise boundary surveys which are laid down in a numeric cadastre.

In a system of registration of deeds it is the deed itself which is registered. In a system of registration of title, however, it is the land parcel itself that is registered, thus effecting "the transference of primary attention from the mobile, mortal, mistakable persons temporarily possessing or claiming rights over patches of the earth's surface, to the immovable, durable, precisely definable units of land affected and the adoption of these as the basis of record instead. "The register itself is proof of title and its correctness at all times is usually guaranteed by the State" (Larsson 1991).

Countries which operate a system of title registration are often divided into three groups, even though this reflects more the differences in land law, than in registration principles:

(a) The English Group
(b) The German/Swiss Group
(c) The Torrens Group

In the English Group use is made of the large scale topographic maps, in the German/Swiss Group use is made of parcel based cadastral maps, and in the Torrens group use is made of isolated survey plans (Grover 2008).

The English group does not involve a cadastre and knows very little boundary surveys. The process does not ask for preparing an index map of the area before registration can start. It uses existing large-scale topographic maps for this. The German/Swiss group uses good cadastral surveys and a well-functioning deeds registry. The title register is called 'land book' and very similar in structure to the land registers. The Torrens system resulted from Sir Torrens' desire to improve on the old English land law system which was very complex, time consuming and expensive. The purpose of the Torrens system is to simplify land transactions and to certify the ownership of an absolute title to realty by having registered plans (Dimopoulou et al. 2003; Elia 2007).

The land registration system in British Rule Cyprus was maintained by English Deeds System. A registered person is considered to be the undisputed owner of the land and his title to ownership is absolute, subject to the Director's power to correct errors or omissions under certain circumstances, and the inherent power of the Courts to order an amendment or cancellation of a registration. Title is not guaranteed by the State and there is no indemnity fund that means that the right holder is protected by 'public faith'. Today every government tries to establish a modern cadastre system for social stability with managing land use information effectively.

After 1974, the island was divided into two political governance as Turkish Republic of Northern Cyprus and Republic of Cyprus. After the separation TRNC government agreed that current cadastral system should be changed and should be modernized. Because the existing cadastral maps are inadequate due to technical reasons, contain boundary errors and insufficient in practice, they need to be renewed. TRNC government aimed to develop a similar approach like the system in Turkey. Although current legislation is based on the old system, new regulations are considered according to the same approach in Turkish system. From this point, we will briefly mention about the system in Turkey:

The system take its source from the Constitution and the Civil Code and the system is based on "State Guarantee". Article 719 (old 645) of the Civil Code states that: "The boundary of immovable is determined with property deeds and boundary marks over the land. If the property deeds and boundary marks do not match, what property deeds say is accepted." In Turkey government is absolute responsible for the regulation of the land registry. If real estate owners and rights holders have been damaged by a change made on the registers without the written permission or a court decision, the losses is compensated by the government. The main difference between the systems is based on that in Turkey there is a cadastral system based on the concept of a central Europe and existing of a legal structure. The structure called "Fix boundary" i.e. precision and accurate measurements are done and connected to the cadastral maps. However in TRNC there is a general understanding of the boundaries called "General Boundary" arising from traditional British system. Cadastral Renovation Project with technical infrastructure is renewed based on the system in Turkey and a new cadaster concept that is compatible with EU aims to be developed (Akyol et al. 1999).

Problems of the Existing System

Modern and sustainable cadaster term is relied on registration with legal assurance. Under most systems of registration of title, the information on the registers is guaranteed so that, in the unlikely event of fraud or error, anyone inadvertently suffering from the incorrectness of the information will be compensated (Dale and McLaughlin 1988). Therefore governments tried to implement the Title registration system for sustainable management of the land information. In TRNC existing deeds system is desired to be converted with title registration.

Deeds registration is a system for registering legal documents, not for registering title to land. A deed does not prove who owns the land; it only records an isolated transaction (Dale and McLaughlin 1988) (Fig. 17.6). Systems of deeds registration are usually described with the main focus on their problems. These problems can be explained by the following defects (Zevenbergen 2002):

- The fact that the deeds merely prove that a transaction took place, without guaranteeing that the intended changes did really occur;
- The fact that it is not compulsory to register (all) changes of ownership, so that a correct impression at one moment may become erroneous later on;
- The fact that the object the deed refers to is not very well described; and
- The fact that the chronologically stored deeds are badly accessible, sometimes only through poorly alphabetized name indexes.

In other words, a deed, in itself, does not prove title. It shows that a transaction took place but does not prove that the parties are legally entitled to carry out the transaction. What is registered is not the title but only the 'evidence' of title, namely instruments purporting to transfer or deal with various interests. Therefore a would-be purchaser has to decide, by examining these instruments, and by inspection of the property, whether or not the vendor is the owner and has the right to sell. Conversion from deeds registration to title registration is not easily succeeded due to different levels of accuracy of parcel plan for title registration as against the plan in deeds. The system is associated with long delays and frustrating processes and procedures leading to loss of public confidence in the system.

Because of the absence of the measurements, calculations, benchmarks and coordinate values existing sheets are accepted as graphic sheets (Fig. 17.7). Nevertheless, this is the only available source; studies like municipal, urban planning, expropriation, etc. are trying to run over according to these sheets. A lot of border violations have occurred due to unhealthy application of the parcels. Land Registry is currently being carried out in unsuitable conditions. The owners registry is keeping recorded; however the information which belongs to the owners cannot be achieved healthily because of the "Surname Law" adopted in 1983. Before that in place of surname for male father's name and for female first her father's name before marriage and after marriage spouse's name is used. Sometimes the chattels belong to the same person is recorded on different pages on owners registry. After cadaster is done, land registers are done according to the demands; however in the

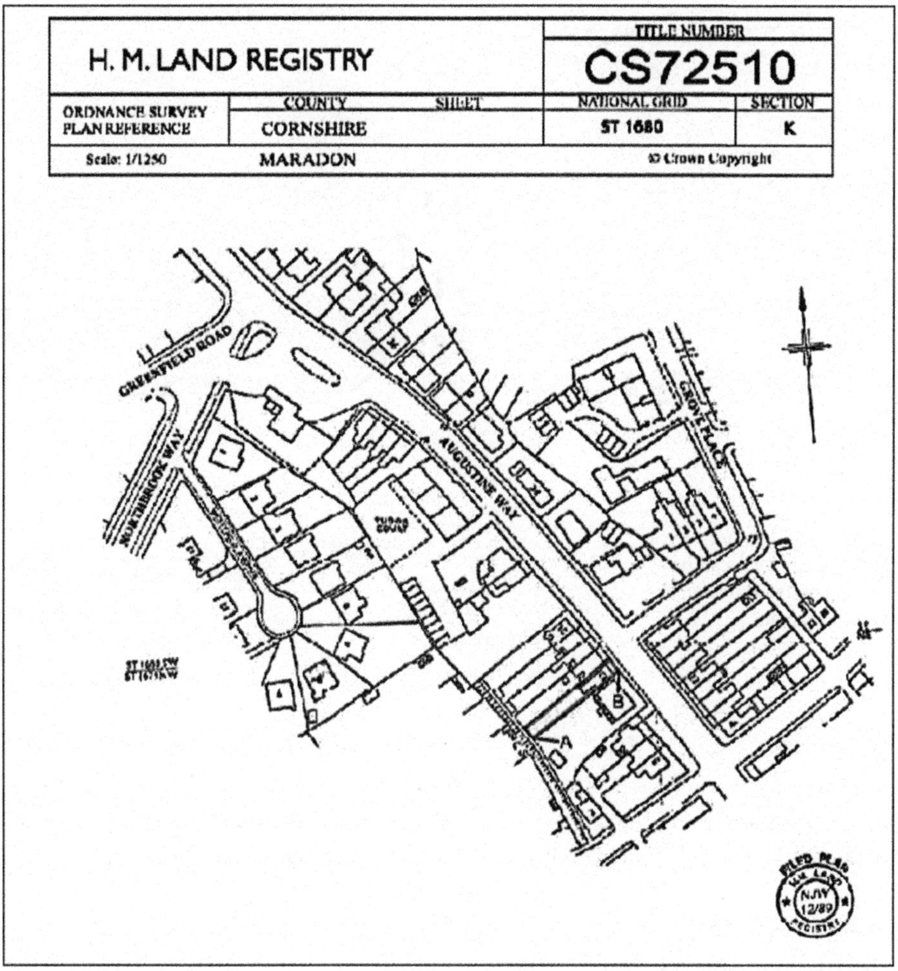

Fig. 17.6 H.M. land registry

old villages before 1983 there is still parcels which are not recorded because of the absence of the majority of the land registry.

The determination of the status of the sheets and records which will be renovated before starting the study is important. It is needed to protect these information and the the rights acquired based on them, also retrieving all these information in accordance with the new and modern technology. Therefore, renovation is a work that requires fund of knowledge. Being knowledgeable of the rights holders about the subject, to understand the nature of the studies, the necessity of working together and maintaining the existing legal rights, announcements, objections and legal procedure are the main processes of this work. Although old maps and records are

Fig. 17.7 Old maps

Table 17.1 Random errors and errors which can be calculated from distance measurements

Length error of steel tape	Length strip counting error
Tilting	Reset (on steel tape start)
Tension and Sag	Reading and spelling errors
Temperature	Personal and other blunders
Deviation from the direction	Obstacle of product size

being accepted as the only legal basis, precision, needs, methods and production technology, when they are produced, need to be understood (Table 17.1).

Studies on the Field in TRNC

Relevant ministries of the TRNC and Turkey signed necessary protocols between the years 1998 and 2008 for renewal studies. In 1998 automating the renewal of cadastral maps and land register of studies in the TRNC which are foreseen in the protocols was started by the personnel assigned from the Republic of Turkey. However, in the past 11 years, only 21 thousand parcels could be able to be renewed. In order to accelerate work with the tender made by the beginning of 2013 TKGM, renewal of the 125 thousand parcels are foreseen to be done in 1-year period as three packages and project cost estimated about 5 million TL. Also it is estimated that in 246 villages there are about 600 thousand parcels and 1.5 million land registry. The result of the tender made by TKGM, contractor company is determined and site delivery is done.

Table 17.2 Boundary definitions

Fixed boundary: Indisputable boundaries which are available on the ground and determined do not change after cadastral works.
Valid boundary: This is the boundary established if the cadastral technical documentation and measurements are free of errors.
Boundary considered valid: This is the boundary created by the owners and other concerned based on the way of usage without any kind of disagreement.
Uncertain boundary: This is the boundary which is not available on the ground but formed by the adjustment plan.
Arguable boundary: This is the border which is available on the ground; however, it's the subject of dispute between the sides.
Exchangeable boundary: This is the boundary which is available on the ground and adjacent places ruled and controlled by the state.

Studies were carried out with:

TKGM: 1 general coordinator, 1 coordinator, 3 engineers, 25 technicians,
TKMD: Advisory Committee (team consisting of TRNC Land Managers, Coordinators, engineers and lawyers)
Contractor: 3 project managers, 15 engineers, 60 technicians (35 technicians from TRNC), 35 GPS and electronic measuring instrument and around 20 terrain vehicles.

In villages and neighbourhoods it is stated that studies will start firstly with the participation of owners then with village headmen and expert witnesses from the region; also announcements about the nature of the work is made. To increase the awareness of the project, it is publicized in the press and the public opinion and given briefings.

Border definitions are the most important factors in these studies (Table 17.2). The result is specified with detailed research on the ground and determined by the headman, local experts and property owners. It must be done by trained technical personnel and very carefully.

Technical Studies

Surface Network Creation Studies

Not only has the scope of the tender, GNSS and Leveling Network that covers all the TRNC surface been established in accordance with the system used in Turkey. All the milestones were built as a "bent bar". There are four CORS (Continuously Operating Reference GPS Stations) stations available and based on these, in total 105 ground control points, 37 C2 stations (SGA-densification GPS Network) as bent bars, and 68 C3 stations (ASN-densification network for detail measurements) were established (Fig. 17.8). Precision of the generated densification network is 1.55 cm in determining the spatial coordinates and 2.08 cm in determining height.

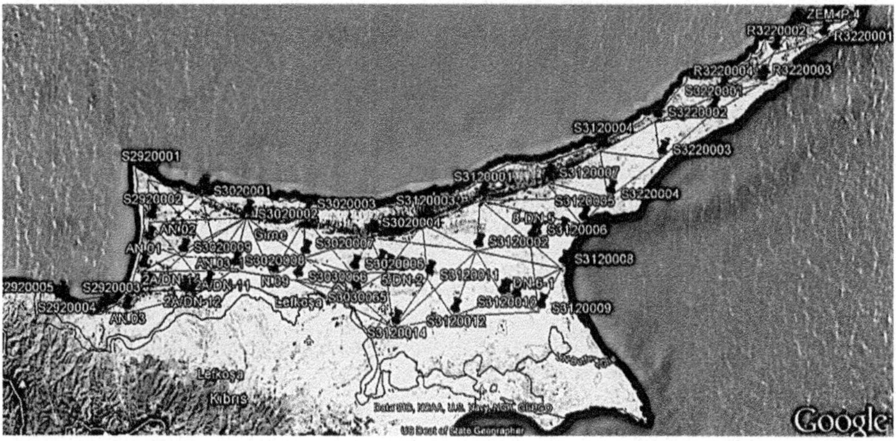

Fig. 17.8 Surface network

Every hour of the day instant height and coordinate information can be determined within the TRNC borders and open for all the public and private sector users.

Firstly the digital data of cadastral offices and the licensed surveying offices have been taken into consideration. The parcel edging have been provided accordingly and the adjacent parcels were evaluated upon this. Even though the former measurement points were present on the units being worked on and these points were measured by Cadastral Registry Directorate and licensed offices in CORS, ITRF and ED50 coordinate system, these have been taken as the same, after controlled (Fig. 17.9). The unmeasured old measurement points (Traverse points) were found on the ground and measured. All the measurements have been converted to CORS system. If there is a parcel-ground point's harmony following the mapping then they have been taken as base during the evaluation. The existing cadastre maps have been made by taking the sets, arches, stone range and layered wall as fixed boundaries as base. The evaluation was made according to the block and parcel communities, the building, wall and the water wells of old times which are shown in the boundaries of parcel taken into consideration.

The boundaries which were used to be the walls in the old parcels were not shown via special mark. Therefore the walls which are thought to be existing in the land previously was taken as notes and considered in the evaluation. In case the parcel and land book show different boundaries the land books and Land Registry Books were decided to be superseding. The facade length written in the land books have been preserved in the evaluation. The breaches and right of ways stated in land records have definitely been observed. The given "Director's Decisions" which had been awarded for the disputed parcels have been investigated and these decisions were observed. When a certain difference have been found to exceed the margin of error in measurements such as wall, fence and building positions, re-measurement or control measurement have been performed. The parcel boundaries have been

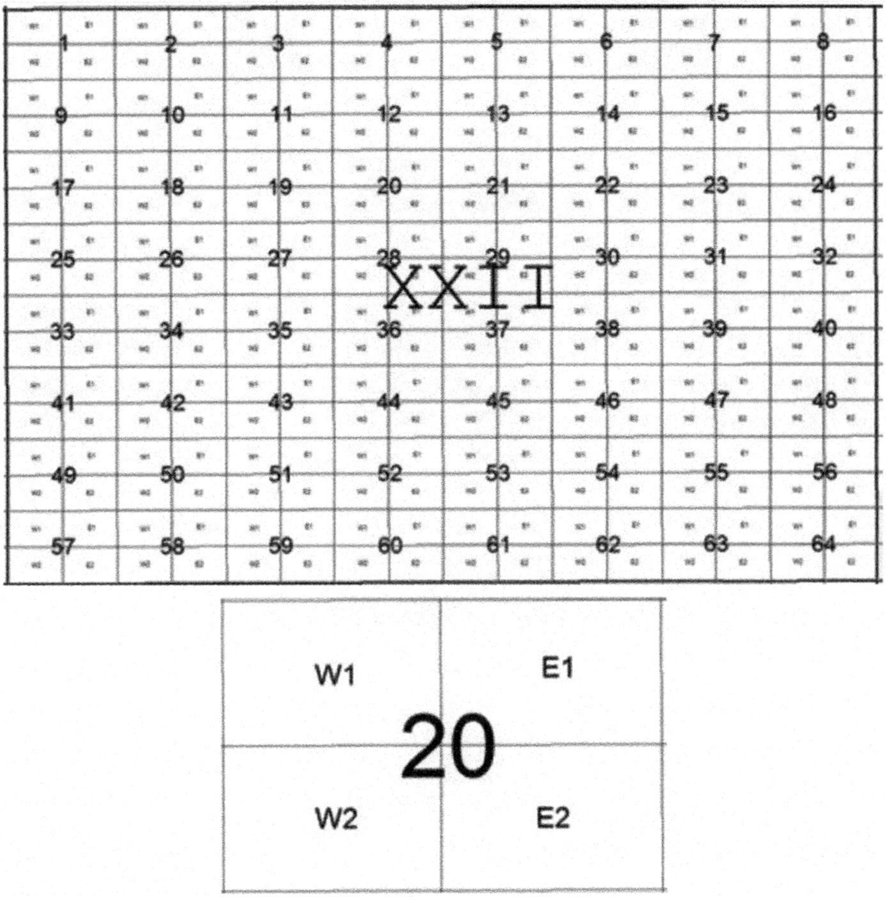

Fig. 17.9 Map index for old maps

determined by taking the road width into the consideration in places where structured and digital data could not be found.

Automation Studies

Automation studies, started for archival purposes, were used for the renovation of the existing documents. These studies are based on the principles of taking the picture of the registers, indexing them in computer and then making data entry (Fig. 17.10).

The first 3/4 months of the work has not been efficient at the desired level due to the adaptation of the study area and TRNC. However; in terms of the subsequent

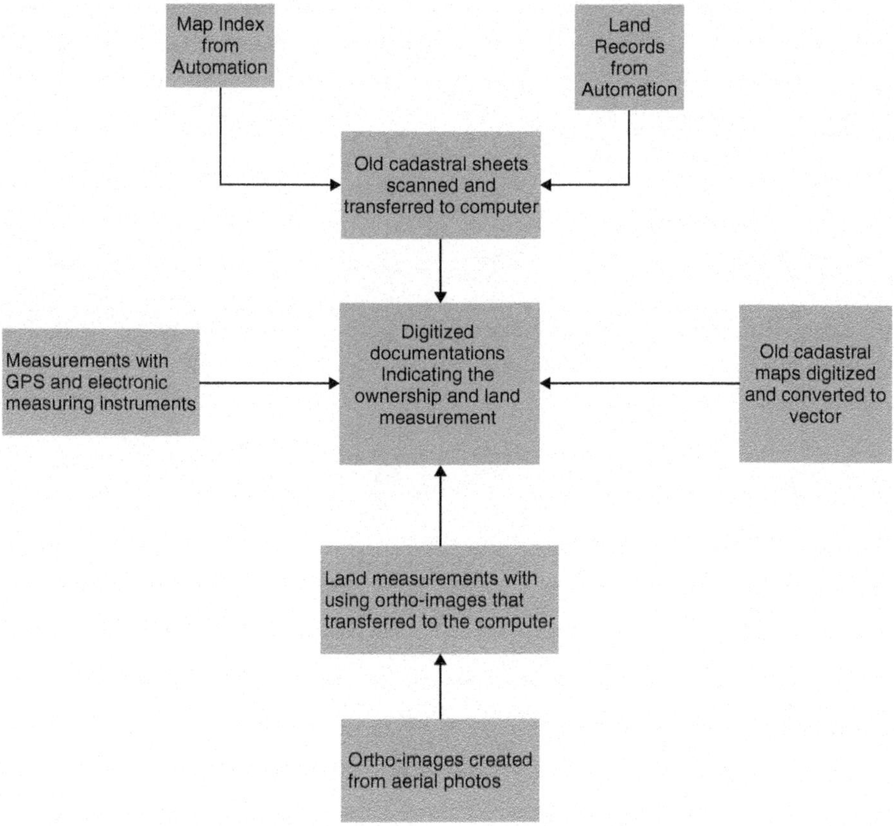

Fig. 17.10 Workflow of the renovation project

process parallel to the appointments, a business plan covering the duration of the project is prepared by TKDM, project coordinator, engineers and technicians and this programme has been strictly obeyed.

The parcel number in the parcel/map system have been formed as 456/4, 456/4/1, … etc., and registered in the land book. These parcel numbers have been accepted as key data in the type of "old parcel no" and it has been re-numbered in a system starting and continuing as the Province, District, Block and Parcel number from to infinitive continuous number. The land books were scanned and loaded to a computer as raster via an index.

The parcel boundaries have been measured literally as to the original boundaries depending on the statements of the experts and owners and have been evaluated as per the border descriptions within the technical regulation. The works made on the land and in the office have been recorded via a report and the mukhtar and witnesses' signatures were taken. The coloured ortho-images of the working places have been taken as basic data during the measurement and evaluation. A new software has been developed in order to match the graphic data and deed registry, and the work has been performed in a more controlled and fast manner without any data loss. Total Parcel 129,962 (there are 1307 sheets with the scale of 1/1000 and 1/2000). Land Registry 1398; Common hold Registry 155 and totally about 1570 new registers were transferred. Commissions examined the objections and they have been concluded.

Conclusions

The cadastral renovation project in TRNC has the characteristics of a property conversion. Cadastral sheets and property records produced according to British imperial system were transformed into a base that is updated, numeric, in the metric system, legally valid and may constitute investments related to the land. Urbanization, construction, expropriation, consolidation and other urban regeneration activities may result in a short time, more economical and effective. Property cases will diminish greatly and renovations also will help the development of relevant legislation in TRNC. To be able to complete the studies carried out laboriously and intensely at the estimated time, has a prestigious importance to both countries.

References

Akyol, N., T. Yomralıoğlu, and C. Bıyık. 1999. KKTC Coğrafi Bilgi Sistemleri Projesi Geliştirme Ön Raporu

Dale, P.F., and J.D. McLaughlin. 1988. *Land Information Management*. New York: Oxford University Press.

Dimopoulou, E., T. Labropoulos, V. Nikolaidou, and P. Zentelis. 2003. *Comparative Analysis of NSDI Policies in Greece & Cyprus Two Different Systems within the EU*.

Elia, E. 2007. Cadastral Template of Cyprus. PCGIAP-Cadastral Template 2003.

Enemark, S. 2006. A cadastral tale. In *Análisis geográficos*. Instituto Geográfico Agustin Codazzi. (Análisis geográficos; No. 30).

———. 2010, April 26–27. *From cadastre to land governance*. Annual World Bank Conference on Land Policy and Administration, Washington, DC.

Grover, R. 2008, September 11–15. *Why the United Kingdom does not have a cadastre – And does it matter?* FIG Commission 7 Annual Meeting Verona, Italy.

Larsson, G. 1991. *Land Registration and Cadastral Systems: Tools for Land Information and Management*. Essex: Bath Press.

Louca, A. and M. Fani. 2008. *The Boundary and the Effects of Cadastral Surveying in Cyprus*.

Zevenbergen, J. 2002. *Systems of Land Registration Aspects and Effects*. Delft: Nederlandse Commissie voor Geodesie (Netherlands Geodetic Commission).

Chapter 18
A Transparent Cadastral System: Fit for Sustainable Development and Legal Security – The Danish Public-Private Cooperation Model

Henning Elmstrøm and Torben Juulsager

Introduction

Maintenance, updating and development of the Danish property cadastre[1] has for more than 250 years been organized in a public-private cooperation model consisting of a central state cadastral authority (today The Danish Geodata Agency) and private licensed chartered surveyors. The licensed chartered surveyors perform the cadastral tasks in the field, prepare the cadastral documents and submit the changes to the central authority that controls, approves and records property changes in the cadastre.

To ensure a continuously "fit for purpose" cadastre has both the cadastral system[2] and the cadastral process[3] adjusted to the needs of a democratic and capitalist society with private landownership and evolved over time in relation to the general societal and technological developments (Williamson et al. 2010).

There is a high focus in the Danish society on managing property rights in a both trustworthy and transparent process to meet the need for legal certainty on real property and consumer protection.

---

[1] "The Danish cadastre is the basis for all land registration in Denmark. Consisting of a country-wide cadastral map, an official register and a cadastral archive" (The Danish Geodata Agency).

[2] The cadastral system includes legislation, institutions, registers, data, IT systems, processes, actors etc. to manage property registration.

[3] The cadastral process includes actors, parties and performing procedures to change the property registration.

H. Elmstrøm (✉)
The Danish Association of Chartered Surveyors (DdL),
Kalvebod Brygge 31-33, 1780 København V, Copenhagen, Denmark
e-mail: elmstroem@plf.dk

T. Juulsager
The Danish Association of Licensed Surveyors (PLF),
Kalvebod Brygge 31-33, 1780 København V, Copenhagen, Denmark

T. Yomralioglu, J. McLaughlin (eds.), *Cadastre: Geo-Information Innovations in Land Administration*, DOI 10.1007/978-3-319-51216-7_18

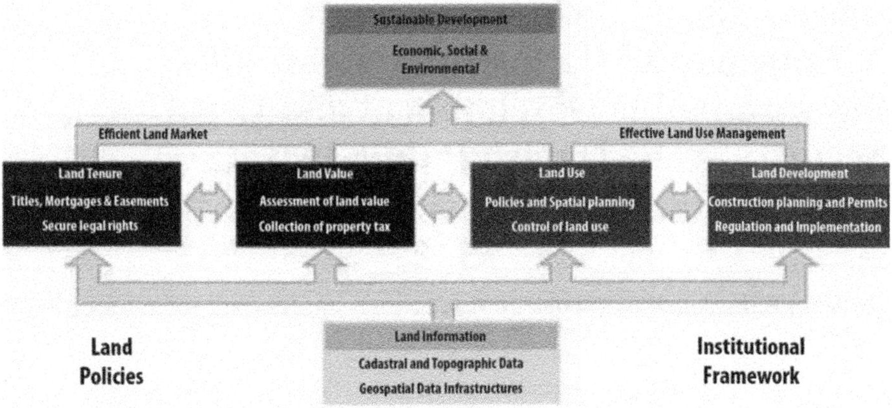

Fig. 18.1 A gobal land management perspective (Enemark 2004)

The underlying philosophy in the Danish cadastral model is to find holistic and balanced solutions that will pave the way for sustainable development of change of property and land use, taking into account "to third parties" – understood broadly as neighbours, the boundary, public regulation, rightsholders in property, the cadastre, the cadastral map etc (Danish Geodata Agency 2006, 2013).

Over time the Danish cadastre has changed from being a primarily fiscal and legal cadastre to be a multipurpose cadastre in to a broader concept of Land Administration Systems. Today the Danish cadastre is a digital basic infrastructural tool in the Danish Land Administration System assisting the functions of land tenure, land value, land use and land development and supporting e-governance by spatial data. A land management design that fits into the conceptual framework described in the global land management perspective is illustrated by the paradigm shown in Fig. 18.1.

This paper presents the Danish cadastral system focusing on basic framework conditions for cadastral activity and professional requirements for the licensed chartered surveyor performing cadastral work:

- Basic framework conditions for the cadastral system
- Professional framework and requirements for licensed chartered surveyors
- The Danish cadastral system

It gives a kaleidoscopic overview of the elements that forms a trustworthy Danish cadastral system and a transparent cadastral process, conducted in a public private cooperation model which ensures a high degree of legal certainty and consumer protection.

Basic Framework Conditions for the Cadastral System

The Danish cadastral system is arising out of private property rights and the inviolability of property, which are basic pillars in the Danish rule of law and economy and sets a constitutional framework for the design of the Danish cadastral system. Furthermore, there is a political consensus on public-private cadastral cooperation model. The Danish cadastral system facilitates an efficient land market, valuation and taxation, supports effective land management and ensures property rights. To ensure legal certainty and consumer protection, the society has established a number of basic framework conditions for the cadastral system.

Institutional Framework

Private property rights are a value bearing foundation of a democratic constitutional state as the Danish. Real property is subject of considerable private and public investment, establishment of rights and serves as the subject of security for raising of loans and for taxation.

It is grounded in the Constitutional Act of Denmark, the framework for Danish democracy, with a set of fundamental principles and rules for society. The Constitutional Act establishes the protection of private property:

> § 73, 1. The right of property shall be inviolable. No person shall be ordered to surrender his property except when required in the public interest. It shall be done only as provided by statute and against full compensation. Legislation, actions and interventions in relation to private property is always taking into account § 73 of The Constitutional Act.

Legal Framework

The legislation in Denmark confirms and consolidates the economic and social significance and value of a secure and trustworthy cadastral system, by setting out a clear framework for responsibility, conduct and performance of cadastral work.

The cadastral system and model in Denmark is basically governed by two main laws: the Subdivision Act and the Chartered Surveyors Act. A series of statutory orders and guidelines sets out more detailed framework for the specific execution of the cadastral work. The Subdivision Act provides the legal basis for the Danish cadastral model and the subdivision requirement. The Chartered Surveyors Act contains professional and ethic requirements and regulations of the licensed chartered surveyors and the chartered surveying companies conducting cadastral work, to ensure legal certainty and consumer protection on defining property boundaries, property creation by an independent, impartial and objective professional surveyor.

The Cadastral Process

The Subdivision Act states that the Cadastral Authority must lead and maintain the land register as a register of all properties and as a cadastral map with associated measuring sheets. The cadastre must include the cadastral designations, area sizes and some listings about the property. The Subdivision Act defines together with the Chartered Surveyors Act that cadastral work may only be performed by practicing chartered surveyors (private licensed chartered surveyors – owners or shareholders of a chartered surveying company) and their assistants who are appointed chartered surveyors.

Subdivision and Land Transfer Requirement

The Subdivision Act states that change in ownership or pledge of an area is only permitted if the area presents a separate real property or an unregistered property. This means that an acquisition of a part area of a real property includes a requirement for subdivision into a separate property or transfer of the part area to another property. Changes must be handled under the rules of subdivision and land transfer. The subdivision and land transfer requirements are supported by the requirement that there must not be established rights of use over a part area of a real property for a longer period than 30 years if the area is part of separate real property.

Subdivision Control

Centrally in the Danish holistic and sustainable cadastral model is that the Subdivision Act states that there must be no subdivision, land transfer or merger if the cadastral change or the intended land use according to the information will result in violation of the another legislation or of planning or other public or private law restrictions on the affected properties.

Sole Purpose

A chartered surveying company must have the sole purpose of performing surveying and cadastral work. And to ensure that customers meet a qualified and professional advisor the Chartered Surveyors Act states that if a surveying company consists of more than one office every office must be managed by a licensed chartered surveyor.

Personal Responsibility

Performing cadastral work is associated with a personal responsibility and ethic rules. The Chartered Surveyors Act states that a licensed chartered surveyor is personally liable, together with the company, for any claims arising in consequence of assistance provided by the chartered licensed surveyor to a client. The licensed chartered surveyor must at the same time behave in a manner consistent with good chartered surveying practice. The surveyors are entrusted with a bonus pater liability.

Ownership and Management Regulations

To ensure surveyor's independence of material and economical interest in relation to the provision of professional responsibility by performing the societal task that lies in performing cadastral work and thereby an objective consideration to third parties interests the Chartered Surveying Act stress ownership and management regulations of chartered surveying companies.

In a chartered surveying company at least 51% of share capital and voting rights shall be owned by practicing chartered surveyors who actively are carrying surveying business in the company. Of the remaining capital and voting rights, other than the mentioned practicing chartered surveyors cannot own more than 15% of share capital each. This limitation does not apply to persons who have their main job in surveying company.

In a chartered surveying company the majority of members in the Supervisory Board and in the Executive Board must be practicing chartered surveyors who actively are carrying surveying business in the company.

Impartiality Requirements

Furthermore to ensure the impartiality of the licensed chartered surveyor conducting cadastral work, there is a requirement that the surveyor may not perform cadastral work concerning a property in cases where the licensed chartered surveyor has a material or any other kind of interests in the property and in the outcome of the case.

Spatial Data Infrastructure Framework

The Danish cadastre is a multipurpose digital/analog register for land registration in Denmark and plays a central role in the Spatial Data Infrastructure (SDI) with a two-part legal and administrative role:

- Authoritative basic data register
- Spatial reference data for eGovernance

The cadastral data are recognized as authoritative basic data – data which may be used without verification. The licensed chartered surveyor is both user of and supplier of cadastral data by performing cadastral work. In this performance the surveyor use many other maps, property data and spatial data.

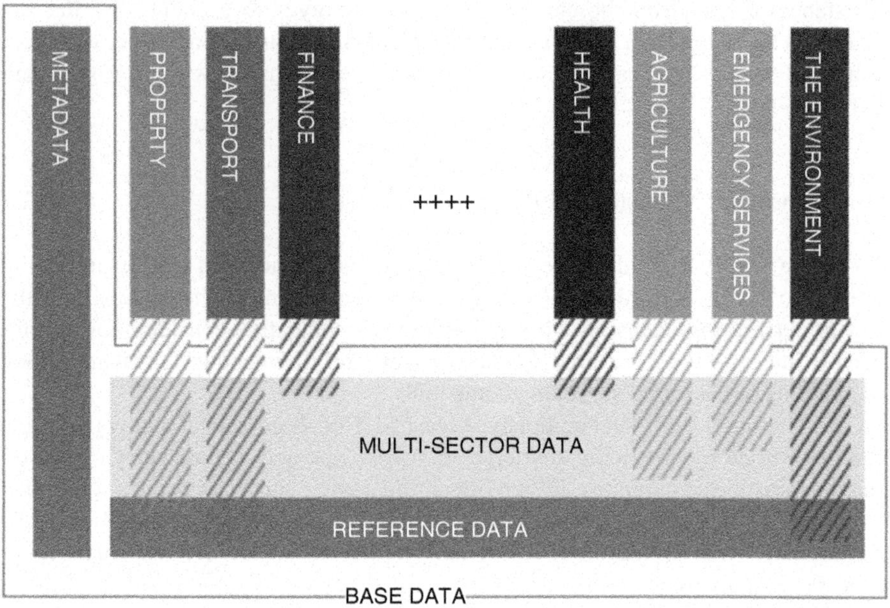

Fig. 18.2 The Danish spatial infrastructure model (The Geodata Agency 2011)

The Danish Spatial Data Infrastructure builds on INSPIRE principles:

- Data should be only collected once.
- Data should be maintained where this can be done most effectively.
- Data should be combinable, regardless of their source.
- It should be easy to get an overview of the available data and internet services.
- There should be clear conditions which assure that data can be utilized by many users in many contexts.

This "philosophy" results in a spatial data infrastructure model based on collections of sector specific geodata themes build up on a "platform" of common multi-sector, reference data and shared infrastructure internet services that all-ow access to documented geodata and to metadata from distributed sources of data (Fig. 18.2). It means that the information in the Danish LIS is sectored in charge of the resort authority, but taking responsibility for cross-agency public sector cooperation.

There exists a unified national collaboration between central, local and regional authorities, universities and private businesses in order to ensure that the national SDI "meets its users' requirements".

Professional Framework and Requirements for Licensed Chartered Surveyors

The property surveyors acting in the cadastral process, are basically entrusted with official authority. They have the professional authority to perform a number of special tasks either representing a public authority, or acting on behalf of a public authority or in cooperation with a public authority.

When the property surveyors represent public authority through their activities, society must set requirements for the professional practice. High professional, business and disciplinary requirements and standards for the profession must be assured through education, consumer protection and claim systems and ethical codes of conduct established by law under national authorities or by self-regulation and self-justice by national surveying associations

Professional Qualifications and Requirements

In order to fulfil societal and consumers expectations regarding the exercise of cadastral work, there is a need for a continuous high level of knowledge and professional skills regarding the licensed chartered surveyor performing cadastral work.

Educational Level

In order to ensure a high professional level of knowledge, the educational background of the licensed chartered surveyor is based on a scientifically high academic education: five years on university level at Aalborg University (campus Aalborg or Copenhagen) consisting of a bachelor degree (3 years) and a master degree (2 years) in surveying and cadastral science – Surveying & Mapping, Geoinformatics, Land Management, Property Economics) (Fig. 18.3). Aalborg University, stands out through a special education model Problem Based Learning (PBL), which shapes and contributes positively to the surveyor's professional DNA.

The "Aalborg Model" is based on regular lectures in the fields of the semester theme combined with problem based work in small groups with real life problems that relates to the semester theme. It provides students with the possibility of acquiring knowledge and skills independently and at a high academic level working analytically and according to interdisciplinary and problem and result oriented methods. During this study model the students will cooperate with the business community on the solution of authentic professional problems.

Besides professional skills the students will develop their abilities within teamwork and will become well prepared for the labour market.

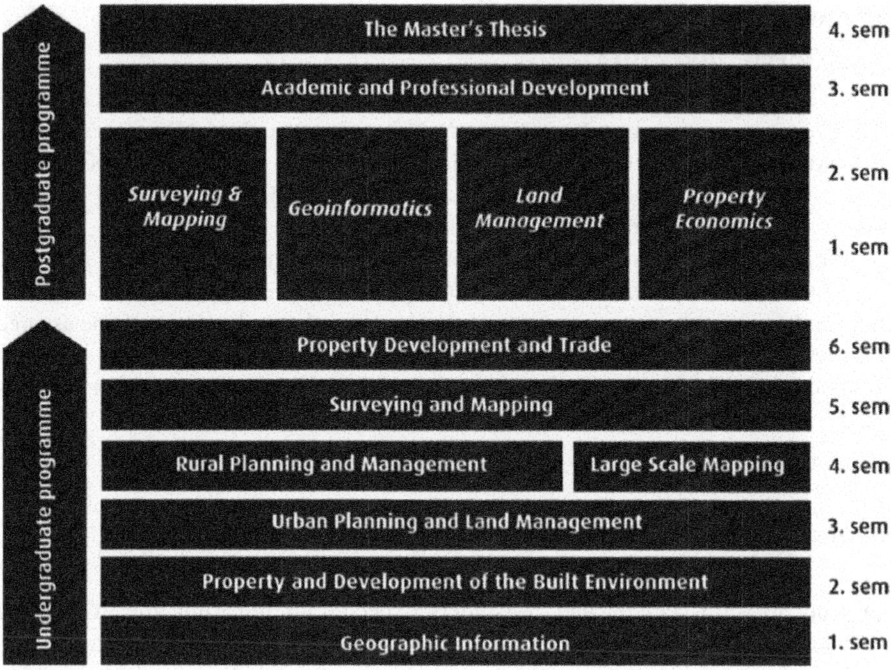

Fig. 18.3 The bachelor and master programme within Surveying, Planning and Land Management (Aalborg University)

Professional Practice

To establish the necessary experience, knowledge and skills to exercise cadastral work it is required that the chartered surveyor must have at least three years of practical supervised experience in performing general cadastral work in a private chartered surveying company or in a cadastral administration doing cadastral work, undergoing further education.

License

Only the one-line combination of the bachelor and master of Science in Surveying, Planning and Land Management can lead to a license as a chartered surveyor in Denmark, in combination with at least three years of professional experience in the employment of a practicing surveyor, doing cadastral work.

The achievement of the license is based on an application to the Danish Geodata Agency with documentation of necessary requirements. The Danish Geodata Agency decides whether to grant applicants licenses as chartered surveyors. The license is personal and can only be used for performing cadastral work in a private chartered surveying company.

Continuing Professional Development

To ensure and maintain high and updated professional skills, development and quality of the professional performance by the chartered surveyor, the Danish Association of Chartered Surveyors (DdL) recommends 37 hours professional training, knowledge and dissemination per year. Continuing Professional Development (CPD) is not compulsory but de facto.

Professional Business and Disciplinary Requirements

In order to fulfil societal and consumers confidence to the licensed chartered surveyor in conducting the professional tasks there is a need for professional liability insurance, claim systems and ethical codes of conduct, with aim of securing high consumer protection.

Insurance Obligation

To protect customers against financial loss that may arise from performance of the cadastral work and other chartered surveying works carried out by practicing chartered surveyors and their employees the Chartered Surveyors Act stress that it shall be mandatory for the practicing chartered surveyor to be covered by professional indemnity insurance, which must be personal. The insurance must cover at least five years after the surveyor permanently ceased company. The Danish Geodata Agency registers and monitors that the demands of compulsory insurance are met.

State Board of Appeal

Any claims raised against a practicing chartered surveyor or licensed chartered surveyor's assistant that they have disregarded the obligations in performing cadastral or surveying works can be brought for the Chartered Surveying Committee. The same applies to claims concerning chartered surveying companies.

The board concerns malpractice complaints in cases of activities which are, by law, performed by licensed chartered surveyors. The board assesses the surveyor's work in order to decide whether he has failed to fulfil his obligations to such an extent that there is reason to impose a penalty in the form of a reprimand or a fine, or if special circumstances justify a withdrawal of the license.

Disciplinary Board

To support the increased professional liability the licensed chartered surveyors are subject to, the Danish Association of Chartered Surveyors (DdL) sets up a Disciplinary Board which considers complaints of all types of work carried out by chartered surveyors or their assistants who are members of the association. It can give an assessment of the quality of the surveyor's work, with an opinion.

The Disciplinary Board has the opportunity to recommend to the surveyor to carry out rectification of work performed. Such a request will usually be followed.

Remuneration Board

Complaints relating to the fee, calculated by a practicing chartered surveyor for any work performance can be brought for the Remuneration Board. The board is established under the Danish Association of Licensed Surveyors.

An opinion of the board will be the basis for an assessment of whether the claim should be brought before the courts. A statement from the board that goes against the surveyor will usually be followed up by the surveyor, and will usually be the basis for any decision if the dispute shall be settled by the court.

Ethical Codes

In order to ensure high ethical standards in the execution of work carried out by practicing chartered surveyors and their employees, their works is based on Statues of Surveyance, a code of conduct developed and enforced by the Danish Association of Chartered Surveyors (DdL). It concerns rules for customer contact and contractual work, including information on the work progress and cost overruns.

In addition the Danish Association of Chartered Surveyors and the Association of Licensed Surveyors (PLF) have joined the "Code of Conduct of European Surveyors" issued by the Council of European Geodetic Surveyors – CLGE.

Professional Representation of Interests

In Denmark there are two organizations that represent and take care of the professional interests of the surveyors – working to develop and strengthen the surveying profession, frameworks for performing surveying activities not least cadastral work and related work for the benefit of its members and for society.

There are two recognized professional bodies in the Danish cadastre sector, actively cooperating and consulting partners in a variety of relevant legal and technical areas both in relation to public and private professional network.

The Danish Association of Chartered Surveyors (DdL)

DdL is responsible for the professional, economic and social interests of the surveying profession as well as member's personal work and professional interests including not least the salary and employment conditions and educational interests. Membership is personal.

General professional orientations communication and knowledge communication is achieved through the association's journal and web site, professional forums and workshops. The association is also responsible for the development and provision of continuing education activities, just as the association is represented in the advisory board for surveyor educational programme at Aalborg University.

The Danish Association of Licensed Surveyors (PLF)

PLF is a professional association of chartered surveying companies performing surveying business under the Chartered Surveyors Act. PLF is a political, professional and employers association carrying out business interests by creating the best possible conditions for exercise of the cadastral work and related business and promoting professional developments.

PLF is included as an active party in legislative development partnerships with various ministries within cadastral work, land management and surveying. PLF also takes a responsibility in the appeal system on cadastral work and other surveying works by being responsible for the operation of the remuneration committee.

Like DdL, PLF also contributes to knowledge transfer and knowledge sharing by conferences, workshops and training activities aimed at cadastral work. PLF is represented in the advisory board for surveyor educational programme at Aalborg University.

The work in PLF is structure around the Board, as the central driver in all business areas and representation of interests. The Board establishes internal committees, working groups and task forces on special work and areas of interest and is represented in relevant external committees and working groups (Fig. 18.4).

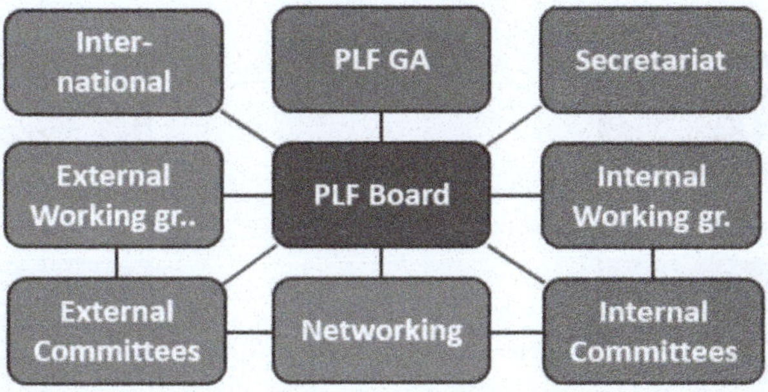

Fig. 18.4 The Association for Licensed Surveyors (PLF) – representation of interests

The Danish Cadastral System

The Danish cadastral system is based on a title system, *"a register of properties presenting 'what is owned by whom'"* (Enemark 2010). Titles are based on the cadastral identification, and the cadastral registration is prior to land registration. The cadastre plays a central role as basis for land and property identification and registration in Denmark, based on registration of property boundaries determined by licensed chartered surveyors.

The Danish cadastral system itself "provide a basic land information infrastructure for running the interrelated systems within the areas of Land Tenure, Land Value and Land Use" (Enemark 2010) (Fig. 18.5) – land administration and land governance.

The Danish cadastre has during time evolved to be a multipurpose cadastre, so in addition to support land administration and land governance the cadastral geo-related authoritative basic data including, not least, the digital cadastral maps is the entrance to eGovernance and digital solutions in the Danish Society.

Due to this central role in the Danish economical and digital infrastructure we will pay attention to the cadastre and the cadastral process and especially the role of the licensed chartered surveyor in execution of the cadastral work in relation to ensuring a credible and sustainable cadastral registration taking into account third party (Fig. 18.6).

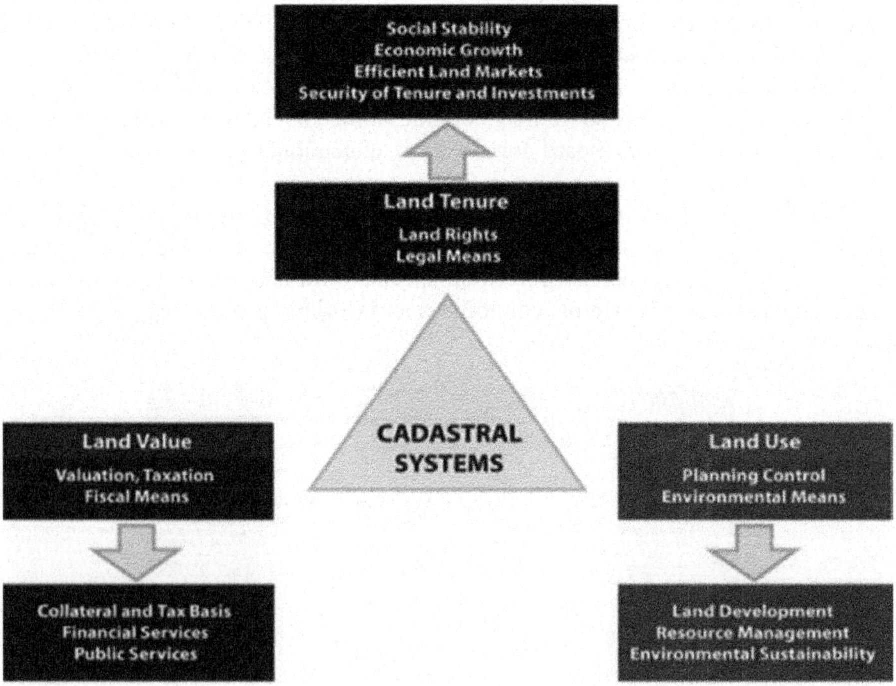

Fig. 18.5 Cadastral system – a basic land information infrastructure (Enemark 2004)

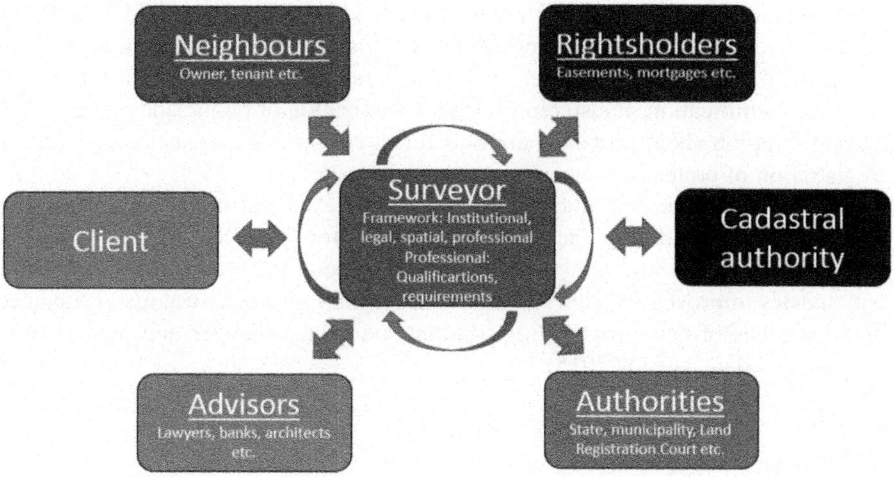

Fig. 18.6 The licensed chartered surveyor – a central objective advisor conducting the cadastral process

The Danish Cadastre

The Danish cadastre is basis for all land registration in Denmark and consists of a cadastral map, a cadastral register and a cadastral archive. The cadastre contains information about properties unique identification, geographical extent in the cadastral map, boundary definition, size of area and certain other property ownership records. The cadastre provides the spatial integrity and unique identification of every land parcel through the cadastral map updated by cadastral surveys carried out by the licensed chartered surveys.

Factually the Danish cadastre consists of approximately 2.5 million parcels. The first cadastre was in force in 1688 and has developed since. In 1997 a digital reform completed the computerizing of approximately 15,000 analogue cadastral maps. In 2001 a digital update system MIA was in place and in 2008 a cadastral digital updating and quality system MiniMAKS was in place. In 2014 the measurement archive became partly digital (information's back to 1950) with digital web based access from the cadastral surveyors. It completed a full digitally cadastral process between the surveyor and the cadastral authority.

The Cadastral Map

The cadastral map is a digital legal cadastral map; however, not to be mistaken for a coordinate cadastre, and is a key component of the Danish cadastre. It will be more indicative to handle the map as a geo-related index map presenting the cadastral register in a visual form, so that individual parcels can be identified and located.

It provides a geometric description of the land parcels represented as a digital closed object by quality labelled coordinates in the UTM-reference system and connected with informative attributes in the cadastral register – first of all the unique property identification. It also represents a visualization of public and private roads and information about protected areas as forest conservation areas, coastal zoning (registration of protection line of coastal and dunes areas).

Due to the original production method of the cadastral map – digitization of analogue cadastral maps – accuracy is variable from few centimetres to several metres (especially in the rural area). New GPS-based measurements of boundaries and updates improve the quality of the cadastral map. The cadastral map is updated daily on basis of new property registrations, boundary changes and field surveys (Danish Geodata Agency 2015).

The Cadastral Register

The cadastral register is a database of property information containing the unique cadastral identification numbers (e.g. 20h Holstebro Markjorder fra Hjerm), property areas, area of private roads and a few public administrative registrations of public rights or restrictions connected to the use of land – e.g. agricultural property (land designated for continued agriculture purposes).

Like the cadastral map, the cadastral register is updated daily. (Danish Geodata Agency 2015).

The Cadastral Archive

The cadastral archive consists of historical information of property registration and includes the documents that have defined property formation through the ages. Not least it includes scanned and digital measuring sheets with field surveys of boundaries which is the basis of the determination and marking of boundaries. The archives also includes scanned historical cadastral maps (Danish Geodata Agency 2015).

Cadastral Process

Cadastral work reflects a change in the property division and often a change in the use of the physical environment. In Denmark the use and development of land/real property is, for natural reasons, a matter not only for private interests but also for public and rightholders interests. Consequently today's legislation on land management in Denmark is trying to find a reasonable regulatory framework for management of the balance between public and private interests.

The consideration for third parties in the property formation and changes process is reflected in the cadastral process. It is set out in the rules of boundary determination, compliance with other laws, mortgagee consultation and allocating easements.

In Denmark the cadastral process is the formal and legal acts and agreements that ensure the implementation of a sustainable and holistic process of project realization. As responsible for the case preparation the licensed chartered surveyor ensures and drives the entire process.

The licensed chartered surveyor is acting in the intersection between public and private interests and as an objective advisor he/she has to represent and manage all the interests related to property change and the future use. The licensed chartered surveyor is popularly said "the ambassador of the boundary". The role as private advisor gives the surveyor a significant legitimacy to be able to handle the balance between private and public interests around the property formation and use, and gives the surveyor the opportunity to take part in both an advisory and design stage.

Simplified the Danish cadastral process includes the following main phases

- Requisition and advisory
- Case preparation
- Approval and registration

A more detailed paradigm of the Danish cadastral process with focus on the practicing chartered surveyors central role in conducting the process is outlined in the following paradigm (Fig. 18.7), which also illustrates potentially involved par-

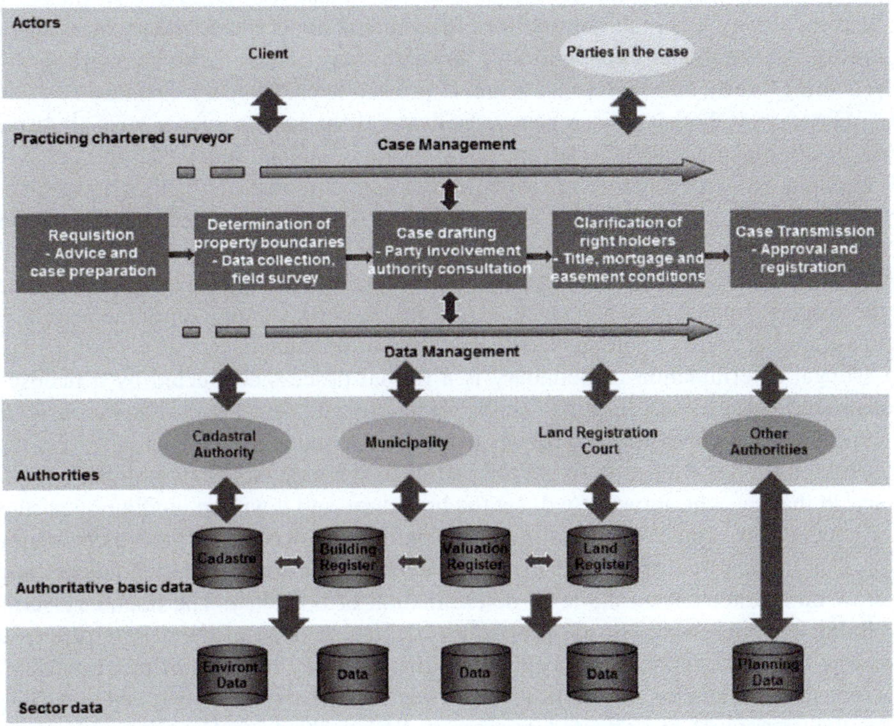

Fig. 18.7 The Danish cadastral process – parties, phases, tasks and information flow

ties, phases and information flow. Based on this paradigm key processes and phases will shortly be described in a licensed chartered surveyors perspective.

Consultancy and Preparation

When the licensed chartered surveyor receives a requisition on a cadastral change he/she makes a screening of property information, public and private legal restrictions. Based on these investigations the surveyor advises and formulates the cadastral case/property changes with the client.

Boundary Determination

The basis for determining existing boundaries are information from the cadastre – measuring sheets or in case of none or insufficient field measuring the cadastral map. The boundary may under Danish law change through prescriptive acquisition. Therefore, when determining and marking the boundary, the licensed chartered surveyor has to investigate whether the property boundary in the field agrees with the information in the cadastre. If there is no deviation the licensed chartered surveyor can mark the boundary in accordance with the cadastral information. At the same time the surveyor is required to inform neighbours about the boundary marking. It must give neighbours the opportunity to submit any comments to the marking for providing a fully informed process and to prevent later boundary disputes.

However, if there is a discrepancy, the licensed chartered surveyor must allow the parties to make a statement before marking the boundary. After this, the licensed chartered surveyor must decide whether the boundary can be fixed in accordance with the cadastre or whether the matter has to be settled in accordance with the rules on rectification of boundaries, transfer of part of property or technical changes. If there seems to exist an irreconcilable disagreement between two neighbours about the property boundary it may be necessary for the party wanting to set the boundary to ask the surveyor to hold legal determination of boundary.

Legal determination of boundary is a formal process conducted by a licensed chartered surveyor given public authority as first instance. It consists of a local inquiry meeting involving the parties to clarify the boundary issue and the licensed chartered surveyor must try to get the parties to agree on the location of the boundary. In that case the surveyor will set the boundary, and it will be binding on parties when they have approved in writing. If the parties do not reach an agreement, on the basis of the information available, the surveyor will set a preliminary boundary and draw up a statement on the legal determination of boundary. The boundary set is binding for the owners of the properties affected, if none of them bring the case before the district court, with a claim that the boundary lies in a different position. When new boundaries are determined and marked, they shall be approved in writing by the owners of the affected properties, before they can be registered in the cadastre.

The general process for boundary determination is conflict prevention which finds concrete expression by performing only about 65 legal determination of boundary held per year and under five of these brought to court.

Clarification and Consultation Authorities

When the cadastral changes are finally decided and the boundaries are determined the licensed chartered surveyor prepares the formal and the technical documents as the basis for approval by the authorities, party involvement, right holders involvement and registration in the cadastre.

The consultation process by the authorities is called "the subdivision control", and this process has to clarify and document that the cadastral changes and the future use of the property is legal with respect to legislation and public regulations. In certain situations, the surveyor has public authority to attest that the cadastral changes do not conflict with public regulations without submitting the case to the authority.

If there is a need for dispensation of laws or public restrictions applying, the surveyor applies for the necessary permission, which can be done in cooperation with the owner or other advisors.

Clarification and Consulting Rightholders

The Land Registration Act and the Subdivision Act ensure protection of mortgages and easement rights by cadastral change of private property.

The Land Registration Act states, that in connection with a land transfer there must be permissions/accept from the mortgagees to the change, as in ordinary cases must be registered as an endorsement of the mortgage in the land register. In special cases when the value of transferred land is insignificant and below a certain threshold value (currently 16,710 Euro), the licensed chartered surveyor has authority to certify, that the land transfer can be done without risk of mortgagee and that the property after the change are still capable of carrying security for the mortgage. This so-called "harmlessness certificate" replaces the registration of mortgages release in the land register, and shall be submitted by the case documents for approval and registration of the cadastral changes.

In relation to rightholders of easements on affected properties, the licensed chartered surveyor must locate the rights and ensure that the easement in future are registered on the right properties after the cadastral change. The requirement applies to both subdivision and land transfer. The land register will be updated by the new registration of location of easements by a so-called "easement statement" from the licensed chartered surveyor.

The Subdivision Act prescribes that land transfer can only be registered in the cadastre, if it is established that the provisions of securing mortgages and easements is observed, which must be documented by a judge certificate.

Approval and Registration of Cadastral Changes

When the surveyor has documented the conditions for registration of the cadastral changes, the documents (registration documents, owner declarations, authority approvals and Certificate of land registration Judge on mortgage, easements and title deeds) will be packed in a digital file package, and submitted through the digital update system MIA to the Danish Geodata Agency for approval and registration. Approval procedures include a technical and legal control of the submitted cadastral case. If case is insufficiently clarified or contains errors, the surveyor will be contacted for additions or corrections. After the control the property changes will be approved and the cadaster will be updated by the new property registrations. When the registration is completely updated, other public records/registers will be updated with the new property information.

Final Remarks

The cadastral system is a part of the Danish societal and economical infrastructure assisting the functions of land tenure, land value, land use and land development. The basic purpose of the system is to ensure a reliable and trustworthy formation, registration and change of property and access to up-to-date authoritative proprietary basic data.

The cadastre plays a central role as a multipurpose register with a two-part primary role legal and administrative as:

- Authoritative proprietary basic data register
- Reference data for spatial eGovernment

The performance of cadastral work – basically determination of boundaries and conducting and preparation of the documents necessary for registration of cadastral changes – is in the Danish context organized in a public private collaboration model with a two shared responsibilities:

- The Danish Geodata Agency (state authority) – approval and registration
- Practicing chartered surveyor (private) – cadastral work is performed by appointed surveyors in private practice and by their appointed assistants

Sustainability

The cadastral work in Denmark is performed in a democratic and constitutional State framework with a strong account of third parties – in this context multifaceted and can be neighbours, mortgagees, rights holders, the community and even the cadastral register. Both the cadastral system and the cadastral process are open and transparent.

In order to manage this complex task the professional identity of the licensed chartered surveyor is holistic and solution oriented. A licensed chartered surveyor has to listen carefully to the clients to know their needs, have a good knowledge in the cadastral legislation and the land management legislation, have to navigate task performance in an intersection of disparate interests (public and private) and finally give the parties a possibility to make an agreement/a sustainable solution. This kind of dialogue based solutions includes a high grade conflict prevention and ensure sustainable solutions.

Due to

- regulations on ownership and management in practicing surveying companies, that ensures the surveyor's independence of economic and specific interests in the cadastral change or the underlying project
- legal framework conditions
- professional and business requirements
- high codes of ethics

the confidence to the cadastral system and the performing professionals is very high in Denmark. The licensed chartered surveyor has a high degree of professional legitimacy. Overall the cadastral system is organized so that it provides a high degree of legal certainty and consumer protection which results in a high degree of credibility in society.

Fit for Purpose

The Danish cadastral system fulfills the constitutional protection of private property and a democratic society's requirement of legal certainty, consumer protection and interests of rightsholders in property.

Over time the Danish cadastral system has managed to evolve from being an analog primarily fiscal oriented system to be a digital multifunctional system. The cadastre has become a digital multipurpose cadastre with a basic role in the Danish spatial infrastructure containing authoritative basic data and is today the entrance to eGovernance and fits in an open digital community structure.

Post Script

PLF Observations on Deregulation Initiatives

The European Commission reached a political agreement in June 2013, which addresses and requires each Member State to perform a review and to modernize their regulations on qualifications governing access to professions or professional titles. The overall purpose is improving access to professions, in particular through a more flexible and transparent regulatory frameworks in Member States, to facilitate the mobility of qualified professionals within the internal market and the cross-border provision of professional services.

This should also have a positive impact on the employment situation and enhance economic growth. In order to boost growth potential and consolidate the way to economic recovery, this review of regulated professions should be a priority. The Commission therefore urges Member States to begin reviewing at national level the qualifications requirements imposed on regulated professions and the scope of reserved activities.

At the same time and in full compliance with EU recommendations, the Danish government launched a study of the abolition of ownership regulation of Danish practicing surveyor companies in order to create growth by

- reducing administrative burdens
- increasing productivity
- increased competition – lower prices.

It is growth initiative no. 49 in a comprehensive growth plan for the Danish society.

PLF Point of View

- Currently there is no workload neither in the Danish Cadastral Agency or in practising surveying companies in reporting and containing changes in ownership and management and supervisory board.
- Property formation and change is nearly 100% dependent on general societal conditions. Pricing on honoraria cannot create growth or increased competition. In a larger ordinary subdivision, surveyors fee represents about one percent of the sales price.
- The competition in the profession is intact. At the moment approximately 75% of all cadastral works is based on a tender, an offer or another tested price.
- The Danish Association of Licensed Surveyors (PLF) has the opinion, that the premises for growth initiative no. 49 are wrong and promoting growth initiative no. 49 can weaken both legal certainty and consumer protection.
- A prerequisite for the legal certainty of the total property formation process is "citizens" access to independent surveyors free of special interests; the surveying company is irrelevant, as it only handles ownership and boundary interests for the benefit of the whole society taking account of third parties.
- The existing ownership and management restrictions, which states that licensed chartered surveyors have a controlling influence in the group of owners in the company and in the business management, ensures the independency and impartiality of the practising chartered company.

References

Danish Geodata Agency, The. 2006. *Property formation in the Nordic countries.* http://eng.gst.dk/media/gst/65040/PropertyformationintheNordiccountries.pdf

———. 2011. *Location – A gateway to eGovernment.* http://eng.gst.dk/media/gst/64820/_Strategi_UK_Final_web.pdf

———. 2013. Statutory order concerning the Chartered Surveyors Act. Consolidation Act no. 680 of 17 June 2013. http://eng.gst.dk/media/gst/2438580/StatutoryorderconcerningtheCharteredSurveyorsActlo.pdf.

———. 2015. http://eng.gst.dk/danish-cadastre/#.VS-Yl8kcS70 (2015.04.16).

Enemark, S. 2004. Building Land Information Policies. In: *Proceedings of Special Forum on Building Land Information Policies in the Americas*, 26–27 October 2004, Aguascalientes, Mexico. http://www.fig.net/pub/mexico/papers_eng/ts2_enemark_eng.pdf.

———. 2010. From Cadastre to Land Governance in Support of the Global Agenda – The Role of Land Professionals and FIG. FIG Article of the Month – December 2010. http://fig.net/pub/monthly_articles/december_2010/december_2010_enemark.html.

Williamson, I.P., S. Enemark, J. Wallace and A. Rajabifard. 2010. *Land Administration for Sustainable Development.* Redlands: ESRI Press Academic. For information. see: http://www.fig.net/news/news_shortstories.htm

Chapter 19
Procedure of Real Estate Acquisition by Foreigners in Turkey

Bayram Uzun, Nida Celik Simsek, and Volkan Yildirim

Introduction

Whether public, private, or something in between, property rights are a social, cultural, *and* economic base laying crucial groundwork for market economies. Within newly minted market economies, it becomes possible – even necessary – to recalibrate intuition towards land and other forms of property as commodities to be bought and sold in an international market (Tess 2004). Also, arising out of technological advances, coupled with the liberalization of capital markets worldwide that allow for efficient movement of capital between countries. Besides every nation, from past to present, has certain restrictions concerning the acquisition of land in the sense of absolute ownership by foreigners (Uzun and Yomralioglu 2007). However the removal of foreign ownership restrictions in many countries, a policy encouraged both by the Organization for Economic Cooperation and Development (OECD) and the European Union, made it easier for people to invest in foreign real estate markets (HPI 2009). Turkey as a developing country, a member of OECD and candidate of EU, has made a series of arrangement in terms of foreign investment on residential property. Turkey is an attractive country in the international land market because of both its geography (with large coastal area and the Mediterranean climate) and its rapidly developing economy. Particularly in recent years, individuals and companies have started to be interested in purchasing real estate in Turkey and this combined with Turkey being accepted as an EU candidate country attention is being paid to the process of real estate purchase by non-Turkish nationals and companies (Uzun and Yavuz 2003). Since 2002, there has been renewed concern in Turkey about purchases of properties by foreign citizens, reflected in numerous amendments and cancellation decisions about such purchases by the Supreme Court and a number of state agencies. Despite the frequent changes, foreign direct

B. Uzun (✉) • N.C. Simsek • V. Yildirim
Geomatics Engineering, Karadeniz Technical University, 61080 Trabzon, Turkey
e-mail: buzun65@hotmail.com

T. Yomralioglu, J. McLaughlin (eds.), *Cadastre: Geo-Information Innovations in Land Administration*, DOI 10.1007/978-3-319-51216-7_19

investment (FDI) rose from just over US$1 billion in the early 2000s to an average of US$13 billion in the 2008–2013 period in Turkey.

Foreign investors have started to play a role in the residential market and Turkey's foreign property market has been growing from strength to strength (WB 2014). Foreign investment in single-family homes, condos, and multifamily apartments, holiday homes, investment and retirement properties have been propping up the real estate market in Turkey. Foreigners have presented demand for real estate, especially along the southern and western coasts of Turkey, and almost 60% of foreign sales countrywide have been in the Aegean and Mediterranean regions. Especially coastal regions have been increasingly popular with British and German buyers over the past 10 years. However Turkish properties have been attracted by various nationals and the distribution of the regions or cities where foreigners dealt with, have been dispersed as heterogeneous. Despite the confusion created by changes made almost every two years to Regulation no. 2644 in section 35 of the Land Registry Law, foreign demand in the Turkish real estate market continues to grow. The current law concerning real estate acquisitions by foreigners shows that Turkey, in the process of harmonization with the European Union, does not have an overly restrictive policy on this issue. The aims of this paper are to examine trends in the foreign real estate market; to analyze the type, quantity, and quality of property acquisitions; and, finally, to investigate the current situation of residential property investment by foreigners in Turkey. To achieve these aims, legal changes have been examined, existing procedures detailed, statistics about real estate purchases and sales collected and analyzed, and basic procedures regarding with foreign property acquisitions compiled. This research has demonstrated that recent legal changes have increased demand in the Turkish foreign real-estate sector and return expectation in the real estate and public corporation sectors.

Legal Aspects of Property Acquisition by Foreigners in Turkey

With reference to the history of foreign investments in Turkey's property sector, it is observed that land legislation and foreign ownerships rules have been amended several times in the last three decades. In other words, the development of policy on foreign acquisition of Turkish properties has been inconsistent.

Juridical Amendments on Title Deed Law (No: 2644)

In our country, the fundamental regulation concerning the real estate acquisition by foreign real persons was made with Land Registry Act numbered 2644, dated December 22nd, 1934. 35th and 36th articles of this act regulates the real estate acquisition by foreigners in our country. But today, these articles have been revoked by the Constitutional Court for different reasons and they have constantly been amended by the legislator.

- The first amendment was made with the Act numbered 3029 and dated June 21st, 1984, but these new items were cancelled with the decision of Constitutional Court dated June 13th, 1985.
- The second amendment was made with the Act numbered 3278 and dated April 22nd, 1986 and two articles were added. However, these new items were cancelled with the decision of Constitutional Court dated October 9th, 1986.
- The third amendment was made with the act numbered 4.4916 which entered into force after being published in Official Gazette dated June 3rd, 2003 and 87th article and 36th article of the village law numbered 442 were repealed. However, this amendment was revoked with the decision of Constitutional Court dated March 14th, 2005.
- The fourth amendment was made with the Act numbered 5444 and dated December 29th, 2005 and a temporary article was added to 35th article. The new modification was partially cancelled with the decision of Constitutional Court dated April 11th, 2007.
- The fifth amendment was made with the Act numbered 5782 and dated July 3rd, 2008. This initiative which created uncertainty in terms of quantity and area of the acquired property was cancelled with the decision of Constitutional Court dated May 12th, 2011.
- The sixth amendment was made with the Act numbered 6302 and dated May 3rd, 2012. 35th article of Land Registry Act (no. 2644) has been rearranged.

Cancellation decisions given almost every two years and amendments made in spite of these decisions indicate that the situation is not legally stable. Nevertheless, the amendment made on 2012 allowed foreign natural or legal persons to acquire immovables in Turkey more easily.

The Title Deed Law in Force (Law No: 2644, Sections 35 and 36)

In accordance with the Article 35 of the Land Registry Law No. 2644, amended by Law No. 6302, which entered into force on 18 May 2012. According to the law:

- The condition of reciprocity for foreigners who wish to buy property in Turkey is abolished, restrictions for foreigners acquiring property in Turkey have been eased with the enactment of the "Reciprocity Law" in 2012.
- The citizens of 183 nations can own property in Turkey. With the new regulation, citizens of neighbouring countries, such as the United Arab Emirates, Middle Eastern countries such as Iraq, Azerbaijan, who were not allowed to acquire real estate in Turkey are now able to do so. However, even though they are on the list,

citizens of countries who share a border with Turkey cannot acquire property in border provinces (GYODER 2014).
- The limit on the size of land able to be bought by foreigners increased to 30 ha.

The law also increases the limit on the size of land foreign buyers can purchase from two-and-a-half hectares of vacant land to 30 ha., and buyers will have to comply with a condition to provide plans for the construction of a house on the land before they make the purchase.

- Foreign individuals and businesses are required to submit their project proposals for vacant plots of land to The Ministry of Environment and Urbanization within two years of purchasing the land. If the ministry approves the project, it will be forwarded to the local land registry office for monitoring.
- The Cabinet will also be able to increase the 30-ha. limit on property purchase to 60 ha. as it deems acceptable.
- The law allows for the purchase of up to 10% of the total area of towns densely populated by foreigners.

The law in force since 2012, has led to a real estate boom. Such that, according to the Real Estate Investing Partners Association (GYODER) said that "96,000 housing units were sold in the first quarter of 2012, 5.5 percent more than a year ago". Following the enactment of the reciprocity law, sales of real estate to foreigners marked an increase in the last two quarters of 2012, and reached USD 2.64 billion (GYODER 2014).

The Ministry of Environment and Urbanization announced that real estate sales to foreigners increased from 2% to 5–6% in the last ten months of 2013. In terms of the real estate prices; while the prices in prime coastal areas typically range €1200–€2600 per sq. metre in 2008, the price has risen up to range $6000–$7000 per sq. metre in 2015.

The Main Process of Foreign-Property Acquisition in Turkey

The responsibility of all types of land registry and cadastre processes concerning the acquisition by foreigners of real estates in Turkey belongs to the Directorship of Foreign Affairs Office of the General Directorate of Land Registry and Cadastre (GDLRC). Foreign individuals and legal entities who want to acquire real estate in Turkey must apply to the Directorship of Land Registry in the location of the mentioned real estate.

Figure 19.1 shows the process of real estate acquisition by a foreigner. The process has been starting with the application phase.

The Council of Ministers may, if the interests of the country requires so, determine or limit the acquisition of real estate and limited real rights by foreign natural persons in terms of country, person, geographic area, time, number, percentage, type, quality, surface area; or stop in whole or in part; or prohibit all together.

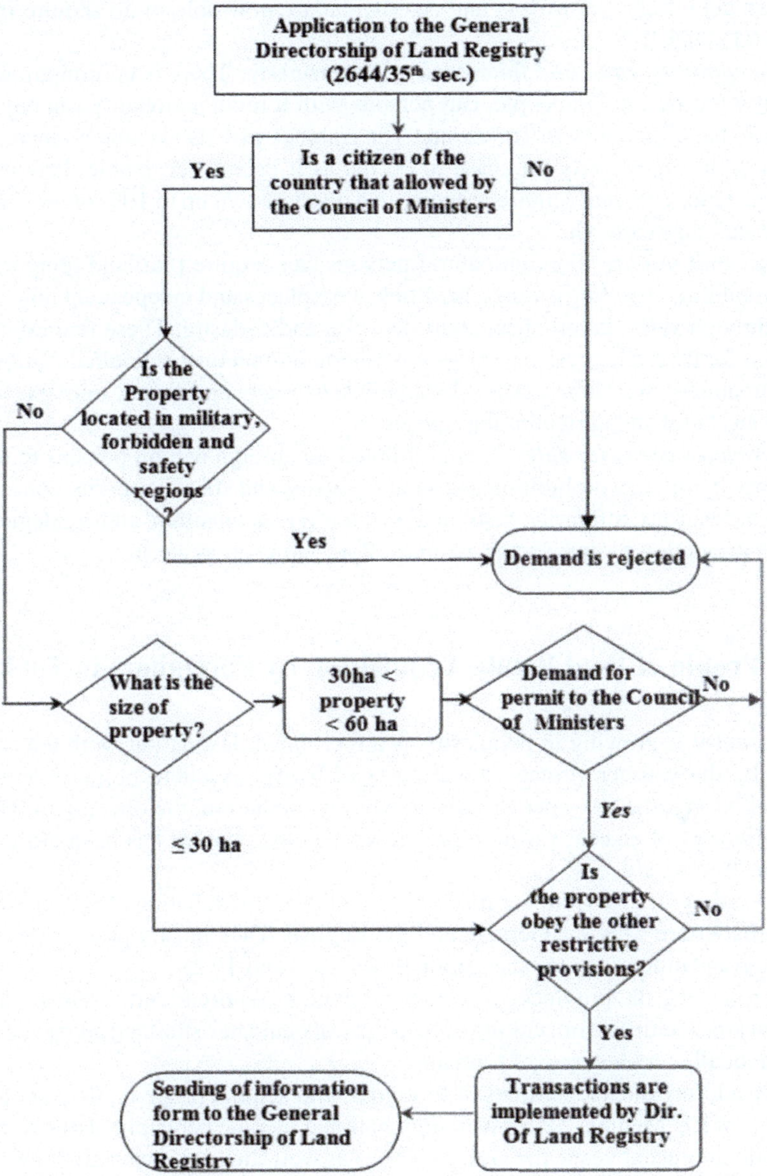

Fig. 19.1 The process of real estate acquisition by a foreigner in Turkey (Adapted from Uzun and Yomralioglu 2007)

The decision on the citizens of what countries can acquire what kind of property is left to the Council of Ministers. The Council of Ministers allow citizens of almost all the United Nations Member States to acquire immovables in Turkey. With the new regulation, citizens of countries, such as the United Arab Emirates, Middle

Eastern countries such as Iraq and Azerbaijan are now able to do acquire immovables (GYODER 2014).

The quantity, ratio and floor area of the property: There is no limitation in the quantity for the foreign people. But persons with foreign nationality can buy maximum 30 ha. of property in Turkey and The Cabinet will also be able to increase the 30-ha. limit on property purchase to 60 ha. as it deems acceptable. Persons with foreign nationality can acquire property in a district/town up to 10% of the total area of the said district/town.

Type and quality: Foreign natural persons can acquire plot/land (empty), plot/land (building, dwelling), tourist facilities, workplaces and independent units (housing) through sales, inheritance, court decision and cadastre. These real estates can be used for dwelling, business/trade, entertainment/meeting, school/education, hospital/treatment, warehouse/bonded warehouse/storage, embassies, military, factory, religious, hotel and agricultural purposes.

Restricted areas for sales: It is prohibited for foreign natural persons to acquire property in military prohibition and security zones and special security zones, independent sections militarily restricted for foreign acquisition and strategic areas (water protection basins, protected areas, flora-fauna areas etc.).

The Profile of Real Estate Acquisition by Foreigners in Turkey

The country is growing in popularity as a retirement destination, with many being lured by the warmer climate, lower costs of living, excellent property value, the natural beauty, the low-cost airlines, its young population, the opening to the East, and offerings of coastal vacation properties. So, the country has been attracted by various foreign citizens.

According to data from the first eight months of 2014, nations who acquired the most real estate were the German and the English. They mostly prefer Aegean and Mediterranean Regions (as seen from Tables 19.1 and 19.2).

Considering the provinces in which foreigners acquired land, these are usually coastal areas with natural, cultural and historical characteristics and advanced socioeconomically and in terms of tourism.

Considering the nationalities of foreigner who acquire property, they are usually citizens of EU Member States with high national income per capita. However, as an exception in recent years, the Gulf countries as well show an interest in immovables in our country. Such that, according to the latest data, the amount of real estate acquired by the Gulf countries showed an increase of 167% per square metre compared to 2013. Investors from the Gulf countries acquired 735 thousand square metres of real estate in 2013, while this number was close to 2 million square metres in 2014. Istanbul, Yalova, Sakarya, Bursa, Kocaeli and Trabzon stand out as the provinces preferred by Arabs. Especially in Trabzon, real estate sales to foreigners showed an increase of 424% compared to the previous year. In this region where

Table 19.1 Acquisition of foreign property according to the cities in Turkey

City name	Not subjected to condominium		Subjected to condominium			Total		
	Master property parcel number	Floor space of the main property (m²)	Condominium ownership parcel number	Condominium ownership number	Floor space of condominium ownership (m²)	Total parcel number	Floor space of total share (m²)	Total person number
Muğla	2.672	3.306.801	3.551	14.572	3.376.262	6.223	6.683.063	22.924
Antalya	2.139	1.803.539	6.400	39.088	4.087.224	8.539	5.890.762	52.212
Hatay	341	1.769.869	7	13	194	348	1.770.063	220
İzmir	1.316	1.264.042	1.497	2.167	279.157	2.813	1.543.199	3.528
İstanbul	2.237	1.249.054	5.821	11.981	828.421	8.058	2.077.475	14.004
Aydin	1.245	1.156.231	2.741	13.016	1.736.984	3.986	2.893.215	19.857
Yalova	545	736.675	499	1.076	89.413	1.044	826.089	1.245
Konya	66	673.220	57	70	8.959	123	682.179	89
Adana	140	590.151	82	116	10.386	222	600.537	167
Amasya	79	509.519	4	4	146	83	509.665	16

Table 19.2 Acquisition of foreign property according to the countries in Turkey

Country name	Not subjected to condominium		Subjected to condominium			Total		
	Master property parcel number	Floor space of the main property (m^2)	Condominium ownership parcel number	Condominium ownership number	Floor space of condominium ownership (m^2)	Total parcel number	Floor space of total share (m^2)	Total person number
Germany	3.523	4.565.431	4.946	8.841	1.091.930	8.469	5.657.361	12.638
England	3.315	2.699.102	6.675	23.322	3.935.932	9.990	6.635.034	36.984
Greece	1.040	305.154	639	1.077	36.211	1.679	341.365	2.277
Holland	770	1.186.053	1.627	3.867	619.276	2.397	1.805.329	5.551
Saudi Arabia	456	1.417.361	589	1.715	202.773	1.045	1.620.134	1.601
France	452	687.827	704	846	103.993	1.156	791.821	1.397
Austria	428	653.084	465	543	55.532	893	708.616	848
Turkish Republic of Northern Cyprus	425	777.033	1.035	1.249	114.198	1.460	891.231	1.513
USA	401	397.412	809	1.000	122.027	1.210	519.439	1249
Italy	292	150.856	492	647	63.282	784	214.138	960

mostly Kuwaitis purchase fields, hazelnut groves and housing, house prices significantly increased.

Conclusions

From the period before the republic up to today, the acquisition of immovable property by foreigners in our country has been allowed with various agreements and laws. Sometimes the rights granted were withdrawn and gaps formed about permissions. That said, in our country, the fundamental regulation concerning the real estate acquisition by foreign real persons was made with Land Registry Act numbered 2644, dated December 22nd, 1934. Over the years, 35th and 36th articles were amended according to needs and sometimes the said articles were revoked by the Constitutional Court. Since 1984, there have been five cancellation decision given in total. In spite of this, 35th and 36th articles were amended each and every time.

While there has not been any changes concerning the military security zones and strategic areas that cannot be acquired by foreigners, the maximum surface area that can be acquired by a foreign real person has been reduced to 2.5 ha. from 30 ha. and then again increased to 30 ha.. In addition, while it was not allowed that the total surface area of property acquired by foreigners exceeded 10% of the total planned area in a province and district, the recent regulation does not allow it to exceed 10% of the total surface area of a province or district. Also, the most important change is the removal of the principle of reciprocity. In this way, citizens of the countries that do not conform to this principle were not able to acquire property in our country; after 2012, citizens of approximately 183 countries were granted real estate acquisition right.

It is observed that different applications were preferred in different periods in Turkey concerning the type, surface area and location of property acquirable by foreigners. The determined acquisition quantity did not have a rational ground, and opinions concerning acquisition restrictions were not created based on impact assessment efforts in our country.

- Legislation and applications concerning real estate acquisition by foreigners is a very dynamic area and changes in this area take place very often.
- It was preferred to facilitate property acquisition for citizens of several countries.
- While there used to be a high demand for the Mediterranean and Aegean coasts by the European Union countries, the Black Sea coasts began to be favoured by the Gulf countries.
- In addition to goals set in development programmes; the removal of restrictions on property acquisition by foreigners and facilities provided to foreigners in real estate market by OECD and EU member states may be shown among the reasons

for constant changes and expansion of acquisition boundaries. As a EU candidate, Turkey has also been affected by these expansions within certain rules.
- Legal regulations made so far show that, as is the case in all the countries in the international unions of which Turkey is a member, foreigners are granted more rights concerning property acquisition and as a result of this, foreign direct real estate investment's contribution to the economy increased to approximately 13.3 billion dollars according to the data from the Central Bank between 2000 and 2013.
- Although these legal regulations have a positive return in terms of economy, international relations and international real estate trade, it is evident that measures related to social, cultural and economical changes caused by foreigners who purchase these properties are not taken. Such that, the increase in the price of a commodity results in a decrease in the demand for it according to the law of demand. However, in cases where the average income is high, for example, for citizens of foreign countries with an annual income of $37.849, this model becomes different. It seems unlikely that Turkish citizens with an average annual income of $13.464 can adopt to these market conditions.

Some suggestions could be made for the solution of above mentioned problems:

- In order to prevent cancellations of the Constitutional Court, the quantity, surface area and location restrictions concerning the real estate acquirable by foreigners must be rationally determined.
- In provinces where the residential area is scarce, such as provinces in the Eastern Black Sea Region, the proportion of real estate to be sold must be as low as possible. In other words, a certain sales limit must be set for each province.
- Due to the absence of a population limit, there is a danger of resident population to become the minority, especially in the tourist settlements. Regional population assessments must be made to prevent this.
- As in other developed countries, in order to prevent the commercial use of real estate sold to foreigners as hotels, hostels and timeshare, it must be compulsory for dwellers to provide identity statement to the local law enforcement. The use of immovable property as off-record commercial property must be thus prevented.
- The ratio of 10% will cause majority of the residential/livable areas to be owned by foreigners, particularly in provinces where they are limited. For example, it has been identified that this was the case for provinces such as Trabzon, Rize, Ordu and Gumushane. Essentially, before including this provision in the law, a technical study must be conducted regarding the relationship between provincial area and residential area. In this case, the ratio in question should be residential area rather than the surface area of the province.
- Formation of a department consisting of experts that will directly examine and follow this issue must be considered on a provincial basis within governorships.
- In the context of land management, an information system integrated with land registry that involves all the national resources related to land, types of land use

based on the development plan, military forbidden zones, military security zones, special security zones that limit the land acquisition and conversion, areas required to be protected due to reasons related to irrigation, energy, agriculture, mining, sit, faith or cultural properties, special conservation areas with flora-fauna characteristics, public real estate presence and real estate owned by foreign natural or legal persons is inevitably needed.

References

GYODER. 2014. Property and Real Estate Investment Trusts Association, the Real Estate Sector Three Quarter Data in 2014, Property and Real Estate Investment Trusts Association of Economics and Business Administration, Istanbul (in Turkish). Available from: http://www.gyoder.org.tr/img/mccontent/20141209154141_25912014_yili_3_ceyrek_verileri_ile_gayrimenkul_sektoru.pdf. Accessed 15 Mar 2015.

HP. 2009. *Housing Price Indices, the World Wide Overview of Real Estate Prices in Recent Years of Research Division of the National Association of Realtors, Washington*. Available from: http://www.realtor.org/sites/default/files/reports/2009/world-wide-overview-real-estate--prices-recent-years-2009-08.pdf. Accessed 22 Feb 2015.

Tess, L.M. 2004. East-Central Europe's new security concern: Foreign land ownership. *Communist and Post-Communist Studies* 37: 213–239.

Uzun, B., and A. Yavuz. 2003. European Union Integration Process of the Foreign Nationals Acquisition of Immovable Property in Turkey. In *9th Turkey Scientific and Technical Conference of Mapping*, Ankara (in Turkish).

Uzun, B., and T. Yomralıoğlu. 2007. Foreigners in Turkey Real Estate Acquisition: A Review in the Context of Land Management, Chamber of Surveying Engineers of Turkey. In *11th Scientific and Technical Conference of Mapping*, Ankara (in Turkish).

WB. 2014. The World Bank Group – Turkey Partnership: Country Program Snapshot. Available from: http://www.worldbank.org/content/dam/worldbank/document/eca/turkey-snapshot.pdf. Accessed 13 Jan 2015.

Chapter 20
Location Profiling in Cadastre for Property Value Intelligence

Mohammed Abdur Razzak

Introduction

Beginning Stories

The situations as depicted in the following three stories represent the realities of land use mechanisms and management issues in relation to modern property value information and its impact on economy and society.

Scenario 1

Around twenty two years ago a study team sought my assistance to guide them and visit the site of devastating river erosion of the Padma, one of the largest rivers of Bangladesh at the Tongibari sub-district, only forty kilometres south to the capital city Dhaka. There I heard the roaring sound and felt stormy river breeze. The sun was still not prepared to set down. There was no crowd near the river bank and an old woman was sitting there alone. When I approached her, I witnessed that she remained unmoved.

As I saw her eyes full of tears, I could not dare to ask her any question at that moment. After a while I asked her somehow the cause of her misery. She took me as someone to help her and turned her eyes with a ray of hope. She narrated the tragic events in the last couple of days and how she lost her only residential house and property and became a helpless beggar overnight. She pointed by her finger a long distance where her house was standing some times earlier. I followed her and

M.A. Razzak (✉)
Registration Department, Government of Bangladesh,
13 Green Corner, Green Road, 1205 Dhaka, Bangladesh
e-mail: mdarajz@gmail.com

T. Yomralioglu, J. McLaughlin (eds.), *Cadastre: Geo-Information Innovations in Land Administration*, DOI 10.1007/978-3-319-51216-7_20

there was nothing but big moving waves and strong river currents of unlimited flows of cruel Padma and she fell in deep silence again.

Scenario 2

The name of the island is Bhola, southest district of Bangladesh, surrounded by the rivers Meghna, Elisha, Tetulia and the Bay of Bengal. I served there for several years. A larger part of Mirjakalu bazar, a rich business centre under Burhanuddin sub-disrict, had fallen into the large river Meghna in the month of April. Parties reported about the sudden fall of property values which became lower than the determined land values based on previous year.

And people were unable to comply with registration requirements and compelled to be reluctant to register their transfer deeds in time as there was no rule or provision to consider such situation. As a result they lost their legal rights and security over their purchased property and the money they spent whereas the government lost its revenue.

Scenario 3

"This is not a road or pukka street, it is an embankment" said the school girl wearing white and blue school dress in reply to my question when I asked her about the location and open green space in front – full of water hyacinth in a deep pond. It was a river, people called it Arial Kha, a large business centre, a river port, police station, hospital, post office, registry office, government rest house and municipality office building – all are there by the river bank. The river has now turned its flow and some char land became visible.

The low land was still not fully usable. But some parties occupied the tiny portion adjacent to the roadside and opened some temporary shops by hanging sheds on bamboo. This made the significant rise of the property value which has to be applicable for the entire area. In this situation, parties have to suffer for several years. They are continuously complaining but no way to escape. Parties of major portion of the area are being deprived year after year from their legal rights of secured tenure and at the same time government is loosing revenue due to lack of timely actions, policy making and implementation of justified values appropriate to property classes, situation and location as well.

Background

According to statistics, in developed countries, the value of land and real estates together with mortgages on properties is about 60–65% of the national asset. The land and property related activities, including property developments, generate about the 30–35% of the GDP. The value of mortgages on properties in developed countries is 30–35% of the GDP (UN 1996, 2001; UN Habitat 2004).

About fifty per cent of the occupation of expanding cities in developing countries is informal, people have no secure tenure (Bathurst Declaration) (United Nations/Federation of International Surveyors 1999). In these countries it is absolutely essential to improve the security of tenure providing appropriate tools for registration of informal or customary tenure.

The implementation of sustainable development (economy, society, and environment) is also one of the main topics worldwide in developed and developing countries as well. There have been many changes related to land and properties during the last decade and resulted in new challenges to be solved. These changes very much affected the developed, transition and developing countries (András Osskó). During these years the concepts, principles and definitions of land, land utilization types, land qualities, land suitability classification and land evaluation procedures were already specified but in some circles the notion of a single, overall "land quality" in the sense of health-of-land has come to the fore.

The total capital we strive to sustain within and between generations consists of separate components:

- the natural capital (the land, the water, the air, genetic material, ecosystems, etc.);
- the human capital (knowledge, science, culture, health, nutrition);
- the institutional capital (schools, universities, research facilities, infrastructure); and
- the social capital (democracy, good governance, civil rights, equity, social harmony).

Philosophy

Philosophers saw nothing less than the establishment of liberty and the abolition of poverty among population by the confirmation of human laws to the natural order intended by the Creator. They saw that there is but one source on which men can draw for all their material needs and that is land and property as the Wealth Generation Cycle – central to the political economy with value for production, wealth generation, capital accumulation and making money for power (shown in Fig. 20.1) and there is one means by which land can be made to yield to their desires – by labour. All real wealth, they therefore saw, is the result or product of the application of labour to land and property.

In order to hoave value, an object must satisfy some human want, and it must exist in a quantity which is sufficient wholly to satisfy all desires for it. In explaining value, economists emphasize on the cost theories of value and the utility theories as classical theory, derived from Adam Smith, logically developed by Ricardo, and substantially completed by Senior, Carey, John Stuart Mill and Cairnes.

Fig. 20.1 The wealth generation cycle

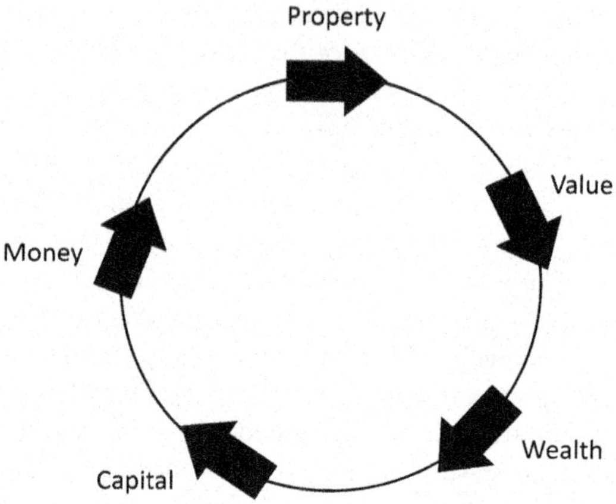

According to this theory, market value is determined by demand and supply, being fixed at the point where the former just equals the latter. Value increases directly with increase in demand, inversely with increase in supply (other conditions remaining the same).

Property

The property as an institution, when limited to its essential elements, consists in the recognition, in each person, of a right to the exclusive disposal of what he or she have produced by their own exertions, or received, either by gift or by fair agreement, without force or fraud, from those who produced it. The foundation of the whole is the right of the producers to what they themselves have produced. Private property as an institution does not owe its origin to any of those considerations of utility, which plead for the maintenance of it when established in economic terms, such as "raw material of the earth" and "gift of nature" for land, "industry" for labour, and "valuable qualities" for productive powers.

Value of Property

Value from valoir, from Latin *valere*, to be strong, able. In political economy, a word that is most commonly used to designate the power of a commodity to command other commodities in exchange. The term is applied, however, to several other conceptions. The potential capacity of an object to meet human needs is sometimes called value – 'value in use', in the terminology of the classical economists. In modern scientific economics, the term 'utility' has for the most use of the word value. Another meaning which the term value conveys is the significance of an object to an individual as the indispensable condition of a certain satisfaction.

Value in this sense of the term is frequently called ‘subjective value’, to distinguish it from ‘objective’ or ‘exchange’ value. Subjective value is of two kinds, ‘subjective use value’, where the importance of an object is gauged by the direct satisfaction to be obtained through its consumption, and ‘subjective exchange value’, where the importance of an object is gauged by the satisfaction it will yield indirectly, through exchange.

Value at a given time represents the monetary worth of property, goods, or services to buyers and sellers. To avoid confusion, appraisers do not use the word value alone, instead they refer to “market value”, “use value”, “investment value”, “assessed value”, or other specific kinds of value. Market value is the focus of most real-property appraisal assignments and its estimation is the purpose of most appraisals (Netzer 1998).

A distinction is usually made between ‘market value’ and ‘normal’ or ‘natural value’. Market value is the purchasing power of a commodity in the open market on a given day; normal or natural value is the value which would prevail if competitive forces worked without friction. Market values fluctuate widely from day to day; normal values change, if at all, only with changes in the fundamental conditions of production and consumption.

The word ‘price’ is often used as synonymous with ‘exchange value’. Economists define price as the power of a commodity to command money in exchange; value is the power of a commodity to command in exchange.

Real Property and Property Markets

Real property is defined by statute to include land, structures and improvements on land, certain mobile homes and machinery and equipment affixed to the land. The constitutional subclasses of real property and their assessed value percentages are as follows: industrial and commercial property, residential property, farm property, and public utility property. Some public utility property is assessed at the lower industrial and commercial percentage pursuant to law. A real-estate market is the interaction of individuals who exchange real-property rights for other assets, such as money. Specific real-estate markets are defined on the basis of property type, location, income-producing potential, typical investor characteristics, typical tenant characteristics, or other attributes recognized by those participating in the exchange of real property.

These real estate-related expenditures are directly linked to the price of goods and services in competitive markets. For example, the costs of roofing materials, masonry, architectural plans, and rented scaffolding are determined by the interaction of supply and demand in specific areas and are subject to the influence of social, economic, governmental, and environmental forces. Totals of real property assessments include assessments of land, structures and improvements along with most mobile homes and machinery and equipment affixed to realty.

Property Profiles

Defining

Property profiles provide a wealth of information useful to understanding a subject property's makeup to research and find properties by address, owner name, legal description or parcel identification number (Stoter 2004).

Property profiles return all pertinent property and owner information for any property. Other reports available include Comparable Market Analysis (CMA), Nearby Neighbours, Subdivision Statistics, Nearby Schools and Businesses, Demographics Sketch Vectors, Aerial Images and Parcel Maps. Wherever in the world businesses, investors and second home buyers look to acquire property, local jurisdictional laws, procedures and risks must be considered. Most active real estate markets have a system for organizing and recording property purchases; however, those systems vary widely.

Combine search includes criteria on location, general property characteristics, mortgage date and amount, sale date and amount, land information (value, acreage, square feet), legal information and miscellaneous characteristics as well as many other fields. All search results may be downloaded or printed directly to labels. Residential and commercial real estate professionals can call up layer picture-quality aerial images, and create tangible property information maps. Whether a company is building a manufacturing plant or a natural gas pipeline, developing an industrial park or resort community, or simply expanding its present facilities, protecting a real estate investment is crucial to a company's financial security and shareholders' peace of mind. No matter what the political or legal climate, property owners around the world benefit from the financial indemnification that a title policy provides.

The Unique Feature of Properties

1. Market position
2. Rental demand
3. Upcoming supply – Research to mitigate the risk of oversupply
4. Growth drivers – Infrastructure and development expenditure
5. Affordability
6. Rental income
7. Property type

8. Quality tenants – Profiles of quality tenants for the area
9. Tenant expectations – Inclusions that attract desirable tenants
10. Property management fees – Local property management fee
11. Most desirable location within the suburb – Locations to target or avoid
12. Screening developers – Identifying quality local developers and builders
13. Product quality
14. Product suitability – Ability to supply product that matches the market demand
15. Property valuation
16. Property inclusions – Property that meets demand expectations
17. Fixed priced contract – Contract terms that mitigate unexpected costs
18. Minimum deposit
19. Settlement terms – Settlement terms that minimise cash commitments
20. Capped holding costs – Contract terms that fix maximum holding costs
21. Passive involvement
22. Community evaluation
23. Site inspection – Personal site inspection
24. Property design
25. Client's expectations

Transportation is an important function of government which would facilitate the creation of a compact city, where people can easily find the facilities they desire for education, commerce, religion and recreation. Good land use, with the freedom of individuals to achieve the highest and best use of land, would ensure a desirable community. A compact city would reduce the need to invade the wilderness and devastate the environment.

Tangible personal property is defined as goods, chattels, and other articles of value which are capable of manual or physical possession and certain machinery and equipment, separate and apart from any real property. For purposes of assessment, tangible personal property has the following three sub classifications and rates of assessment: industrial and commercial property, public utility property, and all other tangible personal property. Some public utility personal property is assessed at the lower industrial and commercial rate.

Indicators and Factors That Influence Land Value

The physical attributes of land include quality of location, fertility and climate; convenience to shopping, schools and parks; availability of water, sewers, utilities and public transportation; absence of bad smells, smoke and noise; and patterns of land use, frontage, depth, topography, streets and lot sizes.

The legal or governmental forces include the type and amount of taxation, zoning and building laws, planning and restrictions. The social factors include population growth or decline, changes in family sizes, typical ages, attitudes toward law and order, prestige and education levels. The economic forces include value and income levels, growth and new construction, vacancy and availability of land. It is the influences of these forces, expressed independently and in relationship to one another, that help the people and the assessor measure value.

The FAO Council definition of sustainable development given in the introduction might be an acceptable starting point to identify issues and indicators. Based on this definition alone (and there are many others), indicators, each of which may integrate more than one variable, would be needed to track:

- the resource endowment, including its abundance, diversity and resilience;
- the environment, for example by reference to its pristine condition;
- the technology in terms of capacity as well as environmental-friendliness;
- the institutions, e.g., fishing rights, enforcement system;
- the human benefits, e.g., food, employment, income;
- the economics of exploitation, e.g., costs, revenues, prices; and
- the social context, e.g., social cohesion, participation, compliance.

Benefits: Governance, Taxation and Justice

Governance and Its Impact on Property Value

Law requires standard valuation methods for industrial and commercial property based on acquisition cost less straight line depreciation. In the absence of better evidence, fixed rates of allowable depreciation must be used depending on how the property is categorized.

Securing Fair Value and Justice by Using Location Profiles

While the major argument for raising public revenue from land rent and natural resources is because it is equitable and fair, it is also the most efficient method of raising the revenue which is needed for public facilities and services. Land is visible, can't be hidden and its valuation is less intrusive than valuations of income and sales. Taxes on labour and capital cause people to consider alternative options, including working with less effort, which produces less real goods.

Adjustments for Use and Location

Adjustments for additional attributes and deficiencies could be made for each individual site, after the base market value had been estimated by the comparative method. The experience from a comparative city could be borrowed and tested in the local area to verify the results.

Land Use and Its Locational Value: Valuation and Evaluation

"Land", the "functions of land", "land evaluation", "land qualities", "sustainability", "resilience", etc. need to be defined carefully to avoid confusion and to assure effective cooperation between international institutions and national planning entities that deal with the assessment of changes in land conditions.

The holistic concept of land was already recognized in the Framework for Land Evaluation, FAO 1995.

- Land is the basis for many life support systems, through production of biomass that provides food, fodder, fibre, fuel, timber and other biotic materials for human use, either directly or through animal husbandry including aquaculture and inland and coastal fishery (the production function).
- A property is an attribute that already gives a degree of information on the value of the land type.
- A land quality (or limitation) is a complex attribute of land which acts in a manner distinct from the actions of other land qualities in its influence on the suitability of land for a specified kind of use.

Framework for Land Evaluation and Land Qualities

The many functions of land:

- production function
- biotic environmental function
- climate-regulative function
- hydrologic function
- storage function
- waste and pollution control function
- living space function
- archive or heritage function
- connective space function

IAS/IFRS Standards for Property Value

Major IAS and IFRS standards and their application areas are as follow: (a) IAS2 applies for Inventories, (b) IAS11applies for Construction Contracts, (c) IAS16 applies for Property, Plant and Equipment, (d) IAS17 applies for Leases, (e) IAS40 applies for Investment Property, (f) IAS41 applies for Agriculture and (g) IFRS6 applies for Exploration for and Evaluation of Mineral Resources.

Property Registration: Central to Property Value Applications

Registration systems fulfill a good variety of human needs such as legal validation of all types of agreed, documented, signed and executed contracts, documents and deeds of any transaction, statement, terms and conditions for making basis of evidence, delivery of information like certificates, licenses, searches, inspections, from the original documents, reports; preservation and protection of databases, administration and supervision of records and registries, archival security management, geographic and land information (GIS and LIS), like parcels (khatians), cartographic maps (manual or digital), ICT and public sector information (PSI); and collection of public revenue e.g., stamp duty, registration fees, gain tax, VAT, local government taxes, court fees, fines etc. (Razzak 2011).

The system also includes development issues like sustainable development, sustainable land management (SLM), regulatory reform, land value taxation (LVT), valuation techniques and approaches, assessment, accounts, and determination mechanisms with a set of technological, procedural and technical arrangements to provide appropriate services in practice.

The conceptual framework of land registration includes a set of complex terms, components, and situations such as laws, systems, institutions and activities; and land revenue, land tenancy, land rights, land valuation, land transfer, land tax, land litigation and land reform etc. To explore such components, their interrelationships and impacts on public finance, a closer look into the definitive structure seems to be more important. Land registration is the "process of determining, recording and disseminating information about the ownership, value and use of land when implementing land management policies" (UNECE Land Administration Guidelines). "Ownership" should be seen as a broad concept of land tenure within various jurisdictions (statutory, customary, informal, etc.), "land" includes constructions at subsurface level, ground level and above land level (e.g. buildings).

The formal registration system has evolved for more than three centuries into an approach that is being used to varying degrees in the modern civilized nations. While registration plays a crucial role in ensuring each country's right to information, public sector information (PSI), property transfer, evidence, transparency and tax system throughout the world. It considers the main segments and registration related issues in international comparisons such as public finance factors like valuation, land value taxation (LVT), income tax, inheritance and gift taxes, capital gains tax, value added tax, stamp duty, registration fees, property transfer tax, wealth tax, records and archives, digital information database like SDI, records of rights (ROR), LIS, GIS, PSI, standards and related issues.

Property registration evolves the central and vital part of the entire system which ensures legal base of ownership with human and judicial support toward making a just, accountable and transparent society. At the same time it contributes a significant revenue share to the public finance and overall economic, social and national development.

Land Registration and Information Systems

The evolution of cadastres, LAS, SDIs and land markets shows that the traditional concept of cadastral parcels representing the built environmental landscape is being replaced by a complex arrangement of over-lapping tenures reflecting a wide range of rights, restrictions and responsibilities, and that a new range of complex commodities, building on this trend is emerging.

Land Registration in Spatially Enabling Government

For modern governments at all stages of development, one question is how best to integrate these processes, especially to offer them in an Internet enabled e-Government environment. In some jurisdictions, title registries may offer some protection to registered owners and/or mortgagees. In others, purchasers may have to rely on legal opinions based on excerpts from official title records. In all cases, for an investor or mortgagor to have true peace of mind about a property acq uisition, a fundamental concern should be to secure "good title" (van der Molen et al. 2004).

E-Government

Technically, digital land-information products offer considerably more possibilities for perfect reproduction and fast, inexpensive and easy distribution. Customers want to be served in a professional way, user-friendly tools, information that is timely, up-to-date, reliable, complete, accurate, relevant, if necessary customised, well-integrated with other relevant data sets of other suppliers, good value for money and systems that are compatible with the customer's working procedures.

Spatial information:

- Cadastre as the fundamental layer of information
- Easy mechanism of spatial enablement
- National land information policy
- Interoperability of spatial information
- Interoperability of all government information
- E-Government service delivery
- Use of "Place" to organise information, services and activities
- Electronic conveyancing

Implementation, Applications and Innovation

Developing Smart Property Value Cadastre

Interactive maps and databases, including Property Value Information Systems (PVIS) create successful and functional platforms of Property Value Intelligence in practice comprised of a set of property value data bases. These are as follow: (i) Market Value Database (MVD), (ii) Assessed and Apprised Value Data (AVD), (iii) Price Value Database (PVD), (iv) Cost Value Data (CVD), (v) Rated Value Databases (RVD), (vi) The Real Property Value Database (RPVD), (vii) Hope or Future Value Database (HVD), (viii) Best and Highest Use Value Database (BUVD), (ix) Assumption and Special Assumption Value Database (ASVD), (x) Alternative Use Value Database (AUVD), (xi) Forced Sale Value Database (FSVD), (xii) Transaction Costs Database (TCD), (xiii) Tax Bases Databases (TBD), (xiv) Synergistic Value or Marriage Value Database (SVD), (xv) Investment Value Database (IVD), (xvi) Mortgage Lending Value Database (MLVD), (xvii) Insurable Security Value Database (ISVD), (xviii) Depreciated Replacement Cost Database (DRCD), (xix) Trade Related Property Value Database (TRPVD), (xx) Development Property Value Database (DPVD), (xxi) Existing Asset Value Database (EAVD), (xxii) Green Value Database (GVD), (xxiii) Degraded Property Value Database (DPVD), (xxiv) Added Value Database etc.

Integrating Profiled Location Data Using Geospatial Business Intelligence with PVI

A data warehouse is a subject oriented, non-volatile, integrated, time variant collection of data in support of management's decisions. Common functions of business intelligence technologies are reporting, online analytical processing, analytics, data mining, process mining, complex event processing, business performance management, benchmarking, text mining and predictive analytics. What is geospatial data and geospatial analytics? When people refer to geospatial data, they are often describing address-related data (a specific address, point of interest, ZIP code, and so on). This data can be matched to a specific latitude and longitude using a process known as geocoding.

Geocodes for addresses and points of interest can also be integrated with other data sources to enhance analysis in dashboards, visualizations, and more advanced modelling. The Geo Database is a collection of geographic datasets, works in concert with ArcGIS software to provide a rich framework for modelling attributes, spatial and temporal relationships, and transactions. Best practices for data modelling and analysis by addressing spatial integrity, attribute integrity, work flow, and scaling. It clarifies geographic data modelling concepts of the geo-database information models.

Geospatial data, sometimes referred to as location data or simply spatial data, is emerging as an important source of information both in traditional and in big data analytics. Geospatial data and geographic information systems (GIS) software are being integrated with other analytics products to enable analytics that utilize location and geographic information. Such analytics are also moving past mapping to more sophisticated use cases such as advanced visualization and predictive analytics. Geospatial data sources include: (a) Global positioning system (GPS) data and (b) Remote sensing data.

Database Deployment Software and Applications

Organizations store feature data in a structured file format such as Autodesk spatial data file (SDF) or SHP. With SDF, organizations benefit from the power of a spatial database without the cost or management overhead. Then organizations can easily extend the reach of their information by using a web mapping application such as Autodesk Map GuideEnterprise to deliver powerful, easy-to-use online maps and related information to audiences of all sizes.

With Topobase and Oracle Spatial, we no longer have to maintain multiple data sets, and we have reduced the risk of data entry errors. Autodesk Geospatial makes it easy for engineers and designers to manage and share mapping data – such as regional scale data sets, cadastral information, and utility network data including pipelines, transformers, and valves. Additionally, teams can import and export data sets from many different CAD and GIS file formats – such as ESRI Shapefiles, MapInfoTAB files, MicroStationDGN, and raster data from multiple coordinate systems – and combine it with DWG files and have the information overlay properly.

Organizations share spatial data with other departments and applications, making spatial data a central part of its IT ecosystem. In this stage, GIS data and functionality get woven into other business systems, integrating with assessor databases, permitting systems, ERP systems, and more. The spatial application server supplies geospatial intelligence and data to these other applications. Autodesk, resellers, partners, and system integrators build powerful solutions to meet the organization's specific business goals and processes.

Autodesk Topobase provides sophisticated solution modules that make it easy for organizations to establish and manage the database deployment. By moving up the geospatial value chain, organizations increasingly leverage their geospatial data for a variety of business functions. Organizations gain the ability to organize data effectively, implement real-world coordinate systems, and work with larger data sets. They deliver increased scalability and security, ability to complete long transactions, and integration with other systems.

Organizations need to move from a file-based environment using DWG, SHP, or SDF to a spatial database environment using the full functionality of a relational database management system (RDBMS). With an RDBMS, hundreds or even thousands of people can create, edit, and manage the same data. With a full RDBMS, organizations get more scalability, as well as added security and the ability to create more sophisticated data models.

Using Data Access Technology, Autodesk Geospatial products work natively with spatial data stored in Oracle, MicrosoftSQL Server and MySQL, as well as with ESRI's ArcSDE middleware. As a result, organizations are able to fully use the security, scalability, sophisticated data models, and multi-user read/write power of an RDBMS. AutoCAD Map 3D provides tools that make data and schema migration from SDF or SHP files to a full-scale RDBMS easy.

Conclusion

This has been devoted to materialize the vision toward achieving the location based property value information intelligence as an easy and accessible platform for prompt decision making at all levels of public, private, business and economic sectors. Incessant research efforts are also vital to enrich the system in practice. Our pragmatic optimism invites the practitioners' communities to move forward as to find the solutions to put in practice with a shared, consorted and a well communicated network in the near future.

References

Netzer, Dick. 1998. *Land Value Taxation: Can It and Will It Work Today?* Cambridge, MA: Lincoln Institute of Land Policy.

Razzak, Mohammed Abdur. 2011. Registration and Information Systems: Toward an Empirical Development in Bangladesh. AHDPH, Dhaka. Amazon.com.

Stoter, J. 2004. 3D Cadastres, PhD TU Delft, the Netherlands.

UN. 1996. *Land Administration Guidelines*. New York/Geneva.

———. 2001. *Inventory of land administration systems in Europe and North America*. London.

UN Habitat. 2004. *Pro Poor Land Management*, Nairobi.
United Nations/Federation of International Surveyors. 1999. Bathurst Declaration on Land Administration for Sustainable Development (1999), Bathurst.
van der Molen, P. et al., 2004. Remarks and observations related to the further development of the core cadastral domain model. In *Proceedings FIG/COST Workshop on 'Standardisation in the Cadastral Domain'*, Bamberg, Germany.

Part IV
Marine Cadastre

Chapter 21
Marine Cadastre Legal Framework for Malaysia

Ashraf Abdullah, Zakaria Mat Arof, Abdullah Hisam Omar, Nazirah Mohamad Abdullah, Chee Hua Teng, Keat Lim Chan, and Hassan Jamil

Introduction

The coastal Malays in particular regarded the seas as natural appurtenances to the land they occupy. Sovereign nations exercised absolute sovereignty and jurisdiction in seas covering the whole of sovereignty area of Malaysia. The seas surrounding the land played a significant role in the defence, economic, and political matters of the nation concerned. The sovereignty of nation depend on concept of unity between land and water expressed by the Malay word *'tanahair'* which literally means land and water is depict to the meaning of native land. Such was already the significance and appreciation of the seas in this region. As early as 1276, during the reign of Sultan Muhammed Shah the first sovereign of the Malacca Sultanate, it was found that the Malays had already designed a set of laws of the sea applicable in sea areas within the jurisdiction of the Malacca Sultanate. These laws were referred to as the *Malaca Code*. Furthermore, introduction of marine cadastre is a new kind of sea uses, spatial extension of ongoing sea uses and the need to better protect and better arrangement of marine parcel for various users, as well as between the users and the environment. Marine cadastre legal framework is the process to define the marine alienation and marine spatial planning to allocate space for specific uses which can

A. Abdullah (✉) • Z.M. Arof
Department of Geomatic Science and Surveying, Faculty of Architecture Planning and Surveying, University Teknologi, MARA (Perlis), 02600 Arau, Perlis, Malaysia
e-mail: ashraf@perlis.uitm.edu.my

A.H. Omar • N.M. Abdullah
Department of Geoinformation, Faculty of Geoinformation and Real Estate, University of Teknologi Malaysia, 81310 Johor Bahru, Johor, Malaysia

C.H. Teng • K.L. Chan • H. Jamil
Department of Surveying and Mapping Malaysia, Jabatan Ukur Dan Pemetaan Malaysia, Wisma JUPEM, Jalan Semarak, 50578 Kuala Lumpur, Malaysia

T. Yomralioglu, J. McLaughlin (eds.), *Cadastre: Geo-Information Innovations in Land Administration*, DOI 10.1007/978-3-319-51216-7_21

help to avoid user conflicts, to improve the management of marine spatial claim and sustain an ecosystem-based management of ocean (Fowler and Treml 2001).

However, the law is less relevant in the recent days. Current practices on handling the marines rights is not much different with on the ground. Marine phenomena in 3D parcel actually differ and should be treated separately. This paper examines the overview of marine cadastre issues in Malaysia related to international law, the governing law, Malaysia law in land matters such as National Land Code 1965 and propose the marine cadastre legal framework for Malaysia and several topics to amendment in NLC1965.

Marine Cadastre in Malaysia

As a maritime country Malaysia had joined as a member in the Law of the Sea Convention of 1982. Marine cadastre in Malaysia was defined as follows: A marine cadastre is a 3D marine parcel administration system with respect to the legal and systematic technical arrangement of marine spatial rights, restrictions and responsibilities for marine space activities covering marine natural resources, marine industry, tourism and national sea park and wildlife conservations (Ashraf 2004). This definition is including the agenda of integrated coastal and marine resources management within the land policy and ocean policy framework (Ng'ang'a et al. 2004). That is why the requirement of legal framework is needed to ensure the ability to achieve goals to be reality. From the study of literature review, the focus of implementation in marine cadastre elements currently practiced and carried out but not in a proprietory way. Besides legal needs; the technical aspects are also limited. Because no specific procedures have been taken into account like transparency, accuracy and clarity on data presentation for parties involved in the marine environment. The implementation of marine cadastre is therefore needed (Sutherland 2011).

In Malaysia territorial water, the main issue is in the territorial water description and the separation of power between Federal and State. In current situation, Malaysia created the new act for territorial water administration and enforced on 22 June 2012 for whole of Malaysia. This issue was continuation from replacement of the Emergency (Essential Powers) Ordinance, No. 7 1969 P.U (A)307A/1969 as mentioned above to make a new jurisdiction and not under the Emergency Proclamation 1969. The Malaysia Government took the decision to scatter the three Emergency Proclamations. The three Emergency Proclamations are The Emergency Proclamation 1966, the Emergency Proclamation 1969 and the Emergency Proclamation 1979. Under this action, all of three Emergency Proclamations were stopped immediately and on the date 24th November 2011 the *Dewan Rakyat* had approved that all the three Proclamations were terminated on 20 December 2011 also done by *Dewan Negara*. From this condition, the new act must be established to conserve Malaysian territorial waters. But from the content of new act, the consideration was similar to (Essential Powers) Ordinance, No. 7 1969 P.U (A)307A/1969 but have some changes about description in term of definition and applications and ensuring parallel to UNCLOS 1982. In this act, the conclusion is a continuation of the Emergency Ordinance 1969 and translated into a new act since the establishment of the Ordinance is not relevant in the present situation and formulated according to the event last year. Here are some excerpts contained in this Act relating to the functions and powers between the federal and state administration.

According to the statement: Paragraph 3 (1) which is related to marine territorial limits of Malaysia is like – Subject to the provisions of this Act, the width of the territorial sea shall be of 12 nautical miles. Statement under 3(2) are mentioned about the baseline issue as stated in the quote: The baselines from which the breadth of that territorial sea is to be measured shall for all purposes be those established in accordance with section 5 of the Baseline of Maritime Zones Act 2006. Regarding Section 5 Baseline of Maritime Zone Act 2006 state as below.

5(1) Subject to subsection (2), the baselines for the purpose of determining the maritime zones of Malaysia shall be:

(a) low-water line along the coast as marked on large scale charts;
(b) seaward low-water line of a reef as shown by the appropriate symbol on charts; or
(c) low-water line on a low-tide elevation that is situated wholly or partly at a distance not exceeding the breadth of the territorial sea from the mainland or an island.

5(2) Not withstanding subsection (1), in respect of any area for which geographical co-ordinates of base points have been declared under section 4, the method of straight baselines interpreted as geodesics joining the consecutive geographical coordinates of base points so declared may be employed for determining the maritime zones of Malaysia.

The above indicative of the position of Malaysia recognizes the low-water line as a line on which the determination of the maritime boundaries with consideration of the matters referred to three situations of low-water line and the first situation is the position of low-water line were identified and marked on the map scale, position seaward low-water line of a reef or authenticity of essays with the symbol shown on the map and the position of the low-water line where the low-tide elevation in whole or in part is not more than 12 nautical miles from the mainland or island position.

In other issues, Malaysia has created the Exclusive Economic Zone 1984 Act which is focusing on determination and declaration of the matters of Malaysia mastery in maritime area.

In EEZ 1984, under the Section 3 (1), the exclusive economic zone of Malaysia as proclaimed by the King of Malaysia vide P.U(A) 115/80 is an area beyond and adjacent to the territorial sea of Malaysia and subject to subsections (2) and (4) extends to a distance of two hundred nautical miles from the baseline from which the breadth of the territorial sea is measured. Subsection (2) mentions that where there is an agreement in force on the matter between Malaysia and a State with an opposite or adjacent coast, questions relating to the delimitation of the exclusive economic zone shall be determined in accordance with the provisions of that agreement. Subsection (4) where having regard to international law, state practice or an agreement referred to in sub-section (2) the King of Malaysia can consider its necessity so to do, he may by order published in the Gazette alter the limits of the exclusive economic zone determined in accordance with subsection (1). Malaysia is a federal state with marine jurisdiction and management responsibility split between the states and the central (federal) government. The amendments to the Emergency (Essential Powers) Ordinance, No. 7 1969 states that territorial water shall be constructed as a reference to such part of the sea adjacent to control the coast thereof not exceeding three nautical miles measured from low water mark. In this situation, the state controls up to three nautical miles from low water mark whilst the federal government has jurisdiction and management responsibility from the said three nautical miles limits to the outer edge of the EEZ and continental shelf. But on 22nd June 2012, the Emergency (Essential Powers) Ordinance, No. 7 1969 (P.U.(A)307 A/1969) had withdrawn power after the Territorial Water Act 2012 was established. This act is to create the Malaysia territorial water to replace the existing. The Emergency (Essential Powers) Ordinance, No. 7 1969 (P.U. (A)307 A/1969 was established under Emergency Proclamation 1969.

In other issues, Malaysia also is lacking due to the LAT demarcation not ready in proper marine laws. Moreover, the legal documentation is important in several issues involved in the marine cadastre elements and aspect, but in Malaysia no provision was established. It is needed to support the policy, rules and regulations for implementation which is important in relating the marine alienation issues. Malaysia has practiced the marine alienation indirectly and that practice has not occurred in suitable marine area and environment under the marine spatial management context. This is lacking due to zoning issue about the marine alienation involved in rights purposes that are applicable.

United Nation Law of the Sea 1982 Related to National Marine Regime

The administration of marine area in Malaysia is governed by legally defined boundaries and follows United Nations Convention on the Law of the Sea (UNCLOS) due to which Malaysia itself can claim, manage and utilize its maritime territories. Based on Country Report on Administering the Marine Environment by Ahmad Fauzi Nordin (2006), Malaysia ratified UNCLOS in October 1996. In line with provisions of UNCLOS:

(i) The Territorial Sea, which is the belt of sea is measured 12 nautical miles (nm) seaward from the territorial sea baseline. On 2nd August 1969, an Ordinance under Article 150(2) of the Constitution known as the Emergency (Essential Powers) Ordinance, No.7, 1969 was promulgated. Under this Ordinance, the territorial waters of Malaysia (except in the Straits of Malacca, the Sulu Sea and the Celebes Sea) were declared as 12 nautical miles from the base line determined in accordance with UNCLOS.

(ii) The contiguous zone which is the belt of sea, contiguous to the territorial sea, measured 24 nm seaward from the Territorial Sea Baseline.

(iii) The Exclusive Economic Zone, which is the area beyond and adjacent to the territorial sea, measured 200 nm seaward from the Territorial Sea Baseline.

This provision is important to guide the marine cadastre implementation in term of legal aspect to ensure that focus of practices must be parallel to UNCLOS 1982 although that law was created focusing on local perspective. Malaysia ratified the UNCLOS 1982 on 14th October 1996 and highlighted the baseline and maritime limitation shown by the big scale chart presented by the list of base point with the geodetic datum. In the context of determination of base point and baseline, Malaysia has practiced the straight base-line approach. Hence, through the endorsement of Baseline of Maritime Zone Act 2006, the Malaysian baselines were established by Malaysian Government and deposit them into United Nation Convention to ensure the sovereignty of Malaysia maritime boundary being declared and avoid the conflict on the next future.

National Land Code 1965 in Malaysia

Until now, Malaysia has established 77 national laws pertaining to diverse marine matters related to marine activities such as shipping, navigation, transportation, petroleum and others. The list of national laws pertaining to marine matters is attached as Appendix. This paper focuses on National Land Code (NLC) 1965 because it was main reference and discusses land matters in Peninsular Malaysia. NLC 1965 came into force from January 1, 1966 and contains 447 sections, 16 schedules, 6 divisions, 35 sub-divisions and have been through about many amendments regarding the current issues. NLC1965 is nearly 48 years old and it is largely a re-enactment of earlier laws. National Land Code 1965 has undergone many changes and revisions and the latest was in February 2012. In regards to delivery system there are more than 40 times the scale of amendments to the National Land Code, which were performed whenever there was an urgent need to address issues of land administration. Latest amendments effected on the National Land Code 2008 can be a launching pad to the evolution of electronic land administration system for the states of Peninsular Malaysia. However, the NLC 1965 governs mainly for land administration purposes and its authority cover until HAT situation and not discussing in details beyond that situation especially on marine environment. The purpose of marine cadastre is not there and need the revision and amendments to answering the issues of marine cadastre. NLC1965 is still relevant to marine cadastre as a new practice on the water issues and limits. It is needed in marine cadastre practices to be amended through a special modification of any statement in related

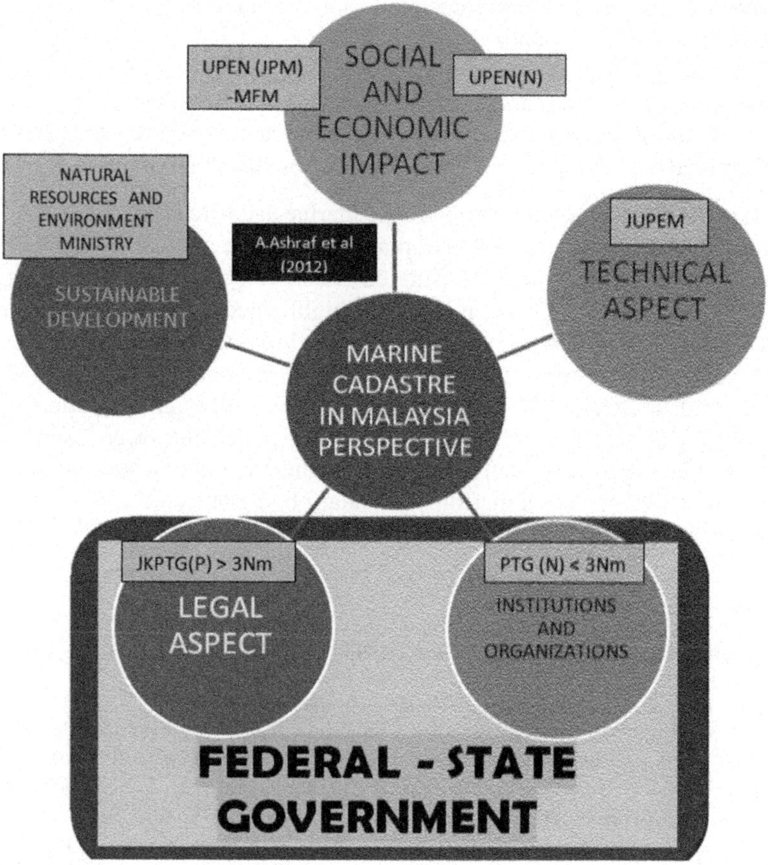

Fig. 21.1 Marine cadastre component and institution involved

section which are connected and can be correlated with the marine cadastre factors.

The Integration of Institution in Marine Cadastre from Malaysia Perspective

However, efforts in establishing the perfect implementation of marine cadastre as a practice have been carried out over the past 5 years. Figure 21.1 shows a resolution issued as a clear picture of the marine cadastre related to general concept from Malaysia perspective to proposed and the connection to main institutions which is suggested to driving the marine cadastre very well (Ashraf and Zakaria 2008).

Figure 21.1 shows five main components that support the general concept of marine cadastre where it covers the legal, institutional and organizational, technical aspects, the social and economic impact and sustainable development. From this aspect of the proposed major institutions involved in marine alienation is where a combination of federal and state governments work in two main aspects related to the law and institutions involved under the law stated. This state of things between federal authority and state where under three nautical miles depend on the power of the state and the State Director of Lands and beyond three nautical miles under the Director General of the Department of Land and Minerals. For technical management, DSMM is the most appropriate institution to highlight the best technical methods to produce a robust procedure in the production of marine title. Marine cadastre also requires that institutions make a detailed study to look at the social and generating profits realized from the implementation of the marine cadastre (Rizqi et al. 2008). In addition, marine environment should always be supervised properly by established ecosystem as well as development activities. Therefore, for sustainable development, it has been proposed for the Ministry of Natural Resources and Environment to put the responsibility for carrying out the specific control activities to ensure that the marine natural resources are not damaged and cause a loss in the country.

The Proposal of Marine Cadastre Legal Framework

For the purposes of marine cadastre in Malaysia, Fig. 21.2 shows a proposed legal framework that are structuring of marine cadastre applications in Malaysia. Figure 21.2 shows the position of the marine cadastre placed in two administrative authorities as clearly stated in the Territorial Water Bill 2012. For the implementation of the marine cadastre in Malaysia, for proposed implementation in their positions, the State Authorities have full power to control only three nautical miles marine area which covers areas of airspace above the water, water column, sea bed and subsea. This framework specifies where the federal government has full administrative powers in the position three nautical miles out to sea whereas it will be subject to federal legal requirements for existing law more than three nautical miles and also in international law such as UNCLOS 1982 or any agreement involving agreements with neighbouring countries. While the position of power under the state government is involved specifically about the implementation of the marine cadastre which include some important aspects that should be implemented in advance like:

(i) Derivation and delineation of lowest astronomical tide
(ii) Large scale mapping produce
(iii) Positioning and delineation of littoral zone
(iv) Beginning of marine cadastre and ending of land cadastre

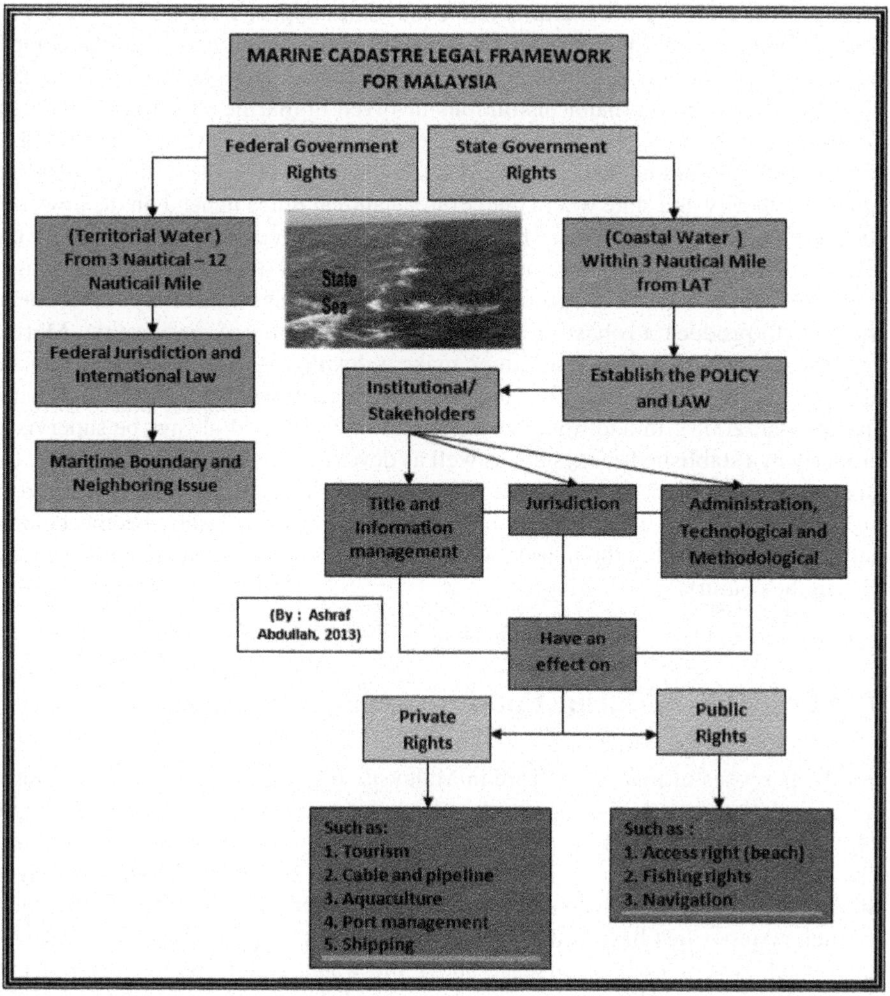

Fig. 21.2 The proposal of marine cadastre legal framework for Malaysia

(v) Technical applications requirement
(vi) Designation of institutions involved
(vii) Standard of procedure for marine alienation
(viii) The guideline and regulation for marine right
(ix) Marine space tax

Once implemented amendments or additions to the existing legislation, the framework emphasizes that institutions will be held accountable by them and the institution in particular will perform in accordance with section provided to them by the amendment of the existing section or in addition to a new section. It is based on their policy regarding marine cadastre practice thought appropriate for the implementation in Malaysia. The development of policy and law are important issues and should be proposed on suitable legislation and viewed in detail and that matter should be concentrated based on current marine environmental factors. The main concern is to be seen in connection with marine elements such as the position of the marine area affected by factors such as time, changes in the vertical aspect of the ever changing natural environment factors of current and waves, the characteristics of marine biodiversity, marine border demarcation and documentation views appropriate to reflect the marine space and given ownership. The important point in the application of marine cadastre implementation is the administrative institutions which have power and can make a decision (Strain et al. 2006). Therefore these institutions should not be burdened with other responsibilities and should have separate scope of their field. For example at present, application management and administration of marine territory under three nautical miles are still under the District Land Office or the State Land and Mines Office, while the agencies added with land administration and management is also a big responsibility. By taking into account these factors, the position of the marine cadastre should be placed under a new institution known as the Department of Marine Space Administration which only focuses on the management of the administrative affairs of the marine regions and functions remain the same as the land office but different arrangement of marines on the issue of alienation and problems of the marine environment (Liu et al. 2011).

For the right to private issue and regarding the concept of alienation, the right to private sector is the administration and management of properties available in all legislation relating to the administration and management of the property. For marine situation, this right is also extended to the concept of land administration of marine space but occur under a number of different factors to grant land ownership. Matters relating to marine alienation is not specifically open to all levels of society, it is where specialization is still a subject under control of the State Government. However, the legal framework of marine cadastral outline some aspects that require priority of alienation aspects of marine related tourism purposes, cables and pipelines, aquaculture, port management and the right part of the cruise shipping sector. In the perspective of the State, marine alienation is a sound investment and basic needs which is on the grounds of social needs and also has a big impact on the economy of the country and not just the state but also all aspects of life.

Discussing on rights to public issue is about the society right and in this aspect; the basic thing is not about the giving the marine alienation but the concept used is related to equal right for society. This aspect is fundamental in ensuring equal right to use the ocean and is still in line with the philosophy that ocean is free and this should be emphasized in the use of marine space in which to put a clear purpose in the legislation so that local and foreign societies can enjoy a marine environment that is free and not bound to any restrictions. As such, the right of access to the beach should not be blocked, and give the right to people to enjoy the beach and also

Table 21.1 Several topics in the proposed amendments in NLC1965

Issues	Topics
1	Review the title name of NLC1965 for application compatibility in marine environment
2	The interpretation issues and concepts of marine alienation
3	The definition of the marine spaces as a title
4	Determination of marine boundary between local states
5	The issue of four dimensional rights in marine spaces
6	Issues in defining coastline boundaries and littoral zone
7	The commencement of marine cadastre applications
8	Display of marine space alienated

as a boat mooring area. Besides fishing rights, priority should also be given to the community and in accordance with the Fisheries Act 1985. The right to sail is also a civil right defined in existing legislation which also should be touched on the matter of international law UNCLOS 1982 and the International Maritime Organization (IMO).

In order to achieve the capability of implementation, several issues addressed in National Land Code 1965 must be proposed to amend and ensure that proposal is under scope of Malaysia perspective. This matter should be covered in social and relationship context with the community and institutions involved; rules and regulations would be suggested to implement with technical approach and politics in land and ocean matters. Table 21.1 shows the several topics suggested as amendments in NLC1965.

Conclusion

Marine cadastre is most important to ponder especially in the legal structure for practices and guidelines. This is a requirement of recent days to ensure the marine resources could be maximizing exploitation to gain the benefit and profit. It should begin from legal aspect as the driver, the great implementation and solution in order to answer all the problems facing and need to update follows the changing of requirements and marine environments itself. This paper is very important and as the starting point to look deeper in terms of laws involving and requires emphasis to ensure marine cadastre as a reality soon. Generally, marine cadastre is the platform to exploit new resource for government and public to share the natural awards to them in creating the better life for future.

Appendix

National laws pertaining to marine matters

No	Category	Laws and regulations
1	Port	Penang Port Commission Act, 1955 (Act 140)
		Port Authorities Act, 1963 (Act 488)
		Port Workers (Regulations of Employment), 2000 (Act 607)
		Sabah Port Authority (Consequential Provisions) Act, 1968 (Act 25)
		Declaration of an Area in Bintulu District to be a Federal Port Act, 1979 (Act 217)
		Bintulu Port Authority Act, 1981 (Act 243)
		Ports (Privatisation) Act, 1990 (Act 422)
2	Shipping	Carriage of Goods by Sea Act, 1950 (Act 527)
		Merchant Shipping Ordinance, 1952
		Merchant Shipping Ordinance, 1960 (Sabah)
		Merchant Shipping Ordinance, 1960 (Sarawak)
		State Boat Rules
3	Light house	Federation Light Dues Act, 1953 (Act 250)
4	Non-living resources	Petroleum Mining Act, 1966 (Act 95)
		Petroleum Development Act, 1974 (Act 144)
		Petroleum and Electricity (Control of Supplies) Act, 1974 (Act 128)
		Petroleum (Safety Measures) Act, 1984 (Act 302)
		Petroleum (Income Tax) Act, 1974 (Act 543)
5	Living resources/Fisheries	Fisheries Act, 1985 (Act 317)
		Fisherman' Association Act, 1971 (Act 44)
		Lembaga Kemajuan Ikan Malaysia Act, 1971 (Act 49)
		National Forestry Act, 1984 (Act 313)
6	Natural resources	Continental Shelf Act, 1966 (Act 83)
		Baseline of Maritime Zone Act, 2006 (Act 660)
		Exclusive Economic Zone Act, 1984 (Act 311)
		Sarawak Natural Resources and Environment (Prescribed Activities) Order, 1994
		Sabah Conservation of Environment (Prescribed Activities) Order, 1999
		National Land Code, 1965
7	Jurisdiction	Emergency (Essential Powers) Ordinance, 1969 (Act 216)
		Extra Territorial Offences Act, 1976 (Act 163)
		State Land Rule

(continued)

No	Category	Laws and regulations
8	Enforcement	Immigration Act 1959/63 (Revised 1975) (Act 155)
		Internal Security Act, 1960 (Act 82)
		Police Act, 1967 (Act 344)
		Malaysian Maritime Enforcement Agency Act, 2004 (Act 633)
		Military Maneuvers Act, 1983 (Act 295)
		Armed Forces Act, 1972 (Act 77)
		Penal Code (Revised 1977) (Act 140)
		Custom Act, 1967 (Revised 1980) (Act 235)
		Evidence Act, 1950 (Revised 1971) (Act 56)
		Poison Act, 1952 (Revised 1989) (Act 366)
9	Tourism	Malaysia Tourism Promotion Board Act, 1992 (Act 481)
		Tourism Industry Act, 1992 (Act 482)
		Tourist Development Corporation of Malaysia Act, 1972 (Act 481)
10	Heritage and antiquity	Antiquities Act, 1976 (Act 168)
		Antiquities and Treasure Ordinance, 1957
11	Telecommunication	Telecommunication Act, 1950 (Act 588)
		Telecommunication Services (Successor Company) Act, 1985 (Act 322)
		Communication and Multimedia Act, 1998 (Act 588)
12	Dispute settlement	Convention on the Settlement of Investment Disputes Act, 1966 (Act 392)
		Arbritration Act, 1952 (Revised 1972) (Act 93)
		Convention on the Recognition and Enforcement of Foreign Arbitral Awards Act, 1985 (Act 320)
13	Forestry/Wildlife	National Forestry Act, 1984 (Act 313)
		Protected Areas and Protected Places Act, 1959 (Act 298)
		Protection of Wild Life Act, 1972 (Act 76)
		Fauna Conservation Ordinance (Sabah), 1963
		Wildlife and Birds Protection Ordinance, 1955
		Forest Enactment 1968 (Sabah)
		Forest Enactment 1954 (Sarawak)
		Planted Forest Rules of Sarawak, 1997
		Land Conservation Act, 1960
		Malaysian Forestry Research and Development Board Act, 1985
		National Parks and Nature Reserves Ordinance, 1998

References

Ashraf, A. 2004. Master thesis: Pengkonsepsualan Kadaster Marine di Malaysia. UTM Skudai.

Ashraf, A., and M.A. Zakaria. 2008. *Development of marine cadastre model and conceptual in Malaysia*. Perlis: UiTM Perlis.

Fowler, C., and E. Treml. 2001. Building a marine cadastral information system for the United States – A case study. *Computers, Environment and Urban Systems* 25(4–5): 493–507.

Kedah State Land Rule 1966.

Liu, W.-H., C.-C. Wu, H.-T. Jhan, and C.-H. Ho. 2011. The role of local Government in marine spatial planning and management in Taiwan. *Marine Policy* 35(2): 105–115.

National Land Code 1965 (Act 56/1965).

National Oceanography Directorate, Malaysia Ocean Policy 2011-2020. 2011. Ministry of Science, Technology and Innovation, Malaysia.

Ng'ang'a, S., M. Sutherland, S. Cockburn, and S. Nichols. 2004. Toward a 3D marine cadastre in support of good ocean governance: A review of the technical framework requirements. *Computers, Environment and Urban Systems* 28(5): 443–470.

Nordin, Ahmad Fauzi. 2006. Country report on marine administration. *JUPEM*, pp. 23–27.

Rizqi, Abdulharis, Eka Djunarsjah and Andri Hernandi. 2008. Stakeholder analysis on implementation of marine cadastre in Indonesia. FIG Working Week 2008, Stockholm, Sweden, 14–19 June 2008.

Strain, L., A. Rajabifard, and I. Williamson. 2006. Marine administration and spatial data infrastructure. *Marine Policy* 30(4): 431–441.

Sutherland, M. 2011. Implementing a marine cadastre. University of the West Indies, St Augustine. Trinidad Tobago. South East Asia Survey Congress.

Chapter 22
The Evaluation of Marine Cadastre Definitions Among Australia, Canada and United States of America Based on Indonesia's Perspective as an Archipelagic State

Yackob Astor, Widyo Nugroho Sulasdi, S. Hendriatiningsih, and Dwi Wisayantono

Introduction

The Unitary State of the Republic of Indonesia both in physical and geographical realities has the natural resources potential much better than other countries. As a tropical country, Indonesia has fertilized land. It is surrounded by some unique species and some variety plants are biodiversity. The location of Indonesia lies along with the pacific track of the fire which led to Indonesia with plenty of minerals and metals such as gold, silver, copper and nickel. Coal, oil and geothermal energy are in high abundance. As an Indonesian island (archipelagic state), it has oceanic territory (6,120,673 km^2) wider than the land area (1,910,931.32 km^2). Approximately 13,466 islands along the coastline of 99,093 km make Indonesia. It has greater marine resources than the natural resources on land. This marine resource condition is seen as an opportunity for Indonesian developing country to build the excellent centre in coastal and marine field.

Some problems arise as Indonesian islands certainly cannot be separated from the conflict or problems that came out both domestic and overseas like the conflict among neighbours related to territorial boundaries. Up-to-now, Indonesian marine

Y. Astor (✉)
Geodesy and Geomatic Engineering Study Program, Faculty of Earth Sciences and Technology, Bandung Institute of Technology (ITB), Bandung, Indonesia
e-mail: yackobastor@yahoo.com

W.N. Sulasdi • D. Wisayantono
Coastal Zone Science and Engineering System Research Group, ITB, Bandung, Indonesia

S. Hendriatiningsih
Surveying and Cadastre Research Group, ITB, Bandung, Indonesia

T. Yomralioglu, J. McLaughlin (eds.), *Cadastre: Geo-Information Innovations in Land Administration*, DOI 10.1007/978-3-319-51216-7_22

setting boundaries among neighbouring countries are still away from being resolved. Among ten countries with marine territorial boundaries, only Australia has sorted the marine boundaries with Indonesia. Meanwhile other neighbouring countries have recently implemented to define the continental shelves limits and some partly marine territorial boundaries as also ZEE (Indonesia's Exclusive Economic Zone). This condition frequently leads to claiming conflicts among the Indonesia and neighbouring countries on marine territorial zone. As a result, it will appear instable and disrupt economic development in those territories (Widodo 2004; Harbimaharani 2010).

Under the 1945 constitution of the Republic of Indonesia, Article 18 verse (1) The Unitary State of Republic of Indonesia shall be divided into provinces and those provinces shall be divided into regencies and municipalities, each of which shall have regional authorities which shall be regulated by law. Article 25 states that The Unitary State of the Republic of Indonesia is an archipelagic state, the boundaries and rights of whose territory shall be established by law. Both articles above declare that each province, district and municipal boundaries and their right must be defined by law. To determine the normatively boundaries that regulated in Act number 2 of 1999 amended Act number 32 of 2004 and then amended by Act number 23 of 2014 on Regional Government, and operationally set out in The Ministry of Home Affair Regulation number 1 of 2006 amended The Ministry of Home Affair Regulation number 76 of 2012 on Guidelines Region Emphasis. Whereas, to determine the rights stipulated in Act number 27 of 2007 amended Act number 1 of 2014 on the Management of Coastal Territories and Small Islands. Issued Act number 32 of 2014 on Marine is one progress in marine resources management managed by sectorial laws and regional regulations.

By the issue of Act number 32 of 2004, the central government gives some authorities to the Regional Government which are not only limited to the government affairs, but also in terms of utilization and management of its wealthy resources including marine resources. This is confirmed that the Indonesian marine managed by some regional governments have the authority to the local marine territorial boundaries including the authority to manage all waters surrounding its islands to 12 nautical miles from the coastline to the open sea and/or in the direction of shores of the province and one third of the jurisdiction of the province to the districts/municipalities (Djunarsjah 2008, 2011; Tamtomo 2004, 2006).

The management of coastal and marine territories in Indonesian archipelago based on the regional autonomy system has a high degree of difficulty. It is because the given number of regencies/municipals in this state are as many as 479 and out of these 324 regencies/municipals have coastal areas. Each part of Indonesian coastal areas has different characteristics from other areas, so it has different ways to manage them. If that is so, the policies and institutional instruments are not the same. As a result, it will also have effect on data provision management information of coastal and marine areas along with each region which will be varied as well.

The national marine resources are not only managed by local but also managed by sector. Based on the identifications result, it is shown that at least 12 ministries are involved in the resource marine management in Indonesia. These ministries (for instance, the Ministry of Maritime Affairs and Fisheries, the Ministry of Energy and Mineral Resources, the Ministry of Environment, the Ministry of Transportation, the Ministry of Culture and Tourism, the Ministry of Commerce, the Ministry of Industry, the Ministry of Home Affairs, the Ministry of National Development Planning, the Ministry of Agriculture, the Ministry of Defence and Security, and the Ministry of Public Works) have their own policies and regulation systems (not integrated among each other's). Their points of views and various management objectives are undirected (without clear management and shared planning together among the ministries) causing the exploitation activities and marine areas functions (marine boundaries) are limited and overlapped among them. For example, marine parcels uses for fish farming overlapping with shipping lanes owing to revenues disruption in fish farming sector, or other example, the areas used for fishing overlapping with Navy space areas, as a result the revenue disruption for fishermen in getting fishes. These uncertain conditions of marine boundary activities in the areas make some barriers among marine economic activities such as fisheries, aquacultures, biotechnology industries, marine tourisms, marine transportations, conservations, explorations and exploitation sectors (Fraser et al. 2003; Waljiyanto 1998).

Based on cultural aspect, the Unitary State of the Republic of Indonesia as an archipelagic state has multicultural ethnicities. There are 10,640 (more than 14%) villages in Indonesia (out of 67,249 villages) as coastal villages with an area of 35,949,021 hectares or 19% of the total area in Indonesian villages. Approximately 92% of coastal villages in Eastern Indonesia are traditional villages who practiced natural resource management based on their own local custom. It is where the implementation of marine management in Eastern Indonesia is frequently confronted to the existence of customary marine management (customary marine law). The problems that occurred are the customary marine exclusivity territorial boundaries which are determined based on their own implementation of custom regulations in their regions. For example, *Haruku* Island (the *Rural Haruku*) in Maluku province has its own customary boundaries among villages determined based on an imaginary line drawn from the land boundary straight out to sea, meanwhile to determine the limits between the boundaries of the village and communal marine public property or common property is by drawing imaginary line between the shallow and the deep seas (Zaenudin 2008). The impacts of customary marine delimitation are conflicting customary boundaries among the traditional villages, the rural sea customary boundary against outsiders and customary marine boundary against local marine authority boundaries. Indigenous and local wisdom issues cannot be avoided being the part of cultural system in Indonesia (Hernandi et al. 2012).

From the above discussion, it can be concluded that the implementation of marine management in Indonesia is strongly influenced by the regional autonomy system, the sectorial system and custom system; this condition is one of the implications as the consequences as an archipelagic state. Discussing on marine management resources in Indonesia, the first step in this research is to perform a comparison of marine resource management through marine cadastre definitions that exist in non-island developed countries such as Australia, Canada and United States. Marine cadastre can be regarded as the application of set principles of cadastre in sea area. Generally, the purpose of marine cadastre is to administer marine space and marine resources including all interests, rights, restrictions and responsibilities that exist in the marine territories (Vaez 2009).

Robertson et al. (1999) defined the marine cadastre as follows "Marine cadastre is a system to enable the boundaries of maritime rights and interests to be recorded, spatially managed and physically defined in relationship to the boundaries of other neighbouring or underlying rights and interests". Then Binns (2004) defined that "Marine cadastre is a spatial boundary management tool which describes, visualizes and realizes legally defined boundaries and associated rights, restrictions, and responsibilities in the marine environment". Marine cadastre in Australia used to create Australian's Marine Management System which then used to regulate some activities such as oil and gas sector, fisheries, aquaculture, shipping, conservation, marine heritage, cable and pipelines, and coastal zone. Australian marine cadastre concept has been implemented in several states such as Queensland and Victoria.

Nichols et al. (2000) conducted good governance of Canada's Oceans to resolve boundary issues as a first step to realize the effective marine management and fair. They defined marine cadastre as follows "Marine cadastre is a marine information system, encompassing both the nature and a spatial extent of the interests and property rights, with respect to ownership and various rights and responsibilities in the marine jurisdiction".

United States Department of Communication – National Oceanic and Atmospheric Administration (NOAA) formulated the marine definition as follows: "The U. S Marine cadastre is an information system, encompassing both nature and spatial extent of interests in property, value and use of marine areas. Marine or maritime boundaries share a common element with their land-based counterparts in that, in order to map a boundary, one must adequately interpret the relevant law and its spatial context. Marine boundaries are delimited, not demarcated, and generally there is no physical evidence of the boundary".

How about Indonesia as the largest archipelagic state in the world? The new concept of marine cadastre in Indonesia has been known for still being introduced, in addition, since the long past the development in Indonesia is largely priority to land area, whereas as the archipelagic state, Indonesia has a wider marine area than the land area. Nevertheless there have been some researches in the field of management and coastal and marine spatial planning which relate to the element of cadastre, the right, restriction, and responsibility based on exploitation and utilization of oceanic spaces (Hasymi 2008). However, from the previous marine cadastre studies that have been conducted in Indonesia, they mostly used marine cadastre definition of non-island countries such as Australia, Canada and United States of America. The existing marine cadastre definitions internationally have been recognized by several countries around the globe (Ng'ang'a et al. 2001; Mårtensson 2004; Ives and Yan 2004; Sook and shin 2010; Srebo et al. 2010).

Giving the definition is very significant in a research or study. The definition is a phrase that expresses the meaning, description, or the main characteristics of the person, object, process or activity. The main role of the definition is to provide limits

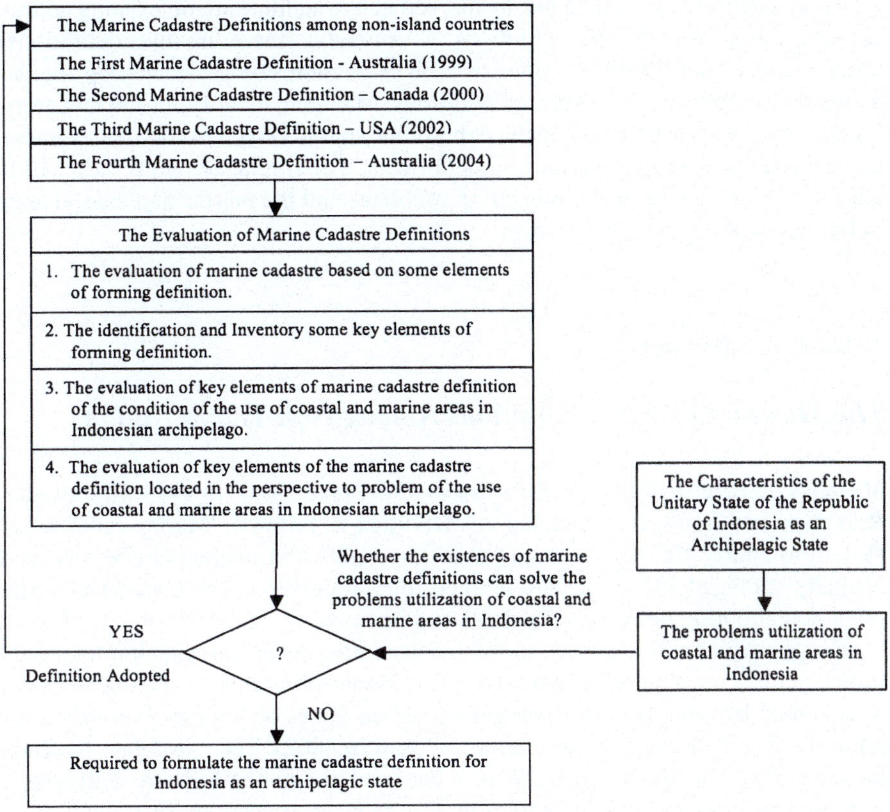

Fig. 22.1 Research methodology

(meaning), the scoop formulation and concept characteristics that become the subject of research. The importance of marine cadastre definitions in this study are the existing definitions of marine cadastre to be used as an approach in building the concept model and marine management in Indonesian archipelago. This study will evaluate the definitions of existing marine cadastre in several non-island countries namely Australia, Canada and United States of America placed in the Indonesian perspective as an archipelagic State.

Methodology

To understand the concept and definition of the marine cadastre in each country, namely Australia, Canada and United States of America, the researchers conducted a literature study from various references to know the important things behind the marine cadastre definitions in each of these non-island nations (Fig. 22.1).

An evaluation was conducted to the respective definitions of existing marine cadastre today, namely the definitions of Australia (there are two definitions), Canada and United States of America. The evaluation was conducted by the elements forming the marine cadastre definitions. And then, identifying and inventorying the key elements that formed the definitions, so it is known in common elements that exist in the four marine cadastre definitions. The similar elements then will be placed into the condition and utilization problems and the marine and coastal areas in Indonesia as an archipelagic state.

Literature Reviews

The Marine Cadastre Definitions Among Non-Island Nations

Marine Cadastre Definition from Australia (1999 and 2004) In 1999, Hoogsteden, Robertson and Benwell formulated the definition as follows: "Marine cadastre is a system to enable the boundaries of maritime rights and interests to be recorded, spatially managed and physically defined in relationship to the boundaries of the other neighbouring or underlying rights and interests".

The above definition is close to the land cadastre definition, which refers to the limits, in this case a maritime boundary (boundaries of maritime). This definition is widely used by other countries, including United States of America (before formulating its own definition of the marine cadastre in 2002). The Australian Research Council (ARC) Marine Cadastre Project has used the marine cadastre definition as a starting point in the development of marine cadastre concept in Australia.

In 2002, Melbourne University in Australia made marine cadastre concept diagram. The concept since then is widely used by various parties and researchers from various countries as a reference (Binns et al. 2004; Sesli and Uslu 2010; Yuwono 2006; Steudler 2004; Rais 2009).

Figure 22.2 shows that the marine cadastre should not be developed separate from the terrestrial environment. Most of the activities occurred in the coastal areas. This region connecting land and sea, is a public access point to the marine environment and is "spilling out" the whole impact of land activities that is wasted or flowing into the sea. Some activities such as Tourism and recreation, Marine protected area, Shipping, Heritage, Cables and pipelines, Aquaculture leases, Mineral and energy, Native title and Ocean waste disposal should be prepared by administrative boundaries and laws that regulate where and when these activities can take place. Rights, restrictions and responsibilities going along with these boundaries should also be noted.

Australian marine is managed by a number of organizations and institutions. Each of them is responsible for the collection, compilation and updating of spatial data related to the organizations or institutions interests (Wisayantono 2009). This condition causes the data becoming of various types and inconsistent, leading marine environment stakeholders difficult to find reliable data. As a consequence,

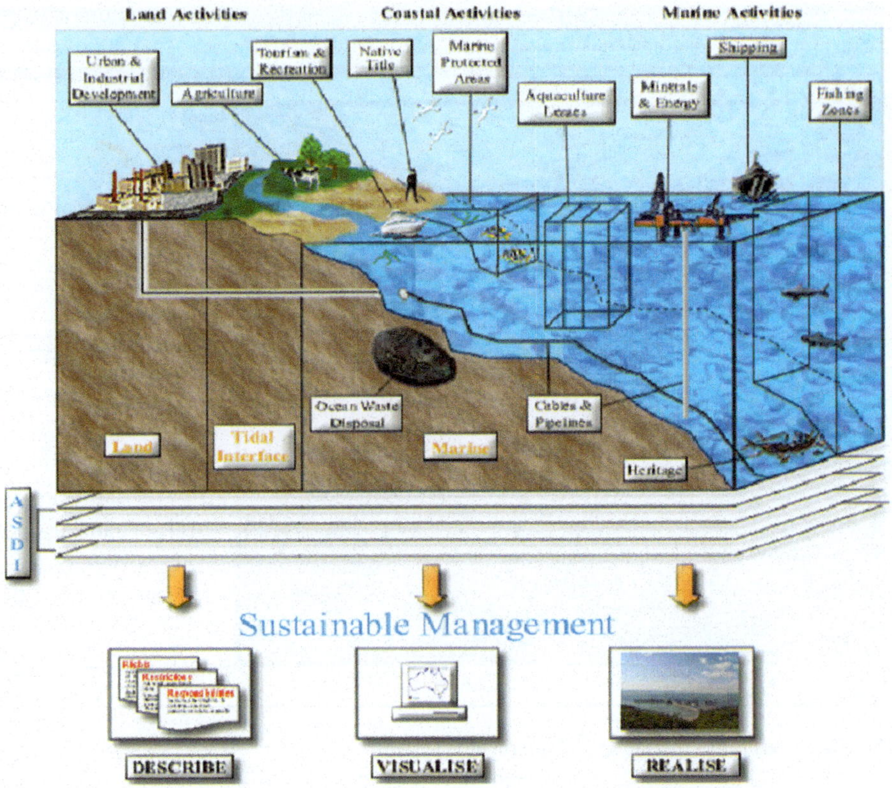

Fig. 22.2 The marine cadastre diagram concept in Australia (Binns 2004)

Australia Spatial Data Infrastructure (ASDI) is formed in the marine environment to support and facilitate the availability and spatial data reliability. This team provides basic realization of the marine integration and terrestrial environment for the sustainable management on natural resources at the whole Australian jurisdictions.

Then Binns (2004) formulated a marine cadastre definition as follows: "Marine cadastre is a spatial boundary management tool which describes, visualizes and realizes legally defined boundaries and associated rights, restrictions and responsibilities in the marine environment".

Binns (2004) gives the marine cadastre definition as more detailed explanation, using the phrase "Marine cadastre is a spatial boundary management tool". It is contrasted to the previous definition which used the phrase "Marine cadastre is a system" (Robertson et al. 1999) or "A marine cadastre is a marine information system" (Nichols et al. 2000) or "Marine cadastre is an information system". The purpose and elements in the marine cadastre definition also stated clearly, that it "describes, visualizes, and realizes legally defined boundaries and associated rights, restrictions and responsibilities in the marine environment".

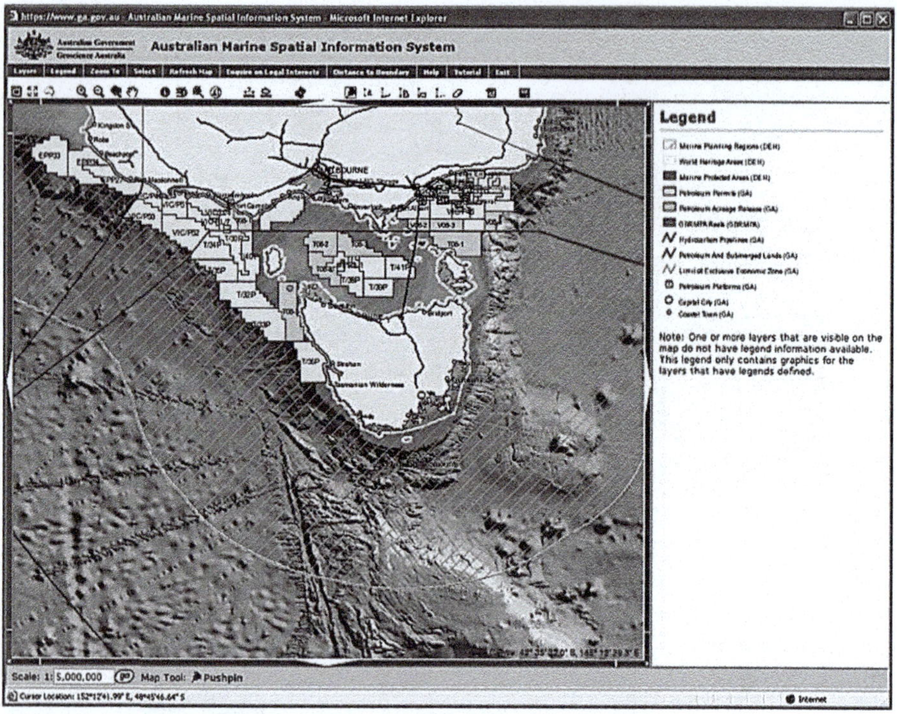

Fig. 22.3 Australian marine spatial information system

Binns definition (2004) serves as a marine cadastre concept development in Australia. The real results of marine cadastre concept in Australia are the user's ability and stakeholders to explain, visualize and realize the "spatial information in the marine environment". Marine cadastre describes the location and extent of rights, restrictions and responsibilities in the marine environment, including the limits and guidelines for the coastal management and marine planning, location, rights, restrictions and responsibilities shall then be visualized through updating spatial data continuously and accurately. The ability to describe and visualize the sea boundaries is an integrated and practical approach to the Australian sea management. Australian marine cadastral concept has been implemented in several states such as Queensland and Victoria (Williamson et al. 2010).

As the marine cadastre concept and development already exist, in 2010 Geoscience Australia, government agencies and private sectors build the Australian Marine Spatial Information System (AMSIS) (Fig. 22.3). It gives information on the Australian marine in holistic and integrated manner. By using this, each user can access a wealth of marine data including jurisdictional boundaries, mining, marine transport activities, fisheries, and other maritime activities.

As for management issues in Australian cadastre elements perspective are marine authority boundary issues among states (0–3 miles) to the federal marine (0–12 miles), and the problem boundaries in marine management activities across

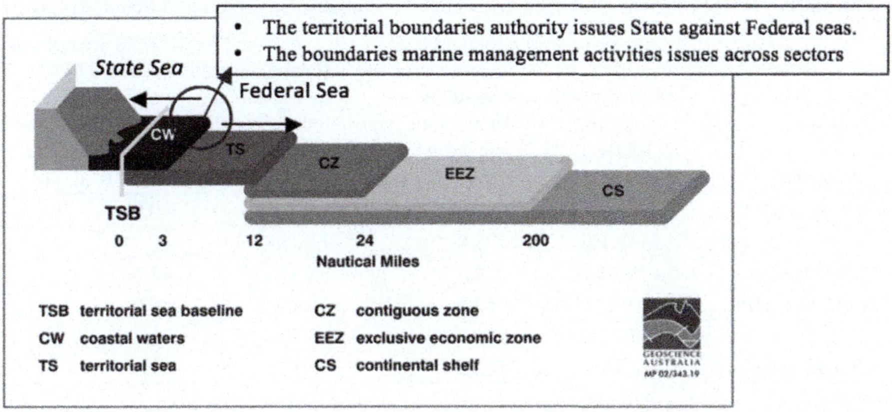

Fig. 22.4 The authority problem in the territorial sea utilization of Australian sea jurisdiction

sectors such as happened in the Great Barrier Reef Marine Park (GBRMP) region (Fig. 22.4). It is one of the largest marine parks in the world to protect coral and other marine biodiversity which is located in the Northwest region of the Queensland state to the federal marine areas.

Concerning to the customary marine management issues, Australia has the National Native Title Tribunal established under the Native Title Act i.e. a board that has an authority to resolve issues and establish indigenous customary rights and interests in Australian land and waters through the Australian federal court decisions.

The existence of indigenous territories in the land and sea areas are recognized and protected by the Australian government. The conflicts arising from the indigenous territories utilization by private or state government as well as the federal government can be minimized by an agreement in the form of land and customary sea use agreement (Table 22.1).

Marine Cadastre Definition from Canada (2000)

Prior to 2000, Canada has defined Multipurpose Cadastre concept that linked law to fiscal cadastre and other spatial information to obtain broader spatial information related to the geological and geophysical data, soils, vegetation, wildlife, hydrology, climate, pollution, health and safety, industry and employment, transport, water and sewerage, gas, electricity, telephone, and emergency services. The implementation of multipurpose cadastre requires cooperation and integration among related institutions. It is realized by the establishment of Spatial Data Infrastructure (SDI) and Marine Geospatial Data Infrastructure as the part of Canadian Geospatial Data Infrastructure which regulated the need for laws, regulations and policies, governance structure, personal arrangements, data management and organization.

Table 22.1 The list of territory indigenous land and customary sea used (up to March 31, 2013)

Jurisdiction	Jurisdiction land area[a] (1000s sq. km)	Total area of land covered by registered indigenous land use agreements (1000s sq. km)	% land covered by registered indigenous land use agreements (1000s sq. km)	Total area of sea covered by registered indigenous land use agreements (1000s sq. km)
New South Wals	800.6	9.8	1.2%	0.0
Northen Territory	1349.1	178.6	13.2%	–
Queensland	1730.6	587.5	33.9%	4.5
South Australia	983.5	387.5	39.4%	0.0
Victoria	227.4	83.4	36.7%	0.3
Western Australia	2529.9	508.0	20.1%	1.1
Tasmania	68.4	–	0.0%	–
Australian Capital Territory	2.4	–	0.0%	–
Commonwealth	n/a	–	n/a	–
Total	7692.0	1754.8	22.8%	5.9

[a]Geospatial Services, National Native Title Tribunal – 11 April 2013

The three dimensional marine cadastre concept are used to represent the rights and interest as the whole that occurred in the ocean; making it easier to determine the rights and interests that exist at the water surface, water column, and subsoil of the bed. It includes the information related to legal tax, environmental and others. The information is then used to determine the laws, regulations and responsibilities of stakeholders.

In 2000, the Canadian government conducted a research under the title: "*Good* Governance *of Canada's Ocean: The Use, Value and Potential of Marine Boundary Data*". The marine research focus is to consider the maritime boundaries issues and the marine boundaries activities. The purpose of this activity is to solve the boundary problem as an initial step to realize the effective marine management and fair. In order to achieve the objective, it requires an understanding of the interaction and spatial relationships among different boundaries types in the sea such as: the boundaries of private and public ownership, the cities, regencies, provinces and territorial jurisdiction and administrative boundaries, national and international, regional environmental protection, military boundaries, pipelines and subsea cables, and limit other activities. The maritime boundaries were identified and subsequently given an evaluation and visualization of each boundary, and then these boundaries serve as a conceptual framework for marine management in Canada.

At the same time (2000), Nichols, Monahan and Sutherland gave Marine Cadastre definition as follows: "A marine cadastre is a marine information system, encompassing both the nature and spatial extent of the interests and property rights, with respect to ownership and various rights and responsibilities in the marine jurisdiction" (Fig. 22.5).

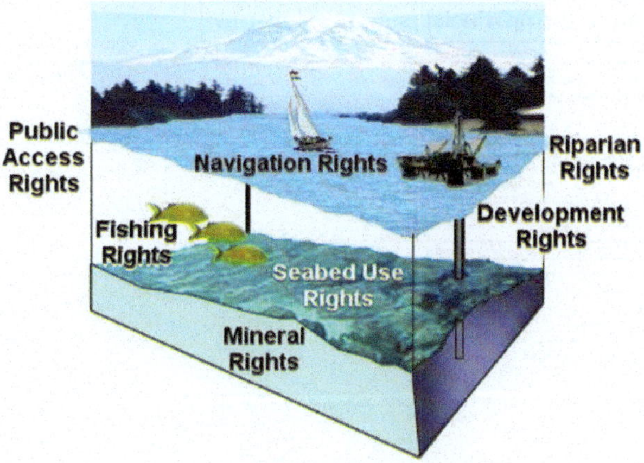

Fig. 22.5 The foundation of the marine cadastre concept in Canada (Nichols et al. 2000)

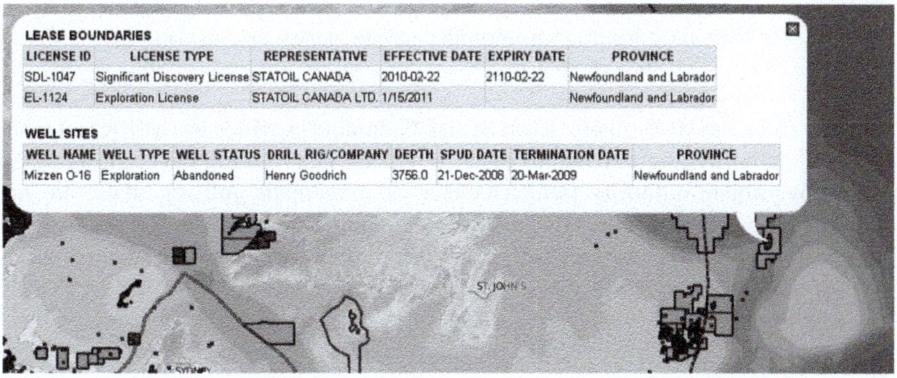

Fig. 22.6 The COIN Atlantic of marine cadastre in Canada

The above definition has varied little understanding of the Good Governance of Canada's Oceans events which focused on the boundaries problem. Nichols et al. introduced the concept of rights and responsibilities into the sea jurisdiction. This marine cadastre definition is widely used as a reference in some countries. In 2008, an advanced development of marine cadastre called *The Coastal and Ocean Information Network for Atlantic Canada* (COIN Atlantic) established a system of marine cadastre application by using St. Margaret's Bay as an area of study (Fig. 22.6).

The main issue of the ocean management in Canada has the similar issues in Australia, which is the authority on sea boundary issue between state and federal governments; there is no single institution (entity) that manages offshore rights and limits, and the right of indigenous people (native) (Fig. 22.7).

In Canadian sea territorial (0–12 miles) are the province authority and federal boundaries are not determined by nautical miles distance as well as in Australian

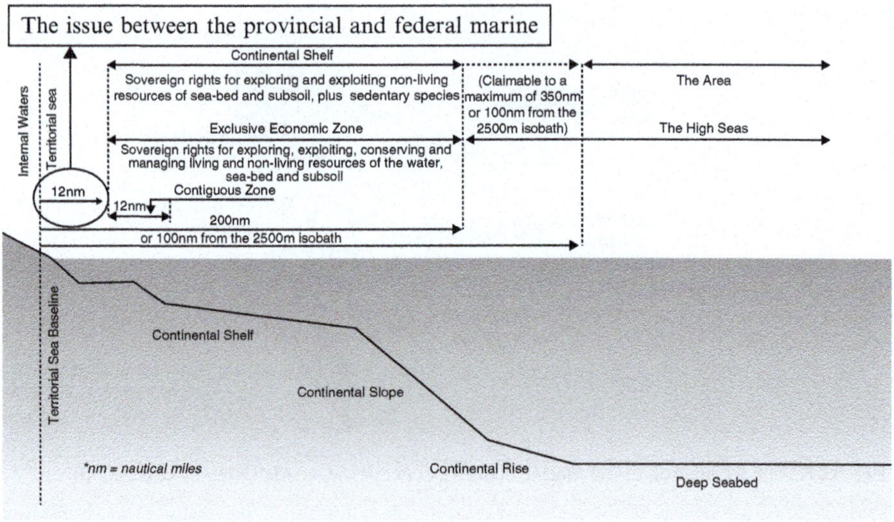

Fig. 22.7 The territorial sea utilization authority problems in the Canadian marine jurisdiction

territory (Sea State 0–3 miles). Even British Columbia province has a different management authority as compared to other provinces, i.e., to bring the sea to limit the local government authority (which consist of municipals and regionals) are not determined by nautical miles distance. This condition often causes the problems in the territorial sea boundary Canadian authorities. Conflicts between private parties against customary rights in the sea utilization, especially fisheries has frequently occurred. Even though the use of customary marine has no exclusivity region, customary law has continuously to be recognized and the existence are protected by the Canadian government.

The Marine Cadastre Definition of United States of America (2002)

In 2002, United States Department of Communication (U.S DOC) – National Oceanic and Atmospheric Administration (NOAA) gave marine cadastre definition as follows: "The U. S Marine Cadastre is an information system, encompassing both nature and spatial extent of interests in property, value and use of marine areas. Marine or maritime boundaries share a common element with their land-based counterparts in that, in order to map a boundary, one must adequately interpret the relevant law and its spatial context. Marine boundaries are delimited, not demarcated, and generally there is no physical evidence of the boundary". This definition is applied in one of instances marine cadastre map (Fig. 22.8).

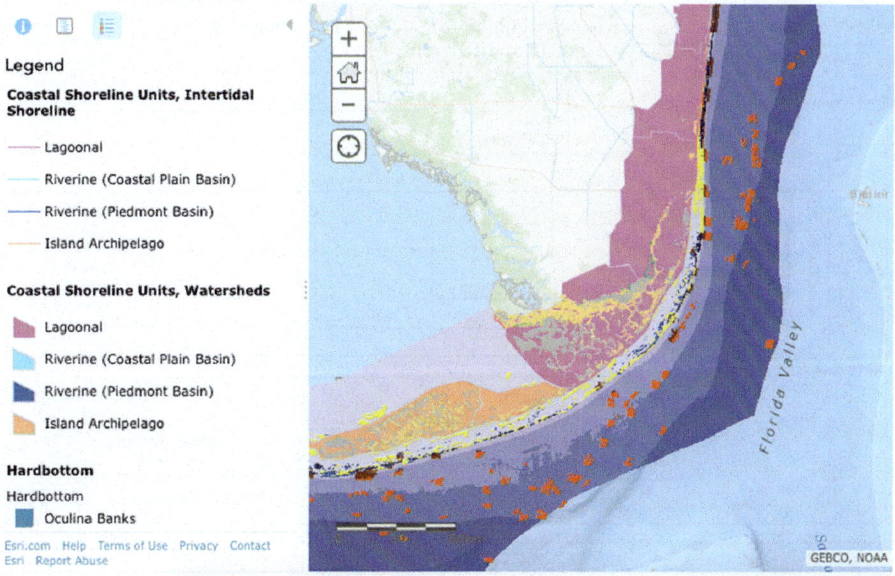

Fig. 22.8 The marine cadastre map in Florida Sanctuary

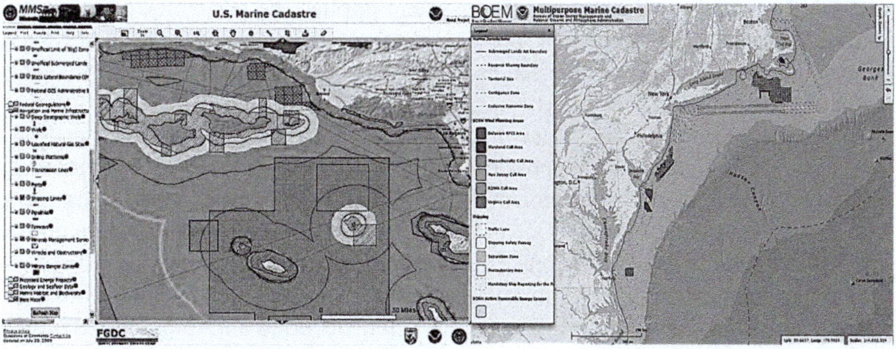

Fig. 22.9 The U.S. marine cadastre

The marine cadastre definition in United States is formulated in a broader perspective, so it does not link the elements of rights, restrictions and responsibilities. It more emphasizes on the determination of the maritime boundaries. American Marine cadastre is largely manifested in the form of *Geographic Information System* (GIS) in *web*-based, based on the data sources authorization that integrate legal, physical data, the data ecology and culture (Fig. 22.9).

Federal Geographic Data Committee (FGDC) and Marine Boundary Working Group (MBWG) developed a web-based marine cadastre. The information displays Jurisdictional boundaries and limits, Federal georegulations, Navigation and marine infrastructure, Proposed energy projects, Geology and seafloor data, Marine habitat

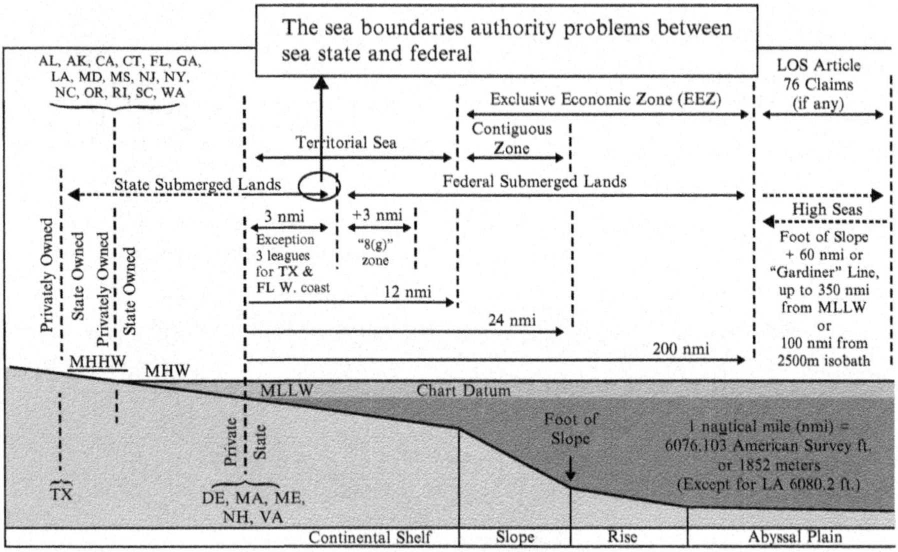

Fig. 22.10 The problems authority of the territorial sea utilization in American jurisdiction sea (Faridha 2010)

and Biodiversity and Base maps. This system applies the principles of integrated marine cadastre for planning by providing an easy and comprehensive access in relevant data information in the sea as an integrated approach to resource management in coastal and marine areas.

Next, the marine cadastre concept developed into a multipurpose marine cadastre by National Oceanic and Atmospheric Administration (NOAA) and Coastal Services Centre and the Mineral Management Service (MMS) which then provide information about Alternative energy, Ocean planning, Habitat conservation, Human use/Recreation, Marine protected areas (MPA), and Aquaculture.

Marine management issues in U. S based on cadastre elements perspective are marine authority boundary problems that often occurred in American territorial zone (0–12 miles) in which there is a sea state authority (0–3 mile) except Texas and the Florida Bay (0–9 miles). All marine waters outside the 3 miles (or 9 miles) are federal and state waters do not have jurisdiction in federal waters. This condition often leads to a marine management authority conflict between the state and the federal governments (Fig. 22.10).

National Oceanic and Atmospheric Administration (NOAA) and other federal agencies have used the Coastal Zone Management Act (CZMA) as a means to overcome the marine management authority problems along with the state and federal programmes to implement coastal and marine management-owned state located in federal waters. CZMA gives the opportunity for states to be able to enter the programme plan into their coastal and marine management in federal waters. If it is approved, the federal government will give license to the states during conducted

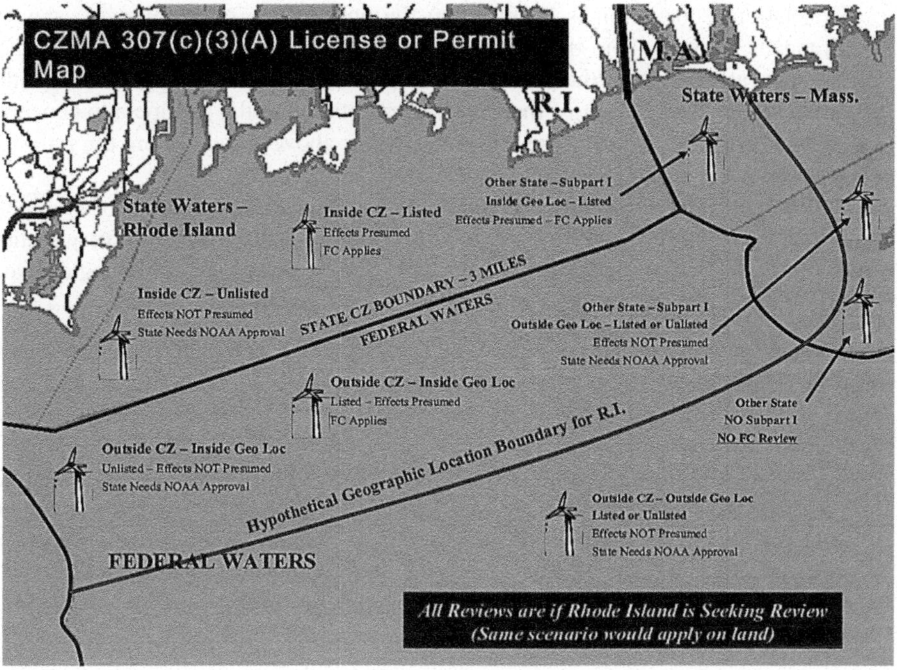

Fig. 22.11 The shared sea management between Rhode Island State and U. S Federal Government

activities in federal waters. For example: the marine resources management for the utilization of alternative energy in the Rhode Island State (Fig. 22.11).

The marine use conflicts among the state or federal government or private parties against indigenous sea rights can be solved by constitutional court decisions; it is where customary law often must give away if faced to the sea use activities for the benefits of the country. There is no customary marine exclusivity area in United States of America.

The Characteristics of Indonesia as an Archipelagic State

The term of Archipelagic State is the decision result of the United Nations Convention on the International Law of Sea in 1982 (United Nations Convention on the Law of the Sea/UNCLOS the 2nd). The concept of an archipelago, set forth in Article 46 (b), is described as a chain of islands, including the parts of island, including waters and other natural beings do to each other so closely that the islands, waters and other natural form is a unitary of geography, economics and principle politics, or which is historically considered so (Abdulharis et al. 2008).

The Nation Conception is based upon the archipelago conception of islands which means the sea where there are many islands. In the archipelago term of the sea or water ratio is greater than the mainland (islands), but both of them are considered as a whole. Thus, the most important interpretation in the archipelago concept is unitary between the sea and the land (and the air above them), where the ratio of the sea area is greater than the ratio of land area.

The geographical conception becomes the basic for the conception of archipelagic state that consisted of one or more archipelagoes, meanwhile not every archipelago classified into archipelagic state or not to be treated as an archipelagic state. Therefore, there are three types of the archipelagoes, namely:

1. *Coastal Archipelago*, which is located along the coast and is generally adjacent to the seashore. It is basically completed by the Geneva Conference in 1958 on the seashore area that allows states draw the basic lines of the archipelago which lies adjacent along the coast.
2. *Mid-ocean Archipelago*, which is located in the middle of the sea off the coast of a country. Basically the International Law Conception concerning *archipelago* comes from the term *archipelago-archipelago*, which generally is an islands cluster that is compact and the distance between the islands is not so wide.
3. *Archipelagic state*, which is an islands cluster or several clusters of islands, became an independent state.

Based on the above explanations, the State island (Archipelagic State) is a country that is supposed to be composed of one or more islands and may include other islands. An archipelago will be considered as a single entity, so that the waters around, surrounding, and connecting the islands (apart from the vast and different dimensions) are the part of internal waters of the state. Forming an island country is determined based upon the archipelagic straight baseline determination and the other baselines as long as not contrary to the 1982 Convention (Article 5, 7 (1) and 47 (1)).

None of the countries which are geographically islands can establish by itself as an archipelagic state legally. At least, there are five sovereign states that obtained an approval in the United Nations Convention on the Law of the Sea (UNCLOS) and qualify as the archipelagic nations: Indonesia, the Philippines, Papua New Guinea, Fuji and the Bahamas. These five Islands countries have to respect the agreement with other countries and should recognize the traditional fishing rights and other legitimate activities of neighbouring countries that directly border in certain areas in the same waters of the archipelago. The terms and conditions for the realization of the rights and further activities are regulated by bilateral agreements between the countries.

In *Djuanda* Declaration, Indonesian government confirms an Archipelagic State principle which viewed at the sea and land areas as a unified and whole in accordance with motherland philosophy (Falah 2010). The Article 46 provisions of the Convention on the Law of the Sea in 1982, if it is linked to the reality in Indonesian characteristics as an archipelagic state, shows that the islands of Unitary State of the Republic of Indonesia are provincial area with some characteristics of:

1. *Coastal archipelago*, it means provincial areas which have islands along with the main seashore, for example: the Jakarta Province, North Sumatra Province, West Sumatra Province, Riau Province, the Papua Province, West Papua Province and other provinces.
2. *Mid-ocean archipelago*, it means provincial areas which have islands in the middle of the sea as the part of the regional territory, for example: East Java Province, South Sulawesi Province, Southeast Sulawesi Province and other provinces.
3. *Archipelagic province* or it is called the Region Islands province which areas have the chain shaped islands; such as Riau Islands Province, Bangka Belitung Islands Province, West Nusa Tenggara Province, East Nusa Tenggara Province, North Sulawesi Province, North Maluku Province and Maluku Islands Province.

Furthermore, as the essence of archipelagic state are a 'unified as a whole territory' (land space, sea space and air space) where boundaries are determined by the sea and inside it there are islands and an islands cluster. The naming of the Unitary State of the Republic of Indonesia as the country still needs to be added to the islands of the archipelago characterized; it is in accordance to what is written in the 1945 constitution, Article 25. It is characterized as archipelago or commonly known as the Archipelago Concept which is archipelagic islands as a single political will, economics, social, cultural, defense and security (Sulasdi 2010).

Based on the above explanation, it is stated that the Unitary State of the Republic of Indonesia is an archipelagic state and has a functional model as follows:

1. Sovereignty = internal waters, archipelagic waters, territorial sea, air space over the territorial sea, the air space over the archipelagic waters, the air space above the waters, seabed, underwater land and natural resources
2. Spatial geographic = land, coastal areas, oceanic areas islands, islands.
3. Governance = central government, provincial governments, local government, municipal government, and regional government
4. Multicultural nationality = tribe, language, religion, culture or customs
5. Natural resources in biodiversity = renewable resources, non-renewable resources, space and region resources, and geographical location
6. Disaster risks = geography of the specific location, types of disasters, the impact or risk mitigation.
7. Defense and security = airspace, land space, sea areas and islands

Analysis and Discussion

The Evaluation of Marine Cadastre Definitions Based on Elements Definition Shaped

Based on the identification and inventory of the key elements of the four marine cadastre definitions above, the next step is to perform a classification based on the similarity of the key elements in order to obtain the nine results of key elements of the marine cadastre as follows: Right, Restriction, Responsibility, Interest, Marine boundaries, Geodetic reference system, Use of marine areas, Marine jurisdiction, and Institution.

The Key Elements Evaluation of Marine Cadastre Definitions

The key elements of the marine cadastre placed in utilization conditions of coastal and marine areas in the waters of Strait Madura East Java Province represents the western part of Indonesia and the waters in Maluku Province represents the eastern part of Indonesia. Organizing the utilization of coastal and marine areas in its western part has a higher level of complexity rather than in its eastern part; it is proven by the number of sectors that get involved in the implementation and utilization of marine diversity conflicts among the sectors and the regions, such as occurred in St. Madura East Java. In contrast to the conditions and problems in the western part of Indonesia, the implementation of sea usage in eastern Indonesia is often confronted to the existence of customary sea management. Maluku Province has an area of 527,191 km^2 and vast seashore 54,185 km^2, which consisted of 559 cays islands. Maluku Province as the largest archipelago in Indonesia has found the existence of indigenous people who use the local customary based archipelagic waters. In Maluku Province, fisheries management practices based on the local wisdom has been long known as *petuanan* and *sasi* (Table 22.2).

The Identification and Inventory of Key Elements in Forming Definition (Table 22.3)

The results evaluation of Table 22.4 is that some of the key elements in the marine cadastre countries Australia, Canada and United States of America have the same elements in the sea use in Indonesia as an archipelagic state (Tables 22.5 and 22.6).

The above results should be completed by the disaster mitigation elements. The Indonesian geographical position is located in the track of fire ring to make this country as a prone to catastrophic volcanic eruptions and earthquakes, coupled with the tsunami disaster and abrasion. The disasters impact that might occur is the loss of rights, restriction and responsibility of a person or parcel on the coastal use and marine resources, so it is needed for reconstruction activities, boundaries utilization and re-recording of the object and the subject of utilization.

Table 22.2 The evaluation of marine cadastre definitions based on the elements forming the definition

No	Country	Marine cadastre definition	Elements of forming definition	Explanation of definition
1.	Australia	Marine cadastre is a system to enable the boundaries of maritime rights and interests to be recorded, spatially managed and physically defined in relationship to the boundaries of other neighbouring or underlying rights and interests	1. System	Marine cadastre is defined as a system.
			2. Boundaries of maritime	
			3. Rights	This definition contains the concept of local autonomy in natural resource management perspective.
			4. Interests	
			5. Recorded	This definition is more focused on the recording, the definition, management and the relationship among the limits in the sea.
			6. Spatially managed	
			7. Physically defined	
			8. Relationship to the boundaries	The only one (from four definitions) that does not mention the legal elements of the sea.
			9. Neighbouring	
			10. Underlying	
2.	Canada	A marine cadastre is a marine information system, encompassing both the nature and spatial extent of the interests and property rights, with respect to ownership, various rights and responsibilities in the marine jurisdiction (Nicolas et al. 2000).	1. Marine information system	Marine cadastre is defined as a system of marine information.
			2. Nature	This definition introduces the concept of the rights associated with ownership of the sea.
			3. Spatial extent	
			4. Interests	
			5. Property rights	
			6. Ownership	
			7. Various rights	
			8. Responsibilities	
			9. Marine jurisdiction	

(continued)

Table 22.2 (continued)

No	Country	Marine cadastre definition	Elements of forming definition	Explanation of definition
3.	United States of America	The U.S. Marine Cadastre is an information system, encompassing both nature and spatial extent of interests in property, value and use of marine areas. Marine or maritime boundaries share a common element with their land-based counterparts in that, in order to map a boundary, one must adequately interpret the relevant law and its spatial context. Marine boundaries are delimited, not demarcated, and generally there is no physical evidence of the boundary.	1. Information system	Marine cadastre is still defined as a system. This definition has little resemblance to the definition of Canada, but it is more focused on defining of the marine boundaries.
			2. Nature	
			3. Spatial extent	
			4. Interests	
			5. Property	This definition contains the concept of the rule of law to resolve conflicts at sea activities.
			6. Value	
			7. Use marine areas	This definition does not completely mention the cadastre elements (rights, restriction and responsibilities).
			8. Marine boundaries	
			9. A common element	
			10. Land	
			11. Based counterparts	
			12. Map a boundary	
			13. Adequately interpret	
			14. Relevant law	
			15. Spatial context	
			16. Delimited	
			17. Not demarcated	
			18. No physical evidence	
4.	Australia	Marine cadastre is a spatial boundary management tool which describes, visualizes and realizes legally defined boundaries and associated rights, restrictions and responsibilities in marine environment (Binns 2004).	1. Spatial boundary	Marine cadastre can be called as a administration tool. This definition contains the concept of cross-sectorial activities in the national perspective.
			2. Management	
			3. Tool	
			4. Describes	
			5. Visualizes	
			6. Realizes	
			7. Legally defined	
			8. Boundaries	
			9. Associated	
			10. Rights	
			11. Restrictions	
			12. Responsibilities	
			13. Marine environment	

Table 22.3 The identification and inventory of the key elements of the marine cadastre definition

The first definition (Australia, 1999)	The second definition (Canada, 2000)	The third definition (United States of America, 2002)	The fourth definition (Australia, 2004)
1. System	1. Marine information system	1. Information system	1. Management tool
2. Marine boundaries:	2.	2. a. Marine boundaries	2. Boundaries
a. The sea boundary adjacent		b. Boundary map	
b. The sea boundary underlying rights or interests		c. Boundary mark	
3. Rights	3. a. Property rights	3.	3. Rights
	b. Ownership rights		
	c. Various rights		
4.	4. Responsibilities	4.	4. Responsibilities
5.	5.	5.	5. Restrictions
6. Interests	6. Interests	6. Interests	6.
7. Recorded	7.	7.	7. a. Describes
			b. Visualizes
			c. Realizes
8. Spatially and physically managed	8. Nature and spatial extents	8. Nature and spatial extents	8. Spatial boundary
9.	9. Marine jurisdiction areas	9. Relevant laws	9. Legally
10.	10.	10. Value and use of marine areas	10. Marine environment

Conclusions and Recommendation

Conclusions

Australian marine cadastre definition builds upon the sea jurisdiction which provides for the authority of state marine use (three miles) and the federal sea. In Indonesian point of view as an archipelagic state, the definition formulated by Hoogsteden, Robertson and Benweel in 1999 contains the concept of local autonomy in natural resource management perspective. While the other marine cadastre definition formulated by Binns in 2004 contains the cross-sectorial concept activities in the national view.

Canadian marine cadastre definition builds upon the sea jurisdiction and is strongly influenced by the authority of sea utilization in federal, province, regency and municipal, even though it does not provide the authority of sea boundary determined by nautical distance. In Indonesian archipelago point of view, this definition

Table 22.4 The evaluation of the key elements of marine cadastre definition to the condition use of coastal and marine areas in Indonesia as an archipelagic state

The key elements of marine cadastre definitions	The marine cadastre definition in non-island countries				The Unitary State of the Republic of Indonesia as an archipelagic state	
	A 1st definition Australia, 1999	B 2nd definition Canada, 2000	C 3rd definition USA, 2002	D 4th definition Australia, 2004	West regional section (Strait Madura, East Java)	East regional section (Waters in Maluku Province)
1. Right	1. Oil and gas development rights.	1. Public access rights	1. Public access rights	1. Oil and gas development rights	-1A1, 1D1 -1A2, 1D2	-1A3, 1D3
	2. Traditional fishing rights.	2. Navigation rights	2. Navigation rights	2. Traditional fishing rights	-1A4, 1D4 -1B1, 1C1	
	3. Aboriginal rights	3. Riparian rights	3. Riparian rights	3. Aboriginal rights	-1B2, 1C2 -1B4, IC4 -1B6, 1C6	
	4. Coastal property rights (including riparian rights): rights for public navigation, recreation and access	4. Fishing rights	4. Fishing rights	4. Coastal property rights (including riparian rights): rights for public navigation, recreation and access	-1B7, 1C7	
		5. Development rights	5. Development rights			
		6. Mineral rights	6. Mineral rights			
		7. Seabed use rights	7. Seabed use rights			
2. Restriction	Based on the federal waters, state waters, and concerning native rights	Based on the federal waters, provincial waters, local government waters and concerning native rights	Based on the federal waters and state waters	Based on the federal waters, state waters, and concerning native rights	2B	2B

3. Responsibility	Under federal law, the states and customs	Under federal law, provincial, local government and native rights	Under federal law and states law	Under federal law and states law	3B	3B
4. Interests	1. Shipping lanes	1. Shipping navigation	1. Alternative energy	1. Tourism and recreation	-4A1, 4B1, 4D3	-4D8
	2. Geophysical exploration.	2. Fishing	2. Ocean planning	2. Marine protected area	-4A3, 4B3, 4D7	
	3. Oil and gas extraction	3. Minerals and energy	3. Habitat conservation	3. Shipping	-4A4	
					-4A5, 4B2	
	4. Defense	4. Development	4. Human use/recreation	4. Heritage	-4A6, 4C5, 4C6, 4D2, 4D6	
	5. Fisheries		5. Marine protected area	5. Cable and pipelines		
	6. Conservation		6. Aquaculture	6. Aquaculture leases	-4D5	
				7. Minerals and energy		
				8. Native title		
				9. Ocean waste disposal		
5. Marine boundaries	1. Boundary jurisdiction:	1. Boundary jurisdiction:	1. Boundary jurisdiction:	1. Boundary jurisdiction:	-5A1a, 5B1a, 5C1a, 5D1a	-5B1b
	a. Federal: territorial Sea, Contiguous Zone, Exclusive Economic Zone, Continental shelf	a. Federal: Territorial Sea, Contiguous Zone, Exclusive Economic Zone, Continental Shelf	a. Federal: Territorial Sea, Contiguous Zone, Exclusive Economic Zone, Continental Shelf	a. Federal: Territorial Sea, Contiguous Zone, Exclusive Economic Zone, Continental Shelf	-5B1b	

(continued)

Table 22.4 (continued)

The key elements of marine cadastre definitions	The marine cadastre definition in non-island countries				The Unitary State of the Republic of Indonesia as an archipelagic state	
	A 1st definition Australia, 1999	B 2nd definition Canada, 2000	C 3rd definition USA, 2002	D 4th definition Australia, 2004	West regional section (Strait Madura, East Java)	East regional section (Waters in Maluku Province)
	b. State: coastal waters (3 miles)	b. Region: province, regency and municipal	b. State: seaward state (3 miles) and revenue sharing (6 miles)	b. State: coastal waters (3 miles)	- 5A2a, 5B2a,	-5A2e, 5D2e
					5C2f, 5D2a	
					-5A2b,	
	2. Boundary activities:	2. Boundary activities:	2. Boundary activities:	2. Boundary activities:	5D2b	
					-5A2c, 5D2c	
	a. Marine protected areas	a. Marine protected areas.	a. Navigation	a. Marine protected areas	-5A2d, 5B2c,	
	b. Fishing zones	b. Defense.	b. Submerged cultural resources	b. Fishing zones	5C2c, 5D2d	
	c. Petroleum exploration and mining	c. Cable and pipeline areas	c. Undersea cables	c. Petroleum exploration and mining	-5B2b, 5C2e	
					-5C2a	
	d. Cable and pipeline areas		d. Offshore aquaculture	d. Cable and pipeline areas		
	e. Native title claims		e. National security	e. Native title claims		
			f. Environmental protection			

6. Geodetic reference system	1. Geodetic and geocentric	1. Geodetic and geocentric coordinates system	1. Geodetic and geocentric coordinates system	1. Geodetic and geocentric coordinates system	-6A1, 6B1, 6C1, 6D1	
					-6A2, 6B2, 6D2	
	2. Projection system: UTM, grid map of Australia	2. Projection system: UTM	2. Projection system: UTM, Outer Continental Shelf (OCS) grid system.	2. Projection system: UTM, Map grid of Australia	-6A3, 6D3	
	3. Horizontal datum: GDA 94, WGS'84.	3. Horizontal datum: NAD27, NAD83, WGS'84	3. Horizontal datum: NAD27, NAD83, WGS'84	3. Horizontal datum: GDA'94, WGS'84		
	4. Vertical datum: Low Water Mark (LWM), Lowest Astronomical Tide (LAT).	4. Vertical datum: Lower Low Water Large Tide (LLWLT) and Lowest Normal Tide (LNT)	4. Vertical datum: MLLW	4. Vertical datum: Low Water Mark (LWM). Lowest Astronomical Tide (LAT)		
	5. Australia Spatial Data Infrastructure (ASDI)	5. Marine geospatial data infrastructure and Canadian geospatial data infrastructure	5. Cadastral data content standard for the national spatial data infra-structure: coastal and marine habitat classification standard	5. Australia Spatial Data Infrastructure (ASDI)		
7. Use of marine areas	1. Water surface	1. Water surface	1. Air column	1. Water surface	-7A1, 7B1, 7C1, 7D1	-7A1, 7B1, 7C1, 7D1
		2. Water column	2. Water surface	2. Water column		
		3. Seabed	3. Water column	3. Seabed	-7B2, 7C3, 7D2	-7B2, 7C3, 7D2
			4. Seabed		-7B3, 7C4, 7D3	-7B3, 7C4, 7D3
			5. Subsurface			

(continued)

Table 22.4 (continued)

The key elements of marine cadastre definitions	The marine cadastre definition in non-island countries				The Unitary State of the Republic of Indonesia as an archipelagic state	
	A 1st definition Australia, 1999	B 2nd definition Canada, 2000	C 3rd definition USA, 2002	D 4th definition Australia, 2004	West regional section (Strait Madura, East Java)	East regional section (Waters in Maluku Province)
8. Marine jurisdiction	1. UNCLOS	1. UNCLOS	1. UNCLOS	1. UNCLOS	-8A1, 8B1,	-8B6, 8A4, 8D4
	2. Federal	2. Federal	2. Federal	2. Federal	8C1, 8D1	
	3. State	3. Province	3. State	3. State	-8B4	
	4. Customs	4. Regency/ municipal		4. Customs	-8B5	
		5. Customs				
9. Institution	Many institutions that managed offshore rights and boundaries.	Many institutions that managed offshore rights and boundaries.	Many institutions get involved. The implementation is coordinated by NOAA.	Many institutions that managed offshore rights and boundaries.	9A, 9B, 9D	

How to read the tables:
1st digit (number) is obtained from the columns of the key elements of the marine cadastre definitions.
2nd digit (upper case) is obtained from the column definition in the marine cadastre in non-islands countries.
3rd digit (number) is obtained from the columns of the key elements one of the marine cadastre definitions.
4th digit (lower case) is obtained from the column of the key elements one of the marine cadastre definitions, is a sub-element.
Example: 5A1a = (5) Marine Boundaries – (A) 3rd definition USA, 2002 – (1). Jurisdiction boundary – (a) Federal: Territorial *sea, Contiguous Zone, Exclusive Economic Zone, Continental Shelf.*

Table 22.5 The key elements of marine cadastre definitions among non-islands countries placed in the perspective of utilization problems in the marine and coastal areas in Indonesia as an archipelagic state

The key elements of the marine cadastre definitions	Marine cadastre definition among non-islands countries				The Unitary State of the Republic of Indonesia as an archipelagic state	
	A 1st definition Australia, 1999	B 2nd definition Canada, 2000	C 3rd definition USA, 2002	D 4th definition Australia, 2004	The marine management problems	The evaluation results
1. Right	1. Oil and gas development rights	1. Public access rights	1. Public access rights	1. Oil and gas development rights	It is still a little kind of rights in the ocean based on the sectorial type's activities, especially since the *HP3* repealed in 2010	The four rights definitions can be used as input for new formulation rights in Indonesia.
	2. Traditional fishing rights	2. Navigation rights	2. Navigation rights	2. Traditional fishing rights		It is not exclusivity applicable area and customary marine tenure in Australia, Canada and USA.
	3. Aboriginal rights	3. Riparian rights	3. Riparian rights	3. Aboriginal rights		
	4. Coastal property rights (including riparian rights): rights for public navigation, recreation, and access	4. Fishing rights	4. Fishing rights	4. Coastal property rights (including riparian rights): rights for public navigation, recreation, and access		
		5. Development rights	5. Development rights			
		6. Mineral rights	6. Minerals rights			
		7. Seabed use rights	7. Seabed use rights			
2. Restriction	Based on federal waters, state waters, and concerning native rights	Based on federal waters, provincial waters, local government waters and concerning native rights	Based on federal waters and state waters	Based on federal waters, state waters and concerning native rights	Overlapping the country authorities among province, municipal/ regency, regional and customs	There are similarities to the Canadian concept. The difference is the determination of provincial and government waters distances are not nautical miles based.

(continued)

Table 22.5 (continued)

The key elements of the marine cadastre definitions	Marine cadastre definition among non-islands countries				The Unitary State of the Republic of Indonesia as an archipelagic state	
	A 1st definition Australia, 1999	B 2nd definition Canada, 2000	C 3rd definition USA, 2002	D 4th definition Australia, 2004	The marine management problems	The evaluation results
3. Responsibility	Under federal laws, states and customs	Under federal laws, provincial, local government, and native rights	Under federal laws and states laws	Under federal laws and states laws	Overlapping and conflicting among the responsibilities: country, province, municipal/regency, regional and customs	The four responsibilities definitions are only divided into the administration area and customary laws. They are not determined on the basis of existing sectorial legislation, so it could not overcome the existing problems in Indonesian seas.
4. Interests	1. Shipping lanes	1. Shipping navigation	1. Alternative energy	1. Tourism and recreation	Overlapping interests'marine management activities across the sectors	Some interests among four definitions can be placed in Indonesia based on the implementation by the state authority, province or municipal/regency. Native title in Australia is not the same as the indigenous sea rights in Indonesia.
	2. Geophysical exploration	2. Fishing.	2. Ocean planning	2. Marine protected area		
	3. Oil and gas extraction	3. Minerals and energy	3. Habitat conservation	3. Shipping		
	4. Defense	4. Development	4. Human use/ recreation	4. Heritage		
	5. Fisheries		5. Marine protected area	5. Cable and pipelines		
	6. Conservation		6. Aquaculture	6. Aquaculture leases		
				7. Mineral and energy		
				8. Native title		
				9. Ocean waste disposal		

5. Marine boundaries	1. Jurisdiction boundary:	1. Jurisdiction boundary:	1. Jurisdiction boundary:	1. Jurisdiction boundary:	Many municipals/ regencies have not been determined and confirmed the regional sea authority. Neither the sectorial boundaries activities in the sea.	There is a similar concept in delimitation by marine authorities in Canada and Indonesia; it is the existence of the province and regency/municipal. The deference is: the determination of the province and regency/ municipal are not determined by nautical miles. Sectorial activities in the marine boundary from each definition can be applied in Indonesia which is concerned with the authority of local and indigenous sea boundary.
	a. Federal: Territorial Sea, Contiguous Zone, Exclusive Economic Zone, Continental Shelf.	a. Federal: Territorial Sea,	a. Federal: Territorial Sea, Contiguous Zone, Exclusive Economic Zone, Continental Shelf.	a. Federal: Territorial Sea, Contiguous Zone, Exclusive Economic Zone, Continental Shelf		
	b. State: coastal waters (3 miles)	b. Contiguous Zone, Exclusive Economic Zone, Continental Shelf.	b. State: seaward states (3 miles), revenue sharing (6 miles)	b. State: coastal waters (3 miles)		
		c. State: province, regency/ municipal				
	2. Activities boundary:	2. Activities boundary:	2. Activities boundary:	2. Activities boundary:		
	a. Marine protected areas	a. Marine protected areas	a. Navigation	a. Marine protected areas		
	b. Fishing zones	b. Defense	b. Submerged cultural resources	b. Fishing zones		
	c. Petroleum exploration and mining.	c. Cable and pipeline areas	c. Undersea cables	c. Petroleum exploration and mining.		
	d. Cable and pipeline areas		d. Offshore aquaculture	d. Cable and pipeline areas		
	e. Native title claims		e. National security	e. Native title claims		
			f. Environmental protection			

(continued)

Table 22.5 (continued)

The key elements of the marine cadastre definitions	Marine cadastre definition among non-islands countries				The Unitary State of the Republic of Indonesia as an archipelagic state	
	A 1st definition Australia, 1999	B 2nd definition Canada, 2000	C 3rd definition USA, 2002	D 4th definition Australia, 2004	The marine management problems	The evaluation results
6. Geodetic reference system	1. Geodetic and geocentric coordinates system	1. Geodetic and geocentric coordinate system	1. Geodetic and geocentric coordinates system	1. Geodetic and geocentric coordinates system	The variety of the geodetic reference system used by each of the sectors has generated geospatial information to be different.	Geodetic reference system in Australia, Canada and United States of America cannot be completely implemented in Indonesia.
	2. Projection system: UTM, Map grid of Australia	2. Projection system: UTM	2. Projection system: UTM, Outer Continental Shelf (OCS) grid system	2. Projection system: UTM, Map grid of Australia		
	3. Horizontal datum: GDA'94, WGS'84.	3. Horizontal datum: NAD27, NAD83, WGS'84	3. Horizontal datum: NAD27, NAD83, WGS'84	3. Horizontal datum: GDA'94, WGS'84		
	4. Vertical datum: Low Water Mark (LWM) Lowest Astronomical Tide (LAT)	4. Vertical datum: Lower Low Water Large Tide (LLWLT) and Lowest Normal Tide (LNT).	4. Vertical datum: MLLW	4. Vertical datum: Low Water Mark (LWM) Lowest Astronomical Tide (LAT)		
	5. Australia Spatial Data Infrastructure (ASDI)	5. Marine geospatial Data infrastructure and Canadian geospatial data infrastructure	5. Cadastral data content standard for the national apatial data infrastructure: coastal and marine habitat classification standard	5. Australia Spatial Data Infrastructure (ASDI)		

7. Use of marine areas	1. Water surface	1. Water surface 2. Water column 3. Seabed	1. Air column 2. Water surface 3. Water column 4. Seabed 5. Subsurface	1. Water surface 2. Water column 3. Seabed	There is no explicit provision related to the implementation of the water surface, water column and seabed activities.	The concept of marine areas utilization in the 2nd, 3rd and 4th definitions can be applied in Indonesia.
8. Marine jurisdiction	1. UNCLOS 2. Federal 3. State 4. Customs	1. UNCLOS 2. Federal 3. Province 4. Municipal 5. Customs	1. UNCLOS 2. Federal 3. State	1. UNCLOS 2. Federal 3. State 4. Customs	The marine boundary setting of the country, province, municipal/regency which are not yet completed.	The Indonesian archipelago has different marine authority than in the non-islands countries.
9. Institution	Many institutions that manage the rights of offshore and seashore boundary.	Some institutions that manage the rights of offshore and seashore boundary.	Many institutions get involved. The implementation is organized by National Oceanic and Atmospheric Administration (NOAA).	Many institutions that manage the rights offshore and seashore boundary.	Many institutions get involved. The implementation of marine management done by sectorial legislation overlapped and contradicted.	The concept of marine cadastre administration in United States of America can be used as an approach to the implementation of marine management solution in Indonesia.

Table 22.6 The evaluation results of marine cadastre elements in Australia, Canada and United States of America to the characteristics of Indonesia as an archipelagic state

The marine cadastre elements	The evaluation results in Australia, Canada and USA to the characteristics of Indonesia as an archipelagic state
1. Marine jurisdictions	Marine cadastre concept to Indonesia should include archipelagic waters elements as distinct to Australia, Canada and United States of America as a coastal state.
2. Authority	The differences in marine management authority boundary among Indonesia and Australia, Canada and United States of America.
	The marine cadastre concept in Indonesia must include marine jurisdictional boundaries elements of province and municipal/regency.
3. Right	The rights in Australia, Canada and United States of America can be used as input to formulate new rights in Indonesia, in conditions of concerning its regional sea boundary authority.
4. Native rights	The Indonesian marine cadastre concept should include the elements of indigenous sea ownership.
5. Interests	The interests in Australia, Canada and United States can be placed in Indonesia with the concerned government marine boundary authority of province, and municipal/regency.
	The Indonesian marine cadastre concept must include Regional Autonomy element.
6. Restriction	The restrictions in Australia, Canada and United States cannot be implemented in Indonesia caused by the elements of the archipelagic state sovereignty, regional autonomy and customary marine laws authority applied in Indonesia.
7. Responsibility	The responsibilities in Australia, Canada and United States of America cannot be implemented in Indonesia caused by the elements of the archipelagic state sovereignty, regional autonomy and customary marine laws authority applied in Indonesia.
8. Marine boundaries	The marine boundaries in Australia, Canada and United States cannot be implemented in Indonesia because of:
	the sovereignty distinction countries,
	the marine boundary authority of the province and municipal/regency
	the prevailing customary sea boundary in Indonesia.
9. Geodetic reference system	Geodetic reference system in Australia, Canada and United States of America cannot be completely implemented in Indonesia.
	Required use of the same geospatial reference system to the variety of marine utilization activities.
10. Institution	The implementation of marine cadastre concept in United States of America can be used as an approach to the application of marine management solution in Indonesia.

contains the regional autonomy concept in natural resource management perspective.

American marine cadastre definition builds upon the sea jurisdiction that sets the authority marine boundary use both federal and state (three miles, except Texas and Florida Bay nine miles). In Indonesian point of view, this definition contains the rule of law concept to resolve conflicts at marine activities.

Thus, the existing marine cadastre definitions among Australia, Canada and United States of America cannot be used in Indonesia. As a highlight, marine cadastre is concerned with how a country, especially Indonesia archipelago, is managing and governing the marine resources administration. This condition causes the marine cadastre definitions among non-islands nations namely Australia, Canada and United States of America cannot be implemented in Indonesian coastal and marine areas.

Recommendation

Required marine cadastre precise definition according to the characteristics of Indonesia as an archipelagic nation. The marine cadastre definition in this country should incorporate jurisdictional boundaries elements of the province, and regency/municipal, the coastal and small islands areas management, and also concern the existence of customary marine law recognized by the government.

References

Abdulharis, R., E. Djunarsjah, and A. Hernandi. 2008. *Stakeholder analysis on implementation of marine cadastre in Indonesia*. In: Proceedings FIG Working Week, Stockholm, Sweden.

Binns, A. 2004. *Defining a marine cadastre: Legal and institutional aspects*. Thesis. Department of Geomatics, The University of Melbourne, Australia.

Binns, A., A. Rajabifard, P.A. Collier, and I. Williamson. 2004. *Developing the concept of a marine cadastre: An Australia case study*. Department of Geomatics, The University of Melbourne, Australia.

Djunarsjah, E. 2008. *The study on the technical and legal aspect of marine cadastre in Indonesia toward natural resources preservation and sustainable development*. Bandung: LPPM – ITB.

———. (2011). *Measuring and mapping technical procedures waters space*. Pilot project in the framework of measurement and mapping services waters activities. The cooperated BPN RI and LPPM ITB.

Falah. 2010. *Assessment of marine cadastre technical aspects of three dimensional (Case study: Coastal Tanjungpinang Riau Island Province)*. Final project, Department of Geodesy and Geomatics, ITB.

Faridha, D. 2010. *Identification of objects cadastre sea waters less than twelve nautical miles from base line islands*. Final project, Department of Geodesy and Geomatics, ITB.

Fraser, R, P. Todd, and P. Collier. 2003. *Issues in the development of a marine cadastre*. Department of Geomatics, The University of Melbourne, Australia.

Harbimaharani, H. 2010. *Study on waters cadastre marine policy*. Final project, Department of Geodesy and Geomatics, ITB.

Hasymi, F. 2008. *Determination limit as supporting implementation of regional sea marine cadastre in Indonesia*. Final project, Department of Geodesy and Geomatics, ITB.

Hernandi, A., R. Abdulharis, S. Hendriatiningsih and M. Ling. 2012. *An institutional analysis of customary marine tenure in Maluku: Towards implementation marine cadastre in Indonesia*. In: Proceedings FIG Working Week, Rome, Italy.

Ives, D. and Y.C. Yan. 2004. *A report on Commonwealth Association for public administration and management*. In: CAPAM's 10th anniversary biennial conference, Singapore.

Mårtensson, N. 2004. *The French ecologic protection zone: Developing marine environmental protection in international law*. Master Thesis, Faculty of Law, University of Lund, Lund, Sweden.

Ng'ang'a, S. M., S. Nichols, M. Sutherland, and S. Cockburn. 2001. *Towards a multidimensional marine cadastre in support of good ocean governance*. In: Proceedings of the international conference on spatial information for sustainable development, FIG, Nairobi, Kenya.

Nichols, S., D. Monahan, and M.D. Sutherland. 2000. Good governance of Canada's offshore and coastal zone: Towards and understanding of the maritime boundary issues. *Geometica* 54(4): 415–424.

Rais, J. 2009. Introduction to marine cadastre in Indonesia. *Journal ISI-UNDIP*. Diponegoro University, Semarang.

Robertson, B., G. Benwell, and C. Hoogsteden. 1999. The marine resource: Administration infrastructure requirements. In: *UN-FIG Conference on Land Tenure and Cadastral Infrastructures for Sustainable Development*, 24–27 October 1999, Melbourne, Australia.

Sesli, F., and G. Uslu. 2010. The importance of marine cadastre for Turkey. *African Journal of Agricultural Research* 5(14): 1749–1758.

Sook, H. L. and D. Shin. 2010. *Issues with building a marine cadastre system in South Korea*. In :Proceedings FIG Working Week, Sydney, Australia.

Srebo, H., I. Fabrikant, and O. Marom. 2010. *Towards a marine cadastre in Israel*. In :Proceedings FIG Working Week, Sydney, Australia.

Steudler, D. 2004. *A framework for the evaluation of land administration systems*. Ph. D Thesis, Department of Geomatics, The University of Melbourne, Victoria 3010, Australia.

Sulasdi, W.N. 2010. *Actual level components integrality mapping in coastal and ocean engineering in Indonesia*. Bandung: Scientific Institute of Technology.

Tamtomo, J. P. 2004. *The needs for building concept and authorizing implementation of marine cadastre in Indonesia*. In: 3rd FIG regional conference TS9.1, 3–7 October, 2004, Jakarta, Indonesia.

———. 2006. *Analysis of land use policies in coastal and marine framework "Marine Cadastre" (Case Study in the Island of Bintan, Riau Islands)*. Dissertation Doctoral Program, IPB.

Vaez, S. 2009. *Marine cadastre and marine administration. Short course on modern cadastre and land administration*. University of Melbourne, Australia.

Waljiyanto. 1998. *Cadastre as based land information system*. In: National seminar dies Natalie XXVIII KMTG FT-UGM, September 28, 1998, Yogyakarta.

Williamson, I., S. Enemark, J. Wallace, ;and A. Rajabifard. 2010. Land administration for sustainable development, Esri, Redlands.

Wisayantono, D. 2009. *Spatial optimization ratio in the management of land resources sustainable coastal zone*. Dissertation Doctoral Program, ITB.

Widodo, S. 2004. *Relationship of marine cadastre and marine spatial planning in Indonesia*. In: 3rd FIG regional conference TS9.2, 3–7 October, 2004, Jakarta, Indonesia

Yuwono, I. R.. 2006. Utilization of marine surveying and mapping for cadastre toward the sea (Marine Cadastre). *Journals*, Geomatics Engineering, Faculty of Civil Engineering and Planning, Institute of Technology of Sepuluf Nopember, Surabaya, Indonesia.

Zaenudin, D. 2008. *Study of the application of legal aspects of marine cadastre in Maluku Province*. Final project, Department of Geodesy and Geomatics, ITB.

Chapter 23
Sustainable Marine Space Managements: Malaysian Perspective

Abdullah Hisham Omar, Nazirah Mohamad Abdullah, Shuib Rambat, Noor Anim Zanariah Yahaya, Rasheila Rahibulsadri, Asraf Abdullah, Rahim Yahya, Hasan Jamil, Teng Chee Hua, and Chan Keat Lim

Introduction

Marine administration has been defined as governing surrounding of the marine space. Governing the surrounding marine space tasks may include sustaining the natural environment, maintaining conservation and managing the resources. In Malaysia, governing such activities involves various departments at government stage as well as the stakeholder. Managing a marine space with approximately 515,000 km^2 area which covered by maritime realm and 4576 km in length by coastline is a tedious task (Teo and Fauzi 2006). Indeed, the maritime adjacent borders with Thailand, Brunei Darussalam, Singapore, Indonesia, Vietnam and the Philippines as shown in Fig. 23.1 mean proper standard of governing the marine space is needed. As part of the South East Asian Region and a founding member of the Association of South East Asian Nations (ASEAN) the relationships with these nations should be important as they are one of the stakeholders in Malaysia marine spaces (Morocco et al. 2003).

A.H. Omar (✉) • N.M. Abdullah • S. Rambat • N.A.Z. Yahaya • R. Rahibulsadri
Department of Geoinformation, Faculty of Geoinformation and Real Estate, University of Teknologi Malaysia, 81310 Johor Bahru, Johor, Malaysia
e-mail: Abdullahhisam@utm.my

A. Abdullah
Department of Geoinformation, Faculty of Geoinformation & Real Estate, UTM, 81310 Skudai, Johor, Malaysia

Faculty of Architecture, Planning and Survey, Universiti Teknologi Mara Malaysia Kampus Perlis, 02600 Arau, Perlis, Malaysia

R. Yahya
SMK Tun Perak, KM 4, Jalan Salleh, 84000 Muar, Johor, Malaysia

H. Jamil • T.C. Hua • C.K. Lim
Cadastral Division, Department of Survey and Mapping Malaysia, Wisma JUPEM, Jalan Semarak, 50578 Kuala Lumpur, Malaysia

T. Yomralioglu, J. McLaughlin (eds.), *Cadastre: Geo-Information Innovations in Land Administration*, DOI 10.1007/978-3-319-51216-7_23

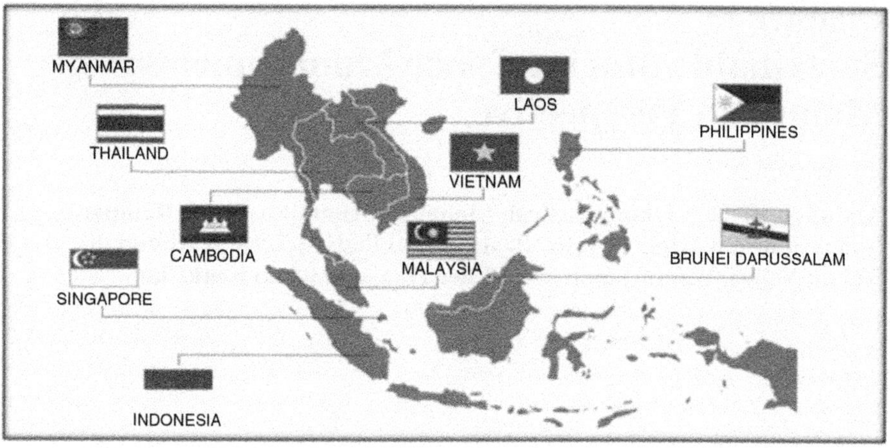

Fig. 23.1 Countries maritime adjacent borders of Malaysia

The Malaysian coastline, which is about 4800 km in length, is rich in coastal resources and has plenty of natural biodiversity. The coastal areas of Malaysia which provision a major portion (70%) of the population, is the navel of socio-economic activities such as urbanisation, agriculture, fisheries, aquaculture, oil and gas exploitation, transportation and communication, tourism, recreation, and others. Indeed, there are numerous industries that are also located in coastal area to facilitate export and to stop the employment mere in urban centre. The expansion of population and the industrialisation are the two core aspects that have contributed to the rapid growth of coastal cities, resulting in an escalation for the demand of coastal land development (Saw et al. 2002). However, the Malaysian marine spaces are not managed by single public institution but it was managed by several departments from the government, the stakeholder and an authorised individual who have interest on the marine spaces. As a result, it creates complex, uncertain and conflict situations in determining the resolution of authority area of true governance.

Based on the above facts, the demand on good governance is one of the main factors that need to be addressed and soon developed by Malaysian marine spaces administrator. It needs to be planned, particularly by means of spatial planning on the level of local, regional and national. As good governance is a term, similar to sustainable development, that can mean many things depending on one's perspective or goals.

In fact, management and governance are the foundation of life and society in order to achieve the human ability to seize the benefits of the natural environment and maintaining a quality resource to be sustained. Importantly, it is also about the decision-making and helm, and the distribution knowledge and influence in an organised entity (e.g., jurisdiction, government departments and others) that pursued goals and objectives quoted from Paquet 1994 and 1997 (Cockburn 2005). In marine spaces, effective management is about covering accuracy, up-to-date, complete and helpful information about the resources that currently exist and the nature

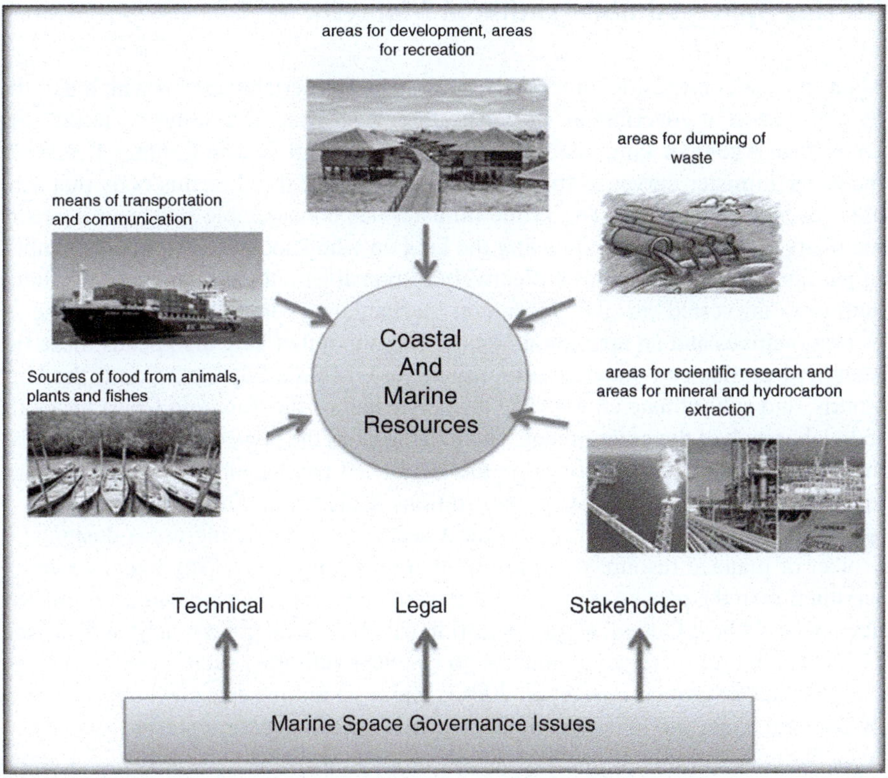

Fig. 23.2 Competing demand for Malaysia coastal and marine resources with marine space governance issues

of the environment in which the resources exist, and also consumers contact for these resources (Ng'ang'a et al. 2004). Furthermore, there are several values of marine space, such as sources of food from animals, plants and fishes, means of transportation, means of communication (subsea cables), areas for development (mineral extraction), areas for recreation, areas for dumping of waste and areas for scientific research. Figure 23.2 clearly shows the Malaysia coastal and marine space that have multiplicity of uses, which often leads to conflict namely technical, legal and stakeholder management. In fact, to avoid conflict, in a multiple use resource there must be rules, hence the importance of institutions and stakeholder frameworks in the administration of coastal and marine space.

Marine Space Administration in Malaysia

Malaysia is a constitutional monarchy (Kerajaan Berperlembagaan) which uses the federal system of government. Its constitution vests executive authority in the Yang Di Pertuan Agong or king, the nominal head of the nation. The Cabinet, headed by the prime minister, serves as the advisory body to the king. It is this body that actually governs the country. The prime minister has considerable power in choosing members of the Cabinet by advising the king on who should be chosen as members of parliament. The Cabinet is collectively responsible to the parliament. Legislative authority is vested in the bicameral parliament, composed of a House of Representatives and an appointed Senate. The appointed king also heads these two houses of parliament. Judicial authority is vested in the Supreme Court, the High Courts, and subordinate courts. The lord president of the Supreme Court heads the judicial branch of the government. The judiciary has the power to deliberate on civil and criminal matters, pronounce on the legality of any legislative or executive act, and to interpret federal and state constitutions (Calestino 2001).

Marine managed areas, in the widest sense, are geographic areas designed to protect or manage resources within the marine environment. Any agency that has jurisdiction in the marine environment can create marine managed areas. A creating agency could be a federal, state, territorial, tribal, or local government and an independent agency, or a regional entity with resource authority, such as a port management council (Suzanne Bass et al. 2006). Malaysia marine space is being managed by national, state and local organizations with various departments and agencies. It is including government, private and educational organizations. Table 23.1 shows the Malaysia Marine Space institutional structure that includes 15 categories of marine space activities, 14 ministries and more than 30 departments/units responsible for the management of the marine space activities.

For these eras, the maritime areas have always been important to Malaysia and based on the maritime sector, economic activities have blossomed which have contributed significantly to Malaysia's economic growth (Saharuddin 2001). However, many economic activities often conflict with the natural environments of coastal areas. The question is why and how this conflict is going on? According to Table 23.1, the majority consent represents the marine activities involves more than one of the different agencies. For example, mixing of authorities between the state, federal and the private sector has resulted in uncoordinated port planning and development (Saharuddin 2001). This creates a conflict jurisdiction and overlapping of functions between several federal ministries, state government and also private sector such as fisheries, environment, state forestry and managing coastal zone which are crucial to the sustainability of marine space administration.

Table 23.1 The Malaysia marine space institutional structure

No	Category	Ministry	Department/Agencies	Division/Council
1	Port	Ministry of Transport	Johor Port Authority	National Shipping Council
			Bintulu Port Authority	
			Klang Port Authority	
		Prime Minister's Department	Kuantan Port Authority	
			Kemaman Port Authority	
			Penang Port Commission	
			Maritime Institute of Malaysia	Malaysian National Shipper's Council
2	Shipping	Ministry of International Trade and Industry	Marine Department of Malaysia	
			Maritime Institute of Malaysia	
3	Light House		Marine Department of Malaysia	
4	Non-living Resources	Ministry of Science and Technology	Department of Standard Malaysia (Standard Malaysia)	National Oil Spill Control Committee
			National Oceanographic Directorate (NOD)	
			Malaysia Remote Sensing Agency (ARSM)	
			Malaysia Meteorological Department	
		Prime Minister's Department	Economic Planning Unit	National Petroleum Advisory Council
		Ministry of Transport	Maritime Institute of Malaysia	
5	Living Resources/ Fisheries	Ministry of Agriculture and Agro-based Industry	Department of Fisheries	National Advisory Council for Marine Park and Marine Reserve
			Fisheries Development Authority of Malaysia (LKIM)	
		Prime Minister's Department	Maritime Institute of Malaysia	
		Ministry of Transport		

(continued)

Table 23.1 (continued)

No	Category	Ministry	Department/Agencies	Division/Council
6	Natural Resources		National Hydraulic Research Institute of Malaysia (NAHRIM)	
			Department of Survey and Mapping Malaysia	
			Department of Director General of Lands and Mines	
		Ministry of Natural Resources and Environment	Department of Irrigation and Drainage Minerals and Geoscience Department	
		Ministry of Transport	Maritime Institute of Malaysia	
7	Forestry/Wildlife		Department of Marine Park Malaysia	
			Department of Environment	
			Forestry Department Peninsular Malaysia	
			Forest Research Institute Malaysia	
			Department of Wildlife and National Park	
			Department of Biosafety	
		Prime Minister's Department	Maritime Institute of Malaysia	
8	Jurisdiction	Ministry of Defence	Royal Malaysia Navy	
		Prime Minister's Department	Hydrographic National Centre	
		Ministry of Transport	Maritime Institute of Malaysia	
9	Enforcement	Ministry of Home Affairs	Royal Malaysian Police	Marine Unit
		Prime Minister's Department	Maritime Enforcement and Coordinating Centre	
			Malaysia Maritime Enforcement Agency	

(continued)

Table 23.1 (continued)

No	Category	Ministry	Department/Agencies	Division/Council
10	Tourism	Ministry of Culture, Arts and Tourism		
11	Heritage and Antiquity	Prime Minister's Department		
		Ministry of Transport	Maritime Institute of Malaysia	
12	Telecommunication	Ministry of Communication and Multimedia		
13	Dispute Settlement		Attorney General's Chamber	Advisory and International Division
		Ministry of Foreign Affairs	Economic Division	Maritime Affairs Units
			Policy and Planning Division	
14	Educations	Ministry of Education		Universiti Teknologi Malaysia (UTM)
				Universiti Malaya (UM)
				Universiti Malaysia Terengganu (UMT)
				Universiti Putra Malaysia (UPM)
				Universiti Kebangsaan Malaysia (UKM)
				Universiti Teknologi Mara (UiTM)
				Politeknik Ungku Omar (PUO)
15	Trade and Service	Ministry of Finance	Internal Tax Division	Secretariat for Cabinet Committee on Trade and Service

Modified after Abdul Hamid Saharuddin (2001)

Role and Responsibilities

Thoughts of environment changes in which civilization operates increasingly determine that there is a need to restructure development practices, in order to ensure the continuity of these practices, or in other words, sustainable development, taking into account the need for harmony between the economic, social and environmental spheres (Cicin-Sain 1993). Aligned with this idea, it becomes essential to examine the responsibilities and roles of different marine space stakeholders in order to ensure that their work in this field is taken into account. This then connects to the idea of marine space sustainable consumption, taken to be an intermediate dynamic feature in the marine space governance development paradigm. Michaelis and Mont and Plepys indicate sustainable consumption as the pattern of consumption resulting from the inter-relation of stakeholders interested in achieving sustainable development.

Hence, with the discussions on sustainable development, it becomes increasingly clear that marine space stakeholders in different fields need to be working and assuming specific roles and responsibilities in this new context. Therefore this study tries to see the role of government, regulators and planning organisations and also role of national policy.

Role of Government, Regulators and Planning Organisations

Marine space administrations will be successful if supported by the appropriate law and regulation on marine management. The law and regulation is dependent on two components: local and international. The local law must be examined under marine cadastre contexts such as National Land Code 1965, Continental Shelf 1966, Territorial Sea Act 2012, State Land Rule and Baselines of Maritime Zones Act 2006. The international law is related to United Nation Convention Law of the Sea 1982 and Convention on the Territorial Sea and the Contiguous Zone 1958.

Management of coastal resources shows that water and land matters fall within the jurisdiction of the State Government, which entails development, planning and zoning powers amongst others (Mokhtar and Ghani Aziz 2003). And yet the living resources are shared between the Federal and State governments. The local authorities, both municipal and district councils, together with relevant government agencies (for example district land offices) act as a channel for both Federal and State governments.

The importance of coastal zone management in Malaysia took place when the Federal Government responded to severe coastal erosion caused by a variety of natural and man-made processes. Malaysian government consequently launched the National Coastal Erosion Study in 1984–1985 and it became a major national concern (Mokhtar and Ghani Aziz 2003). This research was under the responsibility of the Environment and Natural Resources Division of the Economic Planning Unit (EPU) in the Office of the Prime Minister. The important results of this research were recommendations for implementing proper long-term planning to prevent coastal erosion and establishing in 1987 two important institutions related to coastal zone management viz. the Coastal Engineering Technical Centre (CETC) and the National Coastal Erosion Control Council (NCECC) which led to development of important guidelines.

The development guidelines encapsulated in the Federal Government's Outline Perspective Plans (which spans a period of at least 10 years) are then interpreted at State Government levels, and government agency levels, who take the directions formulated for sectors that they represent, and develop specific policy documents, programmes and action plans. State government will be the actual local government of an area for all intents and purposes. They also have access and capability to raise funds, promulgate enactments and regulations, and develop development plans for areas within their boundary (Mokhtar and Ghani Aziz 2003).

There is large volume of published studies describing Malaysia as a plethora of maritime and ocean laws. With respect to the concept of unity between land and water expressed by the Malay word 'tanahair' which literally means, 'land and water' depict the embodiment of the unity assumed simultaneously with the native land. In 1999, Juita Ramli described that as early as 1276 during the reign of Sultan Muhammed Shah – the first sovereign of the Malacca Sultanate – it was found that the Malays has had already designed a set of laws of the sea applicable in sea areas within the jurisdiction of the Malacca Sultanate. These laws were referred to as the Malacca Code.

Malacca Code is all about the laws designed significantly related to the trading activities within the region, which thrived for centuries under the reign of the Malacca Sultanate (Ramli 1999). It is also important to highlight that during this ancient time of living where daily lives were easily entertained, it is notable that such law and order governing matters both on land and at sea had been well administered. Juita Ramli also pointed the nature of victorious civilisation and it is regrettable that as beneficiaries we have failed in perpetuating or pursued to develop, in the least, the codified Malacca laws of the sea.

Today, Malaysia maritime laws consist of multiple meanings and purposes established in specific needs such as to solve any dispute or issue. However, Malaysia's marine legislation does not focus on laws related to the functions of marine space and its characteristics compared to the terrestrial. To ensure success in marine space governance, Malaysian marine laws have to be first examined to have a clear understanding of the Malaysian maritime regime scenario under the scope of marine spaces.

The governance of Malaysia's maritime territory is controlled by legally defined boundaries same as on land (Fauzi 2006). The United Nations Convention on the Law of the Sea (UNCLOS) establishes a jurisdictional regime under which Malaysia itself can claim, manage and utilise its maritime territories. In this regard, Malaysia ratified UNCLOS in October 1996, and in line with provisions of UNCLOS, is entitled to:

(i) The Territorial Sea, which is the belt of sea measured 12 nautical miles (nm) seaward from the territorial sea baseline (Malaysia uses the straight base line approach). On 2 August 1969, an Ordinance under Article 150(2) of the Constitution known as the Emergency (Essential Powers) Ordinance, No.7, 1969 was promulgated. Under this Ordinance, the territorial waters of Malaysia (except in the Straits of Malacca, the Sulu Sea and the Celebes Sea) was declared as 12 nautical miles from the base line determined in accordance with UNCLOS.
(ii) The Contiguous Zone, which is the belt of sea, contiguous to the territorial sea, measured 24 nm seaward from the Territorial Sea Baseline.
(iii) The Exclusive Economic Zone, which is the area beyond and adjacent to the territorial sea, measured 200 nm seaward from the Territorial Sea Baseline.

Confusion occurred until the establishment of a mix of national and international legislations in Malaysia precipitated. It is affected by the development of world-wide laws of the sea since the advent of Western European dominance in international trading (Ramli 1999).

Domestic Laws

There have been several studies in the literature reporting about Malaysia's government being modelled after the British system, somewhat modified because Malaysia's federal structure incorporates 13 states and three federal territories. Nine of those states have rulers or sultans and they elect a monarch, the supreme ruler, every 5 years. The government is based on a parliamentary system, headed by an elected Prime Minister. The Parliament consists of a partially appointed senate and a house of representatives whose members are elected by universal adult suffrage.

The Federal Government has powers such as over external affairs, defence, internal security, civil and criminal law, federal citizenship and naturalization, finance, trade, commerce and industry, taxation, customs and excise duties, shipping, navigation and fisheries, communications and transport, federal works and power, education, medicine and health, social security and tourism. The States' powers include over land and its administration, Islamic law, Malay customs, permits and licenses for mines prospecting, agriculture, forests, local government, states works and water, and riverside fishing. It is essential to heed at this juncture that all of the pre-Federation of Malaysia laws were derived from British domestic laws. It has conclusively been shown from paragraph below.

Malaysia's earliest recorded twentieth century national law – considered remotely relating to management of maritime matters – is the Waters Act, 1920 enacted to provide for the control of rivers and streams. It was not until 20 years later when the Federation of Malaya became an active rubber producer in the region that the Rubber Shipping and Packing Control Ordinance, 1949 was promulgated for the purposes of regulating shipping and packing of rubber for export. In the following years we may observe that domestic laws pertinent to shipping, navigation and port were duly promulgated and enforced. These included Carriage of Goods by Sea Act, 1950; Merchant Shipping Ordinance, 1952; Federation Light Dues Act, 1953; Penang Port Commission Act, 1955; Port Authorities Act, 1963 and so on. This trend was consistent with pre-Merdeka and pre-Federation of Malaysia days when the ruling British were active in pursuing interests in maritime trade arising from an abundance of agricultural produce in the Malay States (Ramli 1999).

International Laws

As highlighted by Aziz Meo Ngah and Nazery Khalid (2014), Malaysia is one of the world's major trading nations and its economic wellbeing depends largely on trade, 95% of which is carried through seaborne means. Malaysia is subjected to international laws in marine matters and various treaties and resolutions have been sealed. The international consultation needs Malaysia to take the relationship with international institutions seriously in deciding to join the international maritime legislation to clarify the rights of marine territory of a country. Malaysia should understand and defend its rights and policies in accordance with the rules of international laws for recognition as a sovereignty of country's maritime. Until now, Malaysia has adopted the international laws in the implementation of all the functions and powers for marine administration and is related to the local legislation as shown in Table 23.2. This information is summarised from Country Report (Fauzi 2006) and Juita Ramli (1999) writing.

To ensure that the Malaysian Marine Space Governance is well managed, country report 2006 also highlight the instrument of Malaysia Governance as shown in Table 23.3. Again it has a number of laws which apply the enforcement of the Malaysian Maritime Zone and seas which cover both the national and international levels and also agreements, circulars and any legal recourse to ensure that the Malaysian sovereignty is safe for the longest time. Eleven provisions among set forth under special laws are as follows:

(i) The Federal Constitution of 1957
(ii) National Land Code [Act 56/65]
(iii) Emergency (Essential Powers) No. 7 [1969] (has been unraveled and replaced)
(iv) Territorial Sea Bill 2012 (replacing the Emergency Ordinance 1969)
(v) Malaysian Maritime Enforcement Agency Act 2004 [Act 633]
(vi) Exclusive Economic Zone Act 1984 [Act 311]
(vii) Continental Shelf Act 1966 [Act 83]
(viii) Fisheries Act 1985 [Act 317]
(ix) The Mutual Assistance in Criminal Article 2002 [Act 621]
(x) The Official Secrets Act 1972 [Act 88]
(xi) Baselines of Maritime Zones Act 2006

Table 23.2 Malaysia international laws

Provision	Relate
Convention on the Law of the Sea, UNCLOS 1982	Maritime and Sovereignty
Collision Regulation (COLREG) Convention 1972	Safety
Safe Manning, Certification, Training & Watchkeeping (SCTW) Convention 1978	Safety
SCTW Convention 1995	Safety
International Maritime Satellite Organization, INMARSAT Convention 1976	Navigation
INMARSAT OA 1976	Navigation
Marine Pollution, MARPOL 73/78 (Annex I/II)	Environment
MARPOL 73/78 (Annex V)	Environment
Convention on the Civil Liability for Oil Pollution Damage, CLC Convention 1969	Environment
International Oil Pollution Compensation, FUND Convention 1971	Environment
Oil Pollution Preparedness, Response and Cooperation, OPRC Convention 1990	Environment
Conference on the Environment and Development, UNCED 1992	Environment
Convention on Facilitation of International Maritime Traffic 1965	Shipping and Transportation
Marine Pollution, MARPOL 73/78 (Annex I/II)	Environment
MARPOL 73/78 (Annex V)	Environment
Convention on the Civil Liability for Oil Pollution Damage, CLC Convention 1969	Environment
International Oil Pollution Compensation, FUND Convention 1971	Environment
Convention on the Control of the Transboundary Movement of Hazardous Wastes and their Disposal (Basel) 1989 (1993).	Environment
Convention on the International Civil Aviation 1964	Airspace
Convention on Psychotropic Substances 1971	Safety

Source: Fauzi (2006) and Ramli (1999)

Malaysia Marine Space Issues

Malaysia's large sea area and its bounty of resources carry immense management responsibilities. These range from ensuring the integrity of its sovereignty over its maritime territories to the sustainable development of marine resources. The country's considerable, strategic stake in the oceans warrants serious, meticulous attention to the governance and administration of its oceanic and maritime affairs. Malaysia marine spaces are many and at times, competing uses and these uses include sources of food from animals, plants and fishes, means of transportation and communication, areas for development, areas for recreation, areas for dumping of waste, areas for scientific research and areas for mineral and hydrocarbon extraction (Teo and Fauzi 2006). Figure 23.3 shows clearly that the Malaysia coastal and

Table 23.3 Instruments of Malaysia governance

Provision	Explanation
Environmental Quality Act, 1974 (Act 127)	An act relating to the prevention, abatement and control of pollution and enhancement of the environment. Act A593 of 1996 provides among others the control of transboundary movement of schedule wastes and their disposal.
Environmental Quality (Amendment) 1985	Section 34A(1) empowers the Minister of Science, Technology and Environment to prescribe the development of Environmental Impact Assessments (EIAs) prior to granting approval to carry out certain activities, and grants control over the approval of projects based upon the results of the EIAs.
Mining Enactment FMS CAP 147	Provides to the states the powers and rights to issue mineral prospecting, exploration licenses, mining leases, and other related matters.
Fisheries Act, 1985 (Act 317)	An act relating to fisheries, including conservation, management and development of maritime and estuarine fishing and fisheries, in Malaysian fisheries waters, to turtles and riverine fishing.
Environmental Quality Act, 1974	Section 29 (1) states that no person shall, unless licensed, discharge wastes into Malaysian waters in contravention of the acceptable conditions stated in Section 21. Section 31(1) states that where any pollutants are being or are likely to be emitted, discharged or deposited, the culprit must install and operate appropriate control equipment. Section 51(1) empowers the Ministry of Science, Technology and Environment to prescribe standards and criteria for the implementation of environmental policy, classification of the environment for protection purposes, prohibit discharge of pollutants into the environment, prohibit the use of equipment that could cause pollution, and, among others, regulate boating and swimming in waters to prevent pollution.
Fisheries Act 1985, Article 9 (1)	For any application that intend to attain the fishing vessel license or permit, any plan, specification or other information regarding the fishing area must be submitted to the Director General of Fisheries.
Fisheries Act 1985 (Marine Parks Malaysia) Regulations 1997	Section 41(1) provides the powers for the Minister to establish any area or part of an area in Malaysian fisheries waters as a marine park or a marine reserve.
Fisheries (Prohibited Areas) Regulations 1994	Fisheries (Prohibited Areas) under section 61 of the Fisheries Act 1985, all forms of fishing and collecting are banned; however no permit is necessary to enter the prohibited area. The Department of Fisheries, Malaysia, controls this regulation.
Marine Parks Malaysia Order 1994	Designated 38 islands as Marine Parks Malaysia. The boundaries of the park extend two nautical miles seaward from the outermost points of the islands measured at low water mark. Within these areas no person shall kill or capture any fish unless he holds a license issued under Section 11.
The Turtles (Prevention of Disturbance) Rules, 1962	The rules state that no vessel, other than a government vessel on official duty, shall enter within half a nautical mile of the low water mark of the above three islands, except with a permit granted by, or on behalf of, the Turtle's Board.

(continued)

Table 23.3 (continued)

Provision	Explanation
Antiquities Act 1976	Provision for the control and preservation of, and research into ancient and historical monuments, archaeological sites and remains, antiquities and historical objects and to regulate dealing in and export of antiquities and historical objects. According to this act, applicants who wish to search for artefacts or shipwrecks, a description of the proposed site, the type and the area of the search and other information must be produced.
Petroleum Act (Safety Measures) 1984	This act was formulated to control activities pertaining to safety aspects in petroleum industry. This act has legal provision to ensure safety of road and rail transportation, sea transportation, air transportation, pipeline transportation, storage and maintenance and use of equipment, building structure and fixation.

Source: Fauzi (2006)

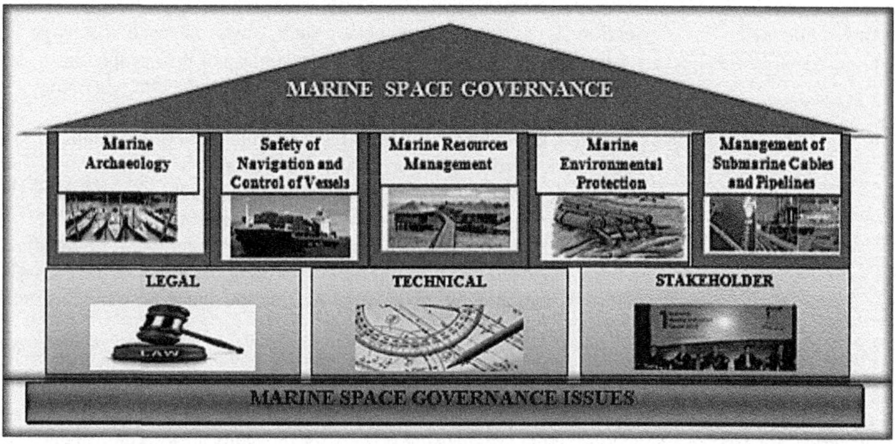

Fig. 23.3 Overview of Malaysia marine space governance

marine space have multiplicity of uses, which often leads to conflict namely technical, legal and stakeholder management. In fact, to avoid conflict, in a multiple use resource there must be rules, hence the importance of institutions and stakeholder frameworks in the administration of coastal and marine space. It is also important to highlight the method of the implanting marine space governance.

As resources are scarce in relation to the demand for it, the scramble for the usage of resources at the coastal and marine space by man is ubiquitous. Accordingly, Table 23.4 exhibits the major issues in administering the rights, restrictions and responsibilities in the marine space environment.

Conclusion from Country Report 2006 in general, state that: since Malaysia has no explicit policies on management or utilisation of marine and coastal resources, the need for coordination between various agencies that manage the marine spaces in different sectors is non-obligatory. A lease issued without consultation with other

Table 23.4 Major issues in administering the rights, restrictions and responsibilities in the marine space environment

Maritime zone	Issues
Coastal Zone	When marine boundaries are not demarcated, there is no physical evidence of the boundary, resulting in disagreement, confusion and conflicting versions of marine boundaries.
	Line of low tide is difficult to determine.
	Natural features like the coastline change over time, so does the marine boundaries.
Territorial Sea	The determination of base points and baselines in accordance to UNCLOS 1982.
	Enforcement agencies operating in the two maritime zones – the 12 nautical mile of territorial sea and the exclusive economic zone. Some enforcement agencies have found it difficult to operate in grey areas i.e. in areas where the territorial waters and EEZ meet at which the demarcation of the boundaries is distinguishable.
	The publication of a chart at a scale adequate for ascertaining the baselines for measuring the breadth of the territorial sea or listing geographical coordinates of these points.
Exclusive Economic Zone	The determination of the outer limits of the continental shelf based on Article 76, UNCLOS 1982, in which coastal states are allowed to claim outer limits of the continental shelf beyond 200 nautical miles, up to a maximum of 350 nautical miles or 2500 metre isobaths plus 100 nautical miles but must submit relevant scientific data to the Commission on the Limits of the Continental Shelf.
	Redelimitation of internal waters, territorial sea, EEZ and continental shelf.
	Updating the Peta Baru Malaysia 1979.

Source: Fauzi 2006

relevant authorities, creates multiple use conflicts and ignorance of the rulings imposed by the other authorities. Moreover, the interstate and inter-district boundaries of marine governance have not been defined, which may lead to confusion to the territorial limits of administration between authorities and causing conflicting maritime claims. In addition, there is a lack of awareness of which ministry/agency issues rights and permits as well as the imposition of conditions and restrictions. As marine spaces are 3D, there are no clear rulings that allow for overlapping rights, for example having petroleum exploration leases overlapping fisheries. It is also important to highlight the integration of data from numerous marine related agencies into the Marine Spatial Data Infrastructure initiative and demarcation of boundaries in the marine spaces.

To design such systems to be useful for managing information on single activities or resource use (e.g., petroleum leases) occurring in marine spaces is uncomplicated. Studies have found that, in order to be of maximum benefit to the governance of marine spaces these information systems will have to be able to manage and visualize information on multiple marine resource interests that overlap in 3-dimensional space and time; these systems should also function in an environment of efficient and effective governance and legal frameworks, and optimal

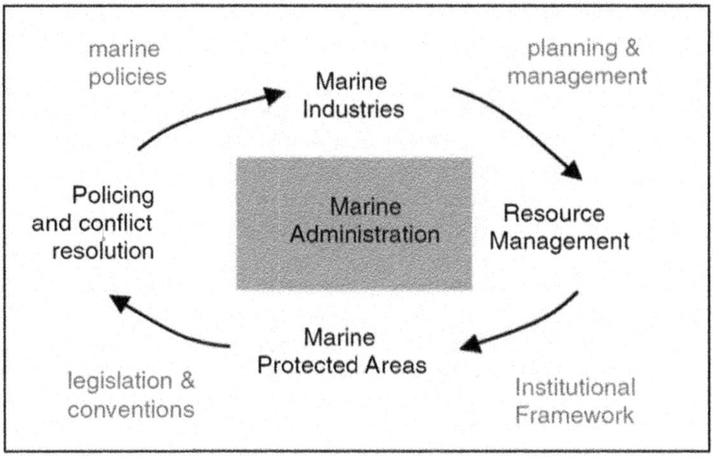

Fig. 23.4 Features of marine administration (Strain et al. 2006)

institutional arrangements that meet the often diverse needs of identified and engaged stakeholders.

Proposed Future Implementation of Malaysia Marine Space Governance

Appropriate administration and management, exploitation, utilisation and conservation of marine resources, economic growth and social values can be improved and sustained (Mukupa 2011), as shown in Fig. 23.4 where coastal and marine activities need administration for marine industries, resources management, marine protected area and policing and conflict resolution. Marine policies, planning and management, institutional framework and legislation and conventions are part of marine administration, to enable sustainability. Furthermore these features will lead to sustainability of marine development, whereas it will balance between social, economic and environmental impacts.

The Marine Policies, Planning and Management

Maritime Institute of Malaysia (MIMA) (National legislation pertaining to maritime management, 1997) has publicized that there are at least 74 national laws at present pertaining to maritime management. This does not include about 35–40 subsidiary legislative items and by-laws which are enforceable with some of the major laws such as the Environmental Quality Act 1984 and Fisheries Act 1985. These laws provide the legal framework for about 15 different aspects ranging from ports, shipping, lighthouse, living resources, non-living resources, environment, telecommunications, trade and education. In the international arena, Malaysia has already ratified at least 21 UN Conventions and 13 IMO Conventions. Three Conventions are subject to ratification, 10 Conventions are under consideration for ratification and another 10 Conventions are recommended for ratification by Malaysia.

As policies are made by two levels of government, they can be either cross sectoral in natural or sectoral. Maritime Institute of Malaysia (MIMA) is the agency directly involved in policy formulation and act. They are specialized in maritime matters and conflict to more versed in maritime transportation regulations, port rules etc. compatible with the MIMA visions to provide maritime-related advice and consultancy services to stakeholders through policy research, training, education and public awareness programmes. The Centre for Ocean Law and Policy (OLAP) is a research unit at MIMA responsible for ocean law and policy issues (see Box 23.1).

Data Management for the Marine Space

The development of a marine space management plan involves a multi-disciplinary approach. It should address issues such as the physical environment, resource inventory, environmental sensitivity, demand or land use projection, socio-economic setting and other factors which are of importance in arriving at the sustainable marine space administration. Abdul Hamid Saharuddin (2001) highlights the importance of quality data and information in sea management as one of the most important components of process to develop management plans and policies.

Box 23.1 The Centre for Ocean Law and Policy (OLAP)
The Centre for Ocean Law and Policy (OLAP) aspires to be Malaysia's national centre of excellence for research in ocean law and policy issues. OLAP aims to provide timely and relevant advice and policy options as well as to identify key areas of interest for Malaysia's multi-disciplinary realm of ocean and maritime law that encompass the Law of the Sea and other related international law, as well as maritime and admiralty law.

OLAP reinforces close working relationships with relevant stakeholders such as the Ministry of Transport, the Marine Department, the National Security Council, the Ministry of Foreign Affairs and the Attorney General's Chambers, as well as fostering links with other local and regional think tanks and international organisations such as the International Maritime Organisation (IMO), Division for Ocean Affairs and the Law of the Sea (DOALOS), International Labour Organisation (ILO) and similar entities.

OLAP undertakes the role of promoting awareness in ocean law and maritime legal aspects to appropriate stakeholders and the public, by conducting seminars, training workshops and conferences.

Source: http://www.mima.gov.my/mima/research

Marine space data management is all about capturing information, analysing, storage and dissemination of the data. Another essential point this paper will highlight is the two important agencies in data management for the marine space which are Department of Survey and Mapping Malaysia (JUPEM) focusing on capturing information and analysing and Malaysia Centre of Geospatial Data Infrastructure (MaCGDI) focusing on data storage and dissemination data.

Department of Survey and Mapping Malaysia (JUPEM)

It is the main Malaysia institution given the responsibilities to technically tackle marine space administration issues. However, that effort must collaborate with academic institution to bring out clearly the theory and methodology to apply and also be suitable for implementation and to be considered in marine environment and factor (see Box 23.2). Below are the functions of JUPEM:

- To advise the government in the field of cadastral survey and mapping along with the state and international boundaries.
- To provide complete and conclusive cadastral information for issuing land, strata and stratum titles.
- To manage efficiently the cadastral and mapping databases.
- To publish photographical, cadastral, thematic and utility maps for the purposes of planning, management of natural environment resources, preservation of environment, development, surveillance and security.
- To provide geodetic infrastructure for the purposes of cadastre survey, mapping, engineering and scientific research.

Box 23.2 JUPEM Corporate Informations
Vision: Making JUPEM an eminent organisation in providing outstanding survey and mapping services as well as geospatial data management towards fulfilling the nation's vision.

Mission: Providing a quality survey and mapping and services and geospatial data management via first-rate system, competent human resource and conducive working environment.

Motto: The catalyst for national development and citizens' prosperity.

Objective: To ensure the products and land survey and mapping services meet the quality acceptance and customer's needs.

To ensure a well maintained, up-to-date cadastral and mapping database to meet needs of the national geospatial infrastructure.

To make JUPEM as an excellent reference centre in the field cadastral survey and mapping.

To survey, determine and demarcate state and international boundaries.

Quality Policy: JUPEM is committed to provide cadastral survey, mapping services and dissemination of high quality geographic information in accordance with established standards and also continuous improvement efforts to ensure customer satisfaction.

Source: https://www.jupem.gov.my

Malaysian Centre for Geospatial Data Infrastructure (MaGDI)

Malaysia Geospatial Data Infrastructure (MaGDI) is an initiative by the government to develop a geospatial data infrastructure to enhance the awareness about data availability and improve access to geospatial information. This can be fulfilled by facilitating data sharing among participating agencies.

MaGDI as the National Spatial Data Infrastructure (NSDI) for Malaysia, is a national infrastructure comprising policy, data, standard, geospatial information and technology, R&D and development of human capital established by MaGDI Circular Letter no. 1 of 2006 – Guidelines for the Implementation of Malaysia Geospatial Data Infrastructure (MaGDI) for the purpose of facilitating the sharing and dissemination of geospatial information amongst government agencies, private sector and the general public. Through this infrastructure, smart partnerships among agencies is continuously being developed to produce and share geospatial information thus providing customer-focused, cost effective and timely delivery of geospatial data.

MaCGDI is a centre established by the government to manage and promote the development of geospatial data infrastructure for Malaysia (MaGDI). MaCGDI is also responsible for coordinating access and delivery of the geospatial information held by all government agencies. MaCGDI was established in December 2002 to

Box 23.3 MaGDI Corporate Informations

Vision: The vision of MaCGDI is to continuously make available and accessible current and accurate geospatial data that promotes a sustainable living environment, economic growth and social progress for all Malaysians.

Mission: "The mission of MaCGDI is to facilitate, coordinate and manage geospatial data infrastructure through the development of policies, standards, data, ICT Technology, R&D and skilled human resources by providing customer-focused, cost effective and timely delivery of geospatial data."

Function: The functions of MaCGDI are:

- To act as an advisor to the Government of Malaysia in the formulation and implementation of policies concerning geospatial data;
- To coordinate activities pertaining to the development of geospatial data and standard for geographic information/geomatics;
- To be a technical reference centre for advisory and consulting services with regard to the development and application of geospatial data;
- To develop and coordinate MaGDI Clearing house activities;
- To plan and conduct human resource development programme in GIS and the related fields;
- To organise various activities in promoting the use of MaGDI throughout the country;
- To become a centre for research and development (R & D) for GIS and the related fields; and
- To represent the public sector in international forum, conferences and meetings involving geospatial data.

Source: http://macgdi.mygeoportal.gov.my

replace the NaLIS Secretariat under the Ministry of Land and Cooperative Development (MLCD). On 27th March 2004, MaCGDI was subsumed under the Ministry of Natural Resources and Environment (NRE).

The main role of the centre is to continuously make available and accessible current and accurate geospatial data that promotes a sustainable living environment, economic growth and social progress for public. MaCGDI is organised with thirteen (13) sections and is set out to carry the following objectives:

- To provide mechanism/infrastructure in supporting the usage and sharing of current, accurate and reliable geospatial information among agencies by employing the latest geospatial techologies; and
- To avoid redundancy of duplicating efforts in collecting, processing, maintaining, providing and dissemination of required geospatial information.

The goal of MaGDI is to enable members of the geospatial communities in Malaysia to share and access geospatial data together seamlessly (see Box 23.3). Through its application MaGDI Explorer, MaGDI facilitates online access to geospatial information as an effort to avoid duplication of effort especially in the collection of geospatial data. It provides a base for geospatial data exploration, evaluation and application for users and data providers within all levels of government, commercial and non-profit sectors as well as the academia and the public.MaGDI governs through its committees and MaCGDI as the coordinator. In the Malaysia contexts, SDI initiatives started from the national level and are expected to filter down to all the states and gradually to all local levels. Like other countries, apart from financial and skill resources which forbid comprehensive big-bang undertaking, knowledge and agreement among agencies on fundamental datasets are required to meet common needs. These issues are usually sorted out by the lead agency, in this case MaCGDI.

Based on the theories, principles and an overview of the literature, we have proposed a framework for future implementation of Malaysia Marine Space Governance such as illustrated in Table 23.5. Accordingly, to identify the marine space data management on utilisations of marine space administration, there are four elements of data management: capturing information, analysing, storage and dissemination of the data. This is consistent with the purpose of sustainable management of marine space where sustainable development involves a continuous process in deciding where certain questions are asked and where the 'right' and the decision were made (Cicin-Sain 1993).

The following eight key categories illustrates the framework of Malaysia Marine Space Governance.

- The *Custodian* of the marine space sector. Prime Ministers Department should be the lead agency to promote the development and oversee the growth of the sector, given its clout that can enable it to gather other agencies and parties involved in the maritime sector to work together (Aziz Meo Ngah and Nazery Khalid 2014).

Table 23.5 Proposed Malaysia marine space governance custodian

Prime Minister's Office
Implementers
Department of Director General of Lands and Mines (Federal)
Technical Support & Geospatial Data Centre
Department of Survey and Mapping Malaysia (JUPEM)
Policies Support
Maritime Institute of Malaysia (Mima)
Spatial Database Infrastructure Coordinator
Malaysia Centre of Geospatial Data Infrastructure (MaCGDI)
Users of Marine Space Services
International and Domestic Import and Export Community
National Security
Oil & Gas Sector
Ports
Freight Forwarders
Supply Chain Managers
Logistics Services Providers
Maritime Support Service Providers
Facilitators
Ministry of Transport
Ministry of International Trade and Industry
Ministry of Finance
Royal Custom and Excise Department
Ministry of Science and Technology
Ministry of Agriculture and Agro-Based Industry
Ministry of Natural Resources and Environment
Ministry of Defence
Ministry of Home Affairs
Ministry of Culture, Arts and Tourism
Ministry of Communication and Multimedia
Ministry of Foreign Affairs
Marine Department
Financial Institutions and Insurers
Legal and Arbitration and Technology Service Providers
Ship Registers
Classification Societies
Suppliers of Marine Space Services
Ship Building and Ship Repairing Yards
Shipbrokers
Ship Management Companies
Ship Owners
Main Engine and Propulsion Manufacturers
Land Transport Service Providers

(continued)

Table 23.5 (continued)

Prime Minister's Office
Warehouse Operators
Aviation Companies
Sources of Human Capital
Maritime Academies and Technical Colleges
Universities and Polytechnics offering courses in Marine Navigation, Engineering Oceanography, Supply Chain Management and Logistics

Modified after Aziz Meo Ngah and Nazery Khalid (2014)

- *The Implementers.* These are parties involved directly in marine space governance. All the activities should be under these recommendations.
- *The Technical Support & Geospatial Data Centre.* Department of Survey and Mapping Malaysia (JUPEM) is the lead agency in terms of record activity above and under, maintaining and updating the marine space data. Again, Department of Survey and Mapping Malaysia (JUPEM) is playing the role of data storage and dissemination. They would operate between the administration/management activities and the data, allowing any user access to appropriate data to support their needs.
- *The Policies Support.* Maritime Institute of Malaysia (MIMA) is an agency directly involved in policy formulation and act. It is specialized in maritime matters and conflict to more educated in maritime transportation regulations, port rules etc.
- *The Spatial Database Infrastructure Coordinator.* Malaysia Centre of Geospatial Data Infrastructure (MaCGDI) for the purpose of facilitating and responsible for coordinating the sharing and dissemination of geospatial information amongst government agencies, private sector and the general public. Through this infrastructure, smart partnerships among agencies is continuously being developed to produce and share geospatial information thus providing customer-focused, cost effective and timely delivery of geospatial data.
- *The Users of Marine Space Services.* These are parties that use services in the marine space sectors such as shipping, port operations and shipyard services. Identifying them is an important step in managing the demand side of the maritime sector (Aziz Meo Ngah and Nazery Khalid 2014).
- The *Facilitators* include government agencies involved in the marine space sector and support services providers in areas such as finance, ICT, legal, tax, consultancy, classification and registry (Aziz Meo Ngah and Nazery Khalid 2014).
- *The Providers of Marine Space Services.* These are parties offering marine space services required by users (Aziz Meo Ngah and Nazery Khalid 2014).
- The *Sources of Human Capital*, include maritime academies, universities, polytechnic and technical colleges providing marine space-related courses and programmes (Aziz Meo Ngah and Nazery Khalid 2014).

This framework shows inter-related with and provides support to one another. According to Aziz Meo Ngah and Nazery Khalid (2014), the custodian of the marine space, the regulatory authorities, industry players and users of marine space services are part of an ecosystem of marine space sector that features stakeholders working in harmony towards attaining common objective in a facilitating and pro-business and pro-investment environment. In such sustainable framework proposed, there is effective management of demand and supply sides of the marine space sector, supported by talented, skilled human capital that matches the marine space governance needs and its rapid development and dynamic operating environment.

Conclusion

This research highlights the need of marine space governance seriously and considers the issues involved in role and responsibilities. By the introduction of the data management into the stakeholder management and organizational performance, the researchers propose to explain how role, responsibilities and data management in marine space administration can lead to successful sustainable marine space governance.

Role, responsibilities and data management – in the broad sense of the process by which stakeholders work together to accomplish a common mission – are essential when stakeholders need to work together closely. Stakeholders must commonly agree on how they will manage the marine space administration functions of marine space tenure, marine space value, marine space use and marine space development, and, equally important, on how they will make this information available to the wider society in order to encourage creativity, efficiency and productive development among citizens and businesses in a sustainable manner. Therefore, role, responsibilities and data management functions is an approach that must be embedded in the marine space governance. A future study investigating marine space stakeholders role, responsibilities and data management would be very interesting.

Acknowledgement The authors would like to express their gratitude to Ministry of Education, Universiti Teknologi Malaysia (UTM), Jabatan Pengajian Politeknik (JPP) and Fundamental Research Grant Scheme (FRGS) – Ministry of Science, Technology and Innovation (MOSTI) for supporting, funding and contribution respectively. This research is funded by FRGS Fund Vot 4F402.

References

Bass, Suzanne, et al. 2006. *Marine managed areas: Best practices for boundary making*, ed. N. C. S. C. Gerald G. Esch. 2234 South Hobson Avenue, Charleston: NOAA Coastal Services Center. 29405–2413 (843) 740–1200.

Calestino, A.B. 2001. Malaysia: Does it really need decentralization? In *Sourcebook on decentralization in Asia. Decentralization and power shift: An imperative for good governance*, ed. A.B. Brillantes Jr. and N.G. Cuachon. Manila: Asian Resource Center for Decentralization.

Cicin-Sain, B. 1993. Sustainable development and integrated coastal management. *Ocean & Coastal Management* 21(1–3): 11–43. doi:10.1016/0964-5691(93)90019-U.

Cockburn, S. 2005. *Developments in Marine Boundary Law to construct a legal framework for offshore and coastal spaces*. M.Sc.E. thesis, Department of Geodesy and Geomatics Engineering Technical Report No. 228. University of New Brunswick, Fredericton, New Brunswick, Canada.

Fauzi, A. 2006. *Country report 2006 (based on the Marine Administration Country Template 2004)*. Malaysia.

Mokhtar, M.B., and S.A.B. Ghani Aziz. 2003. Integrated coastal zone management using the ecosystems approach: Some perspectives in Malaysia. *Ocean & Coastal Management* 46(5): 407–419. doi:10.1016/S0964-5691(03)00015-2.

Morocco, C., S.B. Thia-Eng, and M. San. 2003. Coastal and ocean governance of the seas of East Asia: Towards an era of new regional cooperation and partnerships. Tropical Coasts.

Mukupa, W. 2011. *Land administration to support sustainable development*. FIG Working Week 2011.

Ng'ang'a, S., M. Sutherland, S. Cockburn, and S. Nichols. 2004. Toward a 3D marine cadastre in support of good ocean governance: A review of the technical framework requirements. *Computers, Environment and Urban Systems* 28(5): 443–470. doi:10.1016/j.compenvurbsys.2003.11.002.

Ngah, Aziz Meor, and Nazery, Khalid. 2014. *MIMA Bulletin*. Maritime Institute of Malaysia (MIMA).

Ramli, J. 1999. *A new maritime legal regime for Malaysia within the context of ocean governance*.

Saharuddin, A.H. 2001. National ocean policy – New opportunities for Malaysian ocean development. *Marine Policy* 25(6): 427–436. doi:10.1016/S0308-597X(01)00027-6.

Saw, I., H. Seang, I. Ziauddin, A. Latif, B.K. Pantai, and J. Pengairan. 2002. *1st annual seminar on sustainability science and management: Issues of the coastal zone*. 6–8 May 2002 Kuala Terengganu Future Management of Shoreline Area Through Integrated Shoreline Management Plan.

Strain, L., A. Rajabifard, and I. Williamson. 2006. Marine administration and spatial data infrastructure. *Marine Policy* 30(4): 431–441. doi:10.1016/j.marpol.2005.03.005.

Teo, C.H., and A. Fauzi. 2006. *A national geocentric datum and the administration of marine spaces in Malaysia 2*. In FIG PUBLICATION NO 36.

Web Page. (URL-1, <: http://www.mima.gov.my/mima/research >, 16.10.2014).

———. (URL-1, <: http://https://www.jupem.gov.my >, 16.10.2014).

———. (URL-1, <: https://www.jupem.gov.my >, 16.10.2014).

Index

T. Yomralioglu, J. McLaughlin (eds.), *Cadastre: Geo-Information Innovations in Land Administration*, DOI 10.1007/978-3-319-51216-7

The manufacturer's authorised representative in the EU is Springer Nature Customer Service Centre GmbH, Europaplatz 3, 69115 Heidelberg, Germany. If you have any concerns regarding our products, please contact ProductSafety@springernature.com

Printed and bound by CPI Group (UK) Ltd, Croydon, CR0 4YY
30/09/2026
02225501-0008